PEDIATRICS - LABORATORY AND CLINICAL RESEARCH

ALTERATIONS IN GLUCOSE HOMEOSTASIS IN CHILDREN, ADOLESCENTS AND YOUNG ADULTS – WHAT'S NEW?

PEDIATRICS - LABORATORY AND CLINICAL RESEARCH

Additional books in this series can be found on Nova's website under the Series tab.

Additional e-books in this series can be found on Nova's website under the e-book tab.

PEDIATRICS - LABORATORY AND CLINICAL RESEARCH

ALTERATIONS IN GLUCOSE HOMEOSTASIS IN CHILDREN, ADOLESCENTS AND YOUNG ADULTS – WHAT'S NEW?

E. OTTO-BUCZKOWSKA
EDITOR

nova publishers
New York

Library of Congress Cataloging-in-Publication Data

ISBN: 978-1-63483-861-0
Library of Congress Control Number: 2015951301

Published by Nova Science Publishers, Inc. † New York

CONTENTS

PREFACE

*E. Otto-Buczkowska**

Medical Specialist Centre in Gliwice, Poland

This book is dedicated to the presentation of the latest research on selected issues associated with type 1 diabetes. The authors are well-known professors of Polish medical universities who have published numerous scientific papers.

Type 1 diabetes is the predominant form of diabetes in young people. It is an autoimmune disease which is particularly aggressive in small children. It is also a pandemic illness characterized by autoimmune, genetic and metabolic abnormalities. It concerns mainly children and young people, and therefore is a condition the patient will have to struggle with for many years hence. This is why it is extremely important to maintain maximum metabolic compensation to delay the occurrence of chronic complications which are an issue not only for the patient, but also for the society as a whole, as they involve a considerable financial burden.

At present, we can observe the development of great advances in the treatment of this form of diabetes. Modern insulin preparations have been introduced as well as improved methods of glucose monitoring and insulin delivery. New ways to delay the autoimmune processes are being studied and implemented. It is becoming more and more realistic to create the "artificial pancreas," which should greatly facilitate the achievement of normal metabolic control. In the long term, a possibility may exist to completely cure diabetes.

A wide dissemination of knowledge concerning the latest research on type 1 diabetes is extremely important – among patients and, above all, among medical practitioners. This book is to serve as an approximation of the latest knowledge for physicians who take care of patients with diabetes on a daily basis.

---

* E-mail: em.buczkowski@pro.onet.pl

In: Alterations in Glucose Homeostasis in Children
Editor: E. Otto-Buczkowska
ISBN: 978-1-63483-861-0

Chapter 1

PATHOGENESIS OF AUTOIMMUNE TYPE 1 DIABETES

Dorota Iwaszkiewicz-Grześ, Mateusz Gliwiński and Piotr Trzonkowski*

[1]Department of Clinical Immunology and Transplantology,
Medical University of Gdańsk,
Gdańsk, Poland

ABSTRACT

Type 1 diabetes (T1D) remains an important health problem in pediatric medicine with increasing incidence around the world [1]. T1D is an autoimmune disease characterized by progressive destruction of pancreatic β-cells and, as a result, insulin deficiency [2, 3]. The destruction of the β-cells is the consequence of a cell-mediated immune response mediated by islet-infiltrating lymphocytes and macrophages (*insulitis*) [1]. Typical targets of this autoimmune attack include glutamic acid decarboxylase, fogrin, insulin, ZnT8 and other antigens specific for pancreatic tissue. T1D are positive for at least one autoantibody against them which is now used in T1D diagnostics and the presence of these autoantibodies identifies also relatives of patients who are at high risk for the disease.

T1D manifests acutely by metabolic decompensation frequently leading to dehydration and ketoacidosis and then progress into end-stage organ insufficiency within years. Complications mainly affect eye, kidney, heart, nervous tissue and vessels. T1D is a multifactorial disease with a genetic background. An important role in its occurrence is attributed to environmental factors such as viral and bacterial infections, viral superantigens, nutrients, drugs and toxins [2, 3]. Nevertheless, the complete list of the risk factors is not yet defined.

Keywords: Type 1 diabetes, molecular mechanisms, genetic factors, immunological factors

---

* E-mail: ptrzon@gumed.edu.pl.

LIST OF ABBREVIATIONS

$(CCL)_5$	chemokine (C-C motif) ligand 5
AIRE	autoimmune regulator
anti-CD28	antibodies anti-CD28
APC	antigen-presenting cell
APS-1	polyendocrine autoimmune syndrome type 1
CTLA4	cytotoxic T lymphocyte antygen 4
CTLs	cytolitic T cells
DCs	dendritic cells
dsRNA	double stranded RNA
FoxP3	forkhead box P3
Fyn	proto-oncogene tyrosine-protein kinase, p59(fyn)
GAD	glutamic acid decarboxylase antibodies
GWA	genome-wide association
HLA	human leukocyte antigens
IA-2	insulinoma-associated antibodies
IAA	insulin autoantibodies
ICA	islet cell antibodies
ICA512	neuroendocrine antigen IA-2
IFIH1	interferon induced with helicase C domain 1
IFN-γ	interferon γ
IL2RA	receptor α of interleukin 2
IL-4, 8, 10, 12, 17, 21, 22	interleukin 4, 8, 10, 12, 17, 21, 22
IPEX	immunodysregulation polyendocrinopathy enteropathy X-linked syndrome
ITAMS	immunoreceptor tyrosine-based activation motif
JAK/STAT	Janus kinase/signal transducer and activator of transcription (STAT)
Lck	lymphocyte-specific protein tyrosine kinase, p56(lck)
LN	lymph nodes
MAP-kinase	mitogen-activated protein kinases
mDCs	myeloid DCs
MHC	major histocompatibility complex
mRNA	messenger RNA
Mϕ	macrophages
NFκB	nuclear factor kappa-light-chain-enhancer of activated B cells
NKT	natural killer T-cells
NOD	non-obese diabetic
pDCs	plasmacytoid DCs
PP	pancreatic polypeptide
PTPN22	protein tyrosine phosphatase non-receptor type 22
SFKs	src-family kinases
SUMO4	small ubiquitin-like modifier 4
T1D	type 1 diabetes

TCR	T cell receptor
TGF-β	transforming growth factor β
Th	T helper cells
TNF	tumor necrosis factor
Treg	regulatory T cell
VNTR	variable number tandem repeats
ZAP-70	zeta-chain-associated protein kinase 70
ZnT8	zinc transporter 8, Slc30A8

INTRODUCTION

Over the last decades, diabetes has been arising as a global problem. Currently 8.3% of the human population suffers from diabetes. At least 10% of this number belongs to T1D and this figure continuously increases, notably in fast-developing economies (Figure 1) [4]. Although the disease onset is strictly dependent on autoimmune activity, initial manifestation does not include immune-related features and it is exclusively related to deficiency of insulin, which causes hyperglycemia, ketoacidosis and finally coma. Chronic damage to certain organs (kidney, eye) or systems (nervous, cardiovascular) and finally premature death depends on insulin deficiency and inappropriate control of blood glucose level, too. Nevertheless, a significant number of T1D patients develops symptoms due to the aberrant activity of the immune system, mainly autoimmune syndromes such as thyroiditis (Hashimoto's disease), celiac sprue and other endocrine diseases.

Although immune system is critical in the development of T1D, it is a multifactorial disease, in which the main role is played by three groups of factors: environmental, genetic and immunological. A current model of the pathogenesis of type 1 diabetes (T1D) was proposed by George Eisenbarth in 1986 (Figure 2) [6].

0-14 years old per 100 000 children per year, 2013 data [4].

Figure 1. New cases of type 1 diabetes around the world.

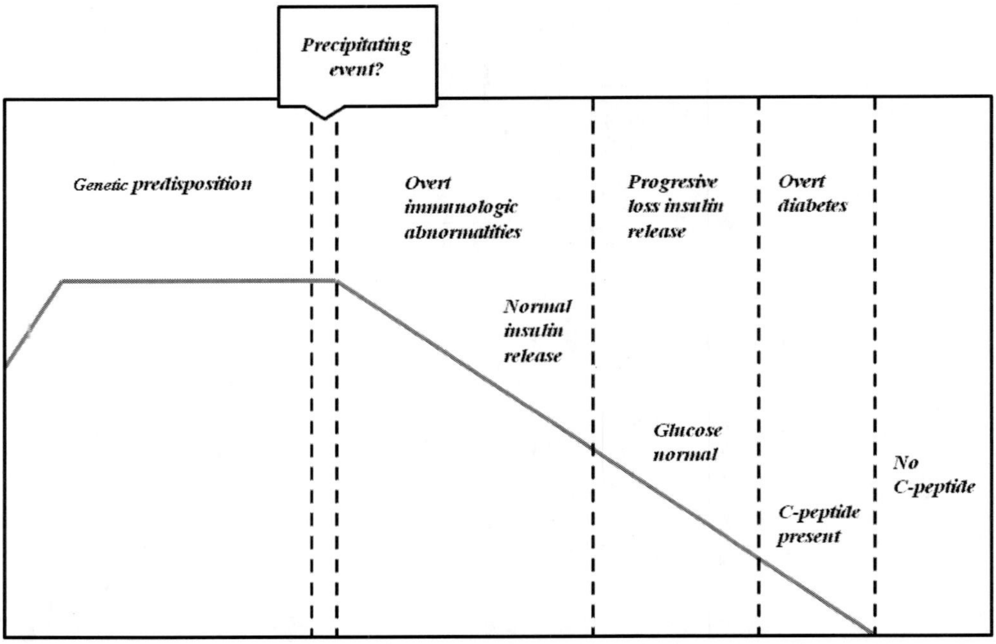

Modified from Eisenbarth GS. Type I diabetes mellitus. A chronic autoimmune disease. N Engl. J. Med. 1986; 314:1360–1368.

Figure 2. Stages in development of T1D [6].

1. Genetic Factors

Type 1 diabetes is a disease with significant genetic background as primarily indicated by inequality of its incidence between general population and families of patients.

Although, the causes of disease on T1D are still incompletely understood, it is certainly associated with genetic factors which predispose to the disease. Many coding regions of genes related to the functioning of immune system cells are associated with the disease. Nevertheless, genes without links to immune system are also involved in the pathogenesis of T1D (e.g., insulin) [7].

1.1. Rare Monogenic Forms

It is rare that T1D is caused by a mutation in a single gene. That rare example is the IPEX syndrome, which symptomatology made also its name: Immune dysregulation, Polyendocrynopathy, Enteropathy, X-linked. The disease is a consequence of the mutation in a gene encoding transcription factor FoxP3, pivotal in the development of regulatory T cells (Tregs). The result is a lack of Tregs function and development of autoimmunity. Another example is polyendocrine autoimmune syndrome type 1 (APS-1), where the mutation in the gene encoding the transcription factor AIRE (autoimmune regulator gene) leads to a lack of insulin presentation in the context of MHC molecules during development of thymocytes. As a result, autoreactive T cells escape deletion in the thymus, cross thymus-blood barrier and traffic to pancreas, but also other sites, where they trigger multiorgan autoimmunity including T1D [8].

1.2. HLA Genes

Genes associated with T1D encode HLA class II receptors which are located on the short arm of human chromosome 6p21. This region contains genes encoding both α and β chains that form the cell surface histocompatibility antigens heterodimer class II. Considering T1D in affected siblings, it can be concluded that the risk increases with the number of common HLA haplotypes of patients (6-10% in siblings and dizygous twins up to 50% in monozygotic twins). The greatest risk of the disease has been found in siblings carrying HLA-DR3 and HLA-DR4 alleles [9].

Molecular genetic analysis allowed for a closer association of the disease with DR and DQ genes, indicating a dominant role of the haplotypes DR3-DQA1*0501-DQB1*0201 and DR4-DQA1*0301-DQB1*0302 in the pathogenesis (gives rise to 30%-50% of genetic T1D risk). The development of the disease is also associated with DRB1*04 alleles, notably with the variants DRB*0405, DRB1*0401, DRB1*0402 and DRB1*0404 increasing the risk. At least one of HLA-DR3 or HLA-DR4 alleles occur in 95% of cases of prediabetic individuals [10, 11]. The involvement of these HLA genes in the predisposition to T1D is indisputable.

Replacement of one or two critical areas in the receptors can significantly increase or decrease the ability to bind specific antigens [11].

The mechanism behind probably involves preferential presentation of peptides that trigger immune reaction against self β cells. In addition, HLA molecules are involved in the selection of the T cell repertoire and elimination of autoreactive lymphocytes. A defect in this process may result in improper selection of T cells responsive to self-tissues including pancreatic islets [10]. Different HLA-DR/DQ genotypes are also associated with other effects in T1D. It is not only an increased risk, but also neutral or protective effects. For example, it has been confirmed that the HLA-DQA1*0102, DQB1*0602, DQB1*0301 and DRB1*0403 haplotypes show protective effect reducing the risk of T1D [11].

Some less stringent associations were observed for HLA class I. This group includes the HLA-A (A24 allele - younger age of onset; A30 - higher risk, A1 - lower risk than haplotypes DR3) and HLA-B (various associations with B18, B39, and B44), and HLA-C (various associations with C3, C8 and C16) [11].

1.3. Non-HLA

Non-HLA genes have lower but significant association with the pathogenesis of T1D. This group includes genes for α receptor of interleukin-2 (IL2RA), cytotoxic T lymphocyte antigen 4 (CTLA4), protein tyrosine phosphatase non-receptor type 22 (lymphoid) (PTPN22), insulin, interferon induced with helicase C domain 1 (IFIH1), small ubiquitin-like modifier 4 (SUMO4) and recently discovered loci from genome-wide association studies (GWA) [12].

a) Insulin (gene located on chromosome 11p15) is a major autoantigen in T1D, so the variants of the gene encoding the hormone are implicated in the pathogenesis of the disease. It concedes about 10% of the genetic predisposition to T1D. Detailed sequence analysis of the gene region shown that there is a correlation between the risk of disease and polymorphism of tandem repeats (VNTR) in the insulin gene promoter. Reduced amount of VNTR increases the risk of T1D development and longer sequence protects the host against T1D. The hypothesis behind regulatory role of VNTR lies in the expression level of insulin in the thymus which affects the binding of transcription factor to the promoter region AIRE [8].

b) PTPN22 (gene is located on chromosome 1p13) is involved in preventing spontaneous T cell activation (by dephosphorylation of Src-family kinases (SFKs), Lck and Fyn, ITAMs, ZAP-70, Vav family proteins and valosin-containing proteins. Hence, it is important negative regulator of receptor signaling. Loss of PTPN22 function by mutation lowers the threshold for activation of autoreactive T cells. On the other hand, suppression of TCR signaling during maturation in the thymus allows autoreactive T lymphocytes escaping thymic selection and therefore their migration to peripheral tissues [13].
c) IL2RA/CD25 –α chain of the interleukin-2 receptor (gene is located on chromosome 10p15) is an essential molecule present on Tregs (IL-2 affects their growth and survival), and activated T lymphocytes. It has been proved that T1D is associated with reduced level of soluble IL2Rα (sIL2Rα). I*n vitro* stimulated peripheral blood mononuclear cells of patients with T1D produce less of sIL2Rα than healthy controls [14].
d) CTLA4 plays an important role in the regulation of T-cell activation and is essential for the proper negative regulation of the immune response. The hypothesis associated with a mutation in the CTLA4 gene (on chromosome 2q33) and its effects on immunopathogenesis T1D is related to the fact that allelic variant lowers the level of mRNA of CTLA4 and reduce control of T cell proliferation. In addition, it influences processing and intracellular trafficking of the CTLA4 protein and may have effect on oligomerization and surface retention [15].
e) *ifih1* (gene is situated on chromosome 2q24) is involved in the recognition of dsRNA viruses derived from picorna. The genetic defect of this molecule can potentially be related to the correct detection and clearance of viral infections and lead to abnormal, diabetogenic immune response [16].
f) *sumo4* is involved in downregulation of NFκB, JAK/STAT pathway and transcription factor activator protein which are involved in the inflammatory reaction. Indirectly, it can induce insulin resistance by tyrosine phosphorylation and suppression of signal transduction of insulin receptors [17].
g) Genome-wide association studies and meta-analyses have identified and confirmed over 40 non-HLA susceptibility loci (including coding regions of immunoregulatory molecules) associated with the risk of T1D. For example, BACH2, PRKCQ, CTSH, C1QTNF6 [18].
h) Mutation in the promoter region of *foxp3/Scurf* gene results in increased risk of T1D. This mutation is less severe than those in the gene itself and does not trigger IPEX [19].
i) Chromosome 18q deletion syndrome (terminal deletion of the long arm) is associated with endocrine disorders. It is related to three genes: *GALR1* (encoding neuropeptide that regulate glucose metabolism and feeding), *CD226* (encoding glycoprotein

expressed on NK cells, naïve and antigen-specific T cells, *TNFRSF11A* (associated with osteoclasts and lymph node development) [20].

2. Immunological Factors

A direct damage to pancreatic β cell is a result of autoimmune reaction in T1D. This process can be detected and followed in the clinic as it is associated with the production of antibodies against islet antigens and accumulation of leukocytes in the islets, so-called *insulitis*. The infiltrations are accompanied by increased apoptosis of β cells in the early stages of the disease. The first theory on pathogenesis of T1D assumed that that decisive role in the destruction of pancreatic islets is played by autoreactive CD8+ T cells that kill the islets in the cytotoxic reaction in which they secrete cytotoxic granules and switch on the apoptosis of β cells (so-called London model) [21]. However, it is well known now that this is a late effector phase which must be preceded by sensitization of the lymphocytes. The process of destruction of the pancreatic cells is initiated by the presentation of processed antigens of β cell antigens in the context of specific variants of MHC class II molecules by antigen presenting cells. The antigens may be directly those from the islets or captured external antigens that resemble islet structures. The latter source is called molecular mimicry and includes viral antigens (e.g., Rubella virus, Coxackie viruses) or environmental ones, like those from the food (e.g., bovine serum albumin from cow milk). Macrophages and dendritic cells (DC) are the main cells that present altered pancreatic antigens [22, 23]. Interestingly, a very narrow repertoire of the autoantigens was reported as a trigger of the disease. In animal studies, it has been observed that of all T cells isolated from pancreatic islets of NOD mice predisposed to develop the disease, 55% showed the presence of a TCR that recognize epitopes of insulin. Over 95% of the clones with insulin binding TCR recognizes a single epitope which comprises amino acids 9-23 of the β chain of insulin [24, 25]. Therefore, based on these observation it is believed that T1D is an autoimmune disease with the dominance of cellular response.

2.1. Cellular Response

Islets of Langerhans are the primary target in T1D. The inflammatory infiltrations in the islets consist mainly of T cells and the intensity of the infiltrations decreases with β cell death and disappears completely when pancreas is finally depleted from β cells. It is the primary proof on the importance of cellular response in the development of T1D. A closer look into the infiltration content in the islets during *insulitis* confirms that, apart from T cells, dendritic cells (DCs) and macrophages can be detected there. This validates an animal model based on NOD mice according to which APC sensitize T cells against the islet antigens and these T cells expand and exert cytotoxic effect on the islets, afterwards. Furthermore, it has been found that adoptive transfer of T cells from the mouse that developed T1D to healthy mice causes rapid development of the disease in the recipients. Furthermore, both in the NOD mice and T1D patients the presence of lymphocytes reactive to the antigens of β cells can be detected. Inappropriate recognition of islet autoantigens can develop in uncontrolled manner and results in clinically manifested T1D mainly when the mechanisms of immune tolerance are ineffective.

This is usually the case of carriers of the susceptibility genes or when environmental antigens mimicking islets induce strong inflammatory immune response.

For T1D to develop, the same for mice and humans, three steps are required (Figure 3) [26]:

i. an immune response needs to be elicited against β cell antigens,
ii. the response needs to induce strong inflammation to destroy β cells,
iii. the regulatory control of the autoimmune responses needs to be ineffective to allow the inflammation in the islets and destruction of β cells.

2.1.1. T and B Cells

Detailed studies proved that both CD4+ and CD8+ T cells need to be present for induction of T1D after transfer of splenocytes into immunocompromised recipients in animal model [26, 27]. Pancreatic β cells express MHC class I, but lack MHC class II proteins, suggesting that direct recognition and cytotoxicity resulting in β cells death can only be mediated by CD8+ CTLs which are the main cells that recognize complexes of peptide antigen presented in the context of MHC class I. CD4+ T cells are assessory in this process providing help to CD8+ T cells and other immune cells via direct cell-to-cell contacts and secreted cytokines [36, 50-54].

Nevertheless, it seems that CD4+ T cells initiate and drive particular responses. CD4+ T cells are mainly involved in driving adoptive immune responses by the activation of CD8+ T cells and B cells [36].

The presence of islet-specific CD4+ T cells that secrete cytokines is a very important for priming the immune response against β cells [26, 33]. To enable an effective CD8+ T cell response, DCs presenting antigens need to be activated, usually through interaction with activated CD4+ T helper (Th) cells [26, 32].

The cytokines secreted by CD4+ T cells license DCs to cross-present exogenous antigens through MHC class I receptors, which normally present endogenous peptides [26, 31]. Activated CD4+ T cells may also trigger inflammation as they are prone to activate macrophages rather than natural killer (NK) cells [34].

Interestingly, antigen-specific CD4+ T cells isolated from blood of T1D patients produce high levels of proinflammatory interferon (IFN) -γ, whereas the CD4+ T cells from healthy donors, produce regulatory cytokine interleukin (IL)-10 rather than IFN-γ. In general, CD4+ T cells are divided according to the pattern of cytokines they secrete. Four subsets: Th1, Th2, Th17 cells and Tregs are of special importance in the pathogenesis of T1D [36].

Th1 cells are responsible for cell-mediated immunity and phagocyte dependent protective responses, which can also destroy the islet β cells and accelerate the course of T1D *via* production of IFN-γ and IL-2 [36]. It seems that this response may be prevalent in T1D patients leading to CD8+ T cell responses and acceleration of macrophage-mediated inflammation.

Th17 cells is a recently discovered subset of CD4+ T cells which produces IL-17. Th17 cells are distinct from Th1 and Th2 cells and play a key role in a variety of infectious diseases, cancer and many autoimmune diseases including T1D. IL-17 produced by these cells is highly proinflammatory and induce both adoptive and innate responses [36, 48].

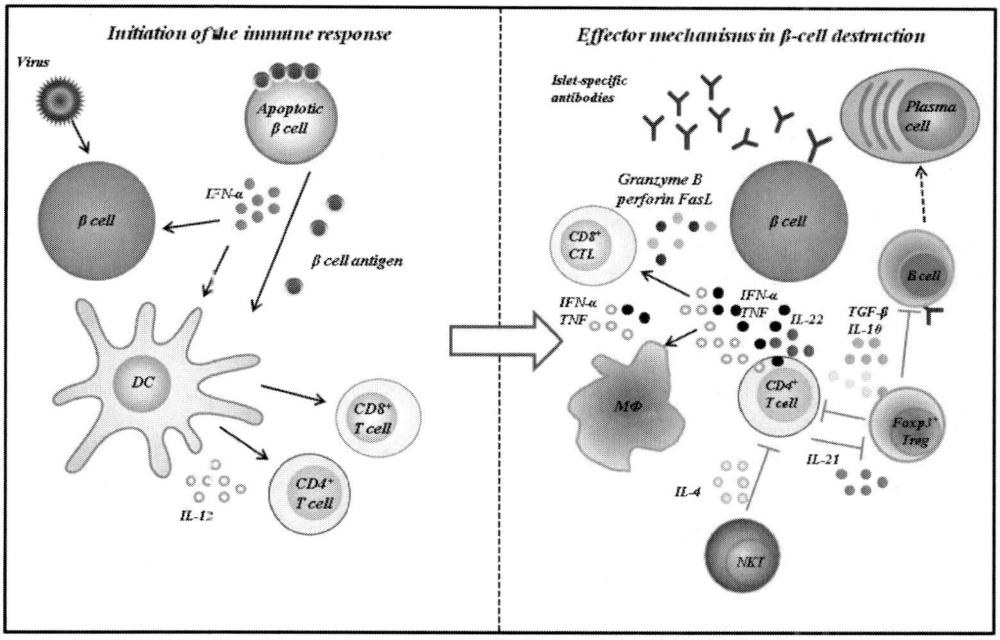

Figure 3. Immune mechanisms involved in β cells destruction. *Initiation:* virus infection can indirectly contribute through increasing expression of MHC class I on β cells and production of proinflammatory cytokines, mainly type I interferons [13, 14]. Exposure to proinflammatory cytokines have effect on β cells themselves, leading to upregulation of expression of chemokines like IL-8, $(CCL)_5$ [26, 29] T cells from recent-onset T1D patients express IL-22 and islet cells express the receptor for IL-22 [26, 30]. Type I interferons activate DCs and promote the presentation of β cell antigens to T cells. DCs can activate CD4+ T cells which promote macrophage-mediated killing through production of cytokines and reactive oxygen species. *Effector mechanisms:* CD4+ T cells can activate islet antigen-specific B cells. They differentiate into antibody-producing plasma cells. Antibodies can bind β cells and mediate complement killing. Activated B cells can be like antigen-specific cells enhancing immune response. Interaction with β cell antigen-specific CD4+ T cells and presence of proinflammatory cytokines can license the DCs to cross-present antigen to β cell antigen-specific CD8+ T cells. Thereby they upregulate their cytotoxic properties. These cytolitic CD8+ T cells (CTLs) can kill β cells through release of granules containing perforin and granzymes. Mechanisms can be stopped by regulatory cell subsets e.g., IL-4-producing NKT cells and Tregs [26].

Increasing evidence has shown that therapeutic agents targeting the IL-17 molecule or directly inhibiting IL-17-producing cells may stop the progression of T1D [36, 49]. Nevertheless, human studies are needed to confirm Th17 involvement in the disease.

Th2 cells mainly produce IL-4 and IL-10, which are responsible for strong antibody production, eosinophil activation and inhibition of several macrophage functions [36, 45, 46]. Immunotherapeutic approaches like anti-CD28 stimulation, which promote and enhance the function of intraislet Th2 cells and secretion of IL-4 by these cells, can effectively prevent the onset of T1D in animal models [47]. Nevertheless, none of the studies in humans confirmed efficacy of this approach.

Tregs are primary controllers of immune responsiveness and peripheral immunological tolerance. This activity is exerted by cell-to-cell inhibition of autoantigen presentation between Tregs and DCs; similar cell-to-cell suppression of autoreactive T cells, suppression via secreted inhibitory IL-10 and TGF-β and killing of autoreactive effectors by cytotoxic granules secreted by Tregs. These cells can also consume high amounts of IL-2 'starving' autoreactive cells which die without this cytokine [26, 35]. Tregs are characterized by the expression of Foxp3 transcription factor and dependency on IL-2. Mutations of *foxp3* gene lead to malfunction of Tregs and therefore severe multiorgan autoimmune and inflammatory disorder called IPEX manifested by early onset of T1D. Adoptive transfer of Tregs in animal model has been shown to offer a protection from T1D, whereas experimental depletion or genetic deficiency in their numbers or activity promotes a more aggressive disease [36, 37-41]. Furthermore, IL-2 administration has been shown to expand and activate Tregs in mice, while a short course of low-dose IL-2 administration at diabetes onset can reverse established disease [36, 42]. Recent studies have also shown that T1D progression in NOD mice is associated with a decrease in numbers and function of Tregs in the inflamed islets, and defects in IL-2 production by effector T cells seem largely responsible [36, 43]. In patients with recent-onset T1D, increased apoptosis and decreased function of Tregs has been described [36]. Expanded autologous Tregs were a safe and well-tolerated therapy in children with T1D, which could inhibit or delay the destruction of pancreatic cells [44]. Nevertheless, the functional defect rather than quantitative defect in human Tregs seems to be a crucial factor in the development of T1D, which requires further investigation [36].

2.1.2. Antigen Presenting Cells (APC)

Antigen presenting cells, such as dendritic cells (DCs) and macrophages (Mϕ), play a crucial role in T cell differentiation providing costimulatory signals and cytokines for T cell priming [36, 58]. DCs in the pancreas take up released β-cell-derived antigens and migrate to the draining lymph nodes (LN) where they interact with islet-specific CD4 and CD8 T cells. Dendritic cells comprise two major classes: plasmacytoid DCs (pDCs) and myeloid DCs (mDCs) [36, 61]. DCs are necessary for the initiation of the anti-islet immune response and if they are eliminated no disease develops [26, 59]. Depending on the signals delivered to T cell by DCs, T cells will differentiate into either inflammatory effector cells such as T helper Th1 cells or anti-inflammatory Th2 or regulatory (Treg) cells [60]. This way DCs are also important in immune tolerance through deletion of self-reactive cells and the induction of Treg [60]. A population of islet-resident DCs has been described which is present even in non-inflamed islets and efficiently present islet antigen to T cells [26, 63]. In T1D, the increased accumulation of DCs along with macrophages in the earliest islet infiltrates of both humans with T1D and NOD mice suggests an important role for these cells in the pathogenesis of T1D [64].

The role of Mϕ in T1D development is more controversial. The studies in NOD mice revealed that diabetogenic CD4+ T cells act primarily through activation of Mϕ and preventing entry of Mϕ into the islets protects from the disease [26, 34, 65], whereas other models have demonstrated that depletion of these cells has no effect on the disease onset [59]. In humans, tumor necrosis factor (TNF) producing Mϕ can be found in the islets of recent-onset T1D patients, but their role is still unclear [26, 66].

A pathogenic role of pDCs in T1D is supported by observations in both humans and rodent models. Type 1 IFN produced in pancreatic islets by these cells could induce or promote the development of the disease [26, 67]. Meyers et al. have further reported that pDCs and mDCs from new-onset patients may have altered TLR7/8 and TLR4 signaling, respectively. It was associated with increased pro-inflammatory cytokines and chemokines expression levels in serum [26, 68].

Cytokines secreted by DCs are considered critical mediators of T1D. For example, IL-1 signaling has roles in β cell dysfunction and destruction *via* NF-κB and MAP-kinase pathways, leading to endoplasmic reticulum and mitochondrial stress and eventually activating the apoptotic machinery in β cell [26, 69]. TNF-α induces maturation of DCs and activates islet-specific pancreatic lymph node T cells, which altogether enhance autoimmune response against β cell in T1D [26, 70].

2.1.3. NK Cells

The role of NK cells in T1D is not clear. It is possible that the presence of NK cells indicates a distinct etiology of disease, because their presence both in humans and mice is correlated with the presence of viral infection. These cells are especially important for pathogenesis in a subgroup of patients in whom viral infection is the initiating event in T1D [26, 74, 75]. Activating receptor NKp46 is considered as the most specific NK cell marker which recognizes mouse and human ligands on pancreatic β cells and triggers degranulation of NK cells [71]. On the other hand, studies have shown a reduction in the frequency of NK cells in the peripheral blood in patients with T1D and reduced surface expression of the activating receptors NKp30 and NKp46 as well as lower mRNA levels of IFN-γ and perforin in NK cells in patients with long-standing T1D when compared to healthy subjects [72, 73].

2.2. Humoral Response

B cells and their products are not directly pathogenic to β cells. Although it seems that B cells are not indispensable in human T1D, these cells might assist in the development of the T1D [36]. Emerging evidence has revealed that they could promote autoimmunity by several mechanisms including: production of autoantibodies with consequent generation of immune complexes, antigen presentation to generate primary autoreactive T cell responses, contribution to the maintenance of CD4+T cell memory and production of pro-inflammatory cytokines [36, 55, 56]. Recent study has demonstrated that B cells are necessary in the islets to promote survival of activated CD8+ T cells accelerating disease progression [36, 57].

Of the highest importance in patients with T1D is the presence of autoantibodies directed against pancreatic islet. The antibodies occur months to years before clinical onset of T1D and then gradually disappear with the duration of the disease. Currently, the presence of the antibody is a gold standard to diagnose autoimmune T1D and monitor the development of autoimmune process in pancreatic islets. While detection of single autoantibody type only is hardly predictive, the presence of two autoantibodies in apparently healthy person confers 25-50% risk of T1D over the next five years. This is even higher if the risk genotype is confirmed.

2.2.1. Islet Cell Antibodies (ICA)

The earliest-known marker were islet cell antibodies (ICA). They appear for many months or even years before detection of diabetes. At the time of diagnosis they are present in approximately 80% of patients, then gradually disappear in a proportion to the duration of the disease. ICA are polyclonal autoantibodies that react with all cells of the islet (e.g., α, β, γ, δ, and PP cells). Lipid and protein autoantigens that are recognized by ICA include sialoglycoconjugate [76, 77], glutamic acid decarboxylase [76, 78] and neuroendocrine antigen IA-2 [76, 79]. ICA do not appear to play an etiological role in β cell destruction but do serve as important markers of β cell autoimmunity. ICA frequency declines following diagnosis, and no more than 5–10% of T1D patients remains ICA-positive 10 years post diagnosis [76].

2.2.2. Insulin Autoantibodies (IAA)

Insulin is a hormone produced by pancreatic β cells, which is not only central to regulating carbohydrate and fat metabolism, but is also central in its pathological role as a T1D autoantigen [81]. At onset of T1D, IAAs occur in 35–60% of children but are decidedly less common in adults [80]. High titer of insulin autoantibodies (IAA) at younger ages is consistent with the concept that these patients develop more aggressive disease. In particular, insulin autoantibodies levels above 2000 nU/mL are almost exclusively found in patients who progress to T1D before age 5 yr, and less than half of individuals developing T1D after age 15 yr carry detectable levels of IAA [81].

2.2.3. Glutamic Acid Decarboxylase Antibodies (GAD)

The antibodies to this antigen were present in about 80% of new onset patients and in pre-diabetics before the appearance of clinical disease. Autoantibodies against GAD are regarded predictor of progression to overt diabetes. When coupled with insulin autoantibodies and islet cell antibodies (ICA), their ability to predict the likelihood of developing T1D in asymptomatic first-degree relatives of T1D patients is very high [81].

2.2.4. Antibodies against the Tyrosine Phosphatase-Like islet antigen IA2/ICA 512

The neuroendocrine antigen IA-2 (ICA512) is a major autoantigen in T1D [81]. It is an enzymatically inactive member of the tyrosine phosphatase family, involved in regulating insulin secretion.

Assessment of the presence of IA-2 autoantibodies contributes to the predictability of the likelihood of T1D development [81]. Like GAD, IA-2 is found in nervous tissue and other endocrine tissues including the pituitary gland. Some IA-2 autoantibody assays detect the entire molecule, whereas others detect portions of the molecule (e.g., IA-2c and ICA512 assays) [76, 82, 83].

2.2.5. ZnT8

Zinc transporter 8 (ZnT8) is specifically and abundantly expressed in islets of Langerhans. Antibodies against it are therefore considered the best marker of T1D. Zn is highly concentrated in β cells and is needed for normal insulin storage.

β-cell specific deletion of ZnT8 in mice results in glucose-intolerance, reduced β-cell zinc accumulation and anomalous insulin granules, as well as blunted first-phase glucose-stimulated insulin secretion, reduced insulin processing enzyme transcripts and increased proinsulin levels [84]. ZnT8 antibodies were found in 26% of patients with T1D who had not shown antibody positivity to other commonly measured autoantigens: GAD, IA2, insulin and, ICA [81].

CONCLUSION

T1D is an autoimmune disease with complex and heterogeneous pathological background. The observations support the hypothesis on a strong genetic background affecting the predisposition to develop the disease. Especially high link is assigned to HLA genes as they present antigens to T cells and to a lesser extent to non-HLA genes associated with the functioning of the immune system, such as a receptor for IL-2 or CTLA4.

Immunopathogenesis of T1D is manifested by *insulitis,* that is, inflammation-related destruction of pancreatic β cells. It is associated with self-sensitization of immune cells which aberrantly react against self pancreatic antigens. It also relates to humoral response, that is, the appearance of antibodies against autologous islet-specific structures. The antibodies, such as anti-ICA, IAA, GAD, IA2 / ICA 512 and ZnT8, are currently used in the routine diagnostics of the disease.

REFERENCES

[1] Herold K. C., Vignali D. A., Cooke A., Bluestone J. A. Type I Diabetes: Translating Mechanistic Observations into Effective Clinical Outcomes. *Nat. Rev. Immunol.*, 13(4): 243–256 (2013).

[2] Gupta S. Immunotherapies in diabetes mellitus type 1. *Med. Clin. North Am.*, 96(3): 621-634 (2012).

[3] Pugliese A. The multiple origins of Type 1 diabetes. *Diabet Med.,* 30(2):135-46 (2013).

[4] International Diabetes Federation, Diabetes Estimates, Diabetes prevalence (%), 2014, http://www.idf.org/atlasmap/atlasmap?indicator=i1&date=2014, (10.04.2015)

[5] WHO: Diabetes. http://www.who.int/mediacentre/factsheets/fs312/en/index.html (01.02.2012)

[6] Eisenbarth G. S. Type I diabetes mellitus. A chronic autoimmune disease. *N Engl. J. Med.,* 314:1360–1368 (1986).

[7] Todd J. A. Etiology of type 1 diabetes. *Immunity,* 32(4), 457-67 (2010).

[8] van Belle T. L., Coppieters K. T., von Herrath M. G. Type 1 diabetes: etiology, immunology, and therapeutic strategies. *Physiol. Rev.,* 91(1), 79-118 (2011).

[9] Redondo M. J , Jeffrey J., Fain P. R., Eisenbarth G. S., Orban T. Concordance for Islet Autoimmunity among Monozygotic Twins. *N Engl. J. Med.,* 359, 2849-2850 (2008).

[10] Skowera A., Ellis R. J., Varela-Calviño R., Arif S., Huang G. C., Van-Krinks C., Zaremba A., Rackham C., Allen J. S., Tree T. I., Zhao M., Dayan C. M., Sewell A. K., Unger W. W., Drijfhout J. W., Ossendorp F., Roep B. O., Peakman M. CTLs are targeted to kill beta cells in patients with type 1 diabetes through recognition of a glucose-regulated preproinsulin epitope. *J. Clin. Invest.*, 118(10), 3390-402 (2008).

[11] Steck A. K., Rewers M. J. Genetics of Type 1 Diabetes. *Clinical Chemistry,* 57(2), 176-185 (2011).

[12] Morran M. P., Vonberg A., Khadra A., Pietropaolo M. Immunogenetics of type 1 diabetes mellitus. *Mol. Aspects Med.*, S0098-2997(14), 00079-X (2015).

[13] Sarmiento J., Wallis R. H., Ning T., Marandi L., Chao G., Veillette A., Lernmark Å., Paterson A. D., Poussier P. A functional polymorphism of Ptpn22 is associated with type 1 diabetes in the BioBreeding rat. *J. Immunol.,* 194(2), 615-29 (2015).

[14] Alcina A., Fedetz M., Ndagire D., Fernández O., Leyva L., Guerrero M., Abad-Grau M. M., Arnal C., Delgado C., Lucas M., Izquierdo G., Matesanz F. IL2RA/CD25 gene polymorphisms: uneven association with multiple sclerosis (MS) and type 1 diabetes (T1D). *PLoS One*, 4(1), e4137 (2009).

[15] Wang J., Liu L., Ma J., Sun F., Zhao Z., Gu M. Common variants on cytotoxic T lymphocyte antigen-4 polymorphisms contributes to type 1 diabetes susceptibility: evidence based on 58 studies. *PLoS One*, 9(1), e85982 (2014).

[16] Fumagalli M., Cagliani R., Riva S., Pozzoli U., Biasin M., Piacentini L., Comi G. P., Bresolin N., Clerici M., Sironi M. Population genetics of IFIH1: ancient population structure, local selection, and implications for susceptibility to type 1 diabetes. *Biol. Evol.*, 27(11), 2555-66 (2010).

[17] Tang S. T., Peng W. J., Wang C. J., Tang H. Q., Zhang Q. Polymorphism M55V in gene encoding small ubiquitin-like modifier 4 (SUMO4) protein associates with susceptibility to type 1 (and type 2) diabetes. *Diabetes Metab. Res. Rev.,* 28(8), 679-87 (2012).

[18] Qu H. Q., Bradfield J. P., Li Q., Kim C., Frackelton E., Grant S. F., Hakonarson H., Polychronakos C. In silico replication of the genome-wide association results of the Type 1 Diabetes Genetics Consortium. *Hum. Mol. Genet.*, 19(12), 2534-8 (2010).

[19] Bassuny W. M., Ihara K., Sasaki Y., Kuromaru R., Kohno H., Matsuura N., Hara T. A functional polymorphism in the promoter/enhancer region of the FOXP3/Scurfin gene associated with type 1 diabetes. *Immunogenetics,* 55(3), 149-56 (2003).

[20] Hogendorf A., Lipska-Zietkiewicz B. S., Szadkowska A., Borowiec M., Koczkowska M., Trzonkowski P., Drozdz I., Wyka K., Limon J., Mlynarski W. Chromosome 18q deletion syndrome with autoimmune diabetes mellitus: putative genomic loci for autoimmunity and immunodeficiency. *Pediatr. Diabetes*, (2014).

[21] Bottazzo G. F. Pujol-Borrell R., Hanafusa T., Feldman M. Role of aberrant HLA-DR expression and antigen presentation in induction of endocrine autoimmunity. *Lancet,* 2(8359): 1115-1119 (1983).

[22] Jasen A. Homo-Delarche R., Hooijkaas H., Leenen P. J., Dardenne M., Drexhage H. A. Immunohistochemical characterization of monocyte-macrophages and dendritic cells involved in the initiation of insulitis and beta-cell de-struction in NOD mice. *Diabetes,* 43(5): 667-675 (1994).

[23] Yoon J. W. Jun H. S., Santamaria P. Cellular and molecular mechanisms for the initiation and progression of beta cell destruction resulting from the collaboration between macrophages and T cells. *Autoimmunity,* 27(2): 109-122 (1998).

[24] Simone E. Daniel D., Schloot N., Gottlieb P., Babu S., Kawasaki E., Wegmann D., Eisenbarth G. S. T cell receptor restriction of diabetogenic autoimmune NOD T cells. *Proc. Natl. Acad. Sci. USA*, 94(6): 2518-2521 (1997).

[25] Wong F. S. Karttunen J., Dumont C., Wen L., Visintin I., Pilip I. M., Shastri N., Pamer E. G., Janeway C. A. Jr. Identification of an MHC class I-restricted autoantigen in type 1 diabetes by screening an organ-specific cDNA library. *Nat. Med.,* 5(9): 1026-1031 (1999).

[26] Wållberg M., Cooke A. Immune mechanisms in type 1 diabetes. *Trends in Immun.*, 34(12): 583-591 (2013).

[27] Phillips J. M., Parish N. M., Raine T., Bland C., Sawyer Y., De La Peña H., Cooke A. Type 1 diabetes development requires both CD4+ and CD8+ T cells and can be reversed by non-depleting antibodies targeting both T cell populations. *Rev. Diabet. Stud.*, 6, 97–103 (2009).

[28] von Herrath, M. Can we learn from viruses how to prevent type 1 diabetes?: the role of viral infections in the pathogenesis of type 1 diabetes and the development of novel combination therapies. *Diabetes,* 58, 2–11 (2009).

[29] Eizirik D. L., Sammeth M., Bouckenooghe T., Bottu G., Sisino G., Igoillo-Esteve M., Ortis F., Santin I., Colli M. L., Barthson J., Bouwens L., Hughes L., Gregory L., Lunter G., Marselli L., Marchetti P., McCarthy M. I., Cnop M. The human pancreatic islet transcriptome: expression of candidate genes for type 1 diabetes and the impact of pro-inflammatory cytokines. *PLoS Genet.,* 8(3):e1002552 (2012).

[30] Arif S., Moore F., Marks K., Bouckenooghe T., Dayan C. M., Planas R., Vives-Pi M., Powrie J., Tree T., Marchetti P., Huang G. C., Gurzov E. N., Pujol-Borrell R., Eizirik D. L., Peakman M. Peripheral and islet interleukin-17 pathway activation characterizes human autoimmune diabetes and promotes cytokine-mediated beta-cell death. *Diabetes,* 60, 2112–2119 (2011).

[31] Joffre O. P., Segura E., Savina A., Amigorena S. Cross-presentation by dendritic cells. *Nat. Rev. Immunol.*, 12, 557–569 (2012).

[32] Smith C. M., Wilson N. S., Waithman J., Villadangos J. A., Carbone F. R., Heath W. R., Belz G. T. Cognate CD4(+) T cell licensing of dendritic cells in CD8(+) T cell immunity. *Nat. Immunol.*, 5, 1143–1148 (2004).

[33] Makhlouf L., Grey S. T., Dong V., Csizmadia E., Arvelo M. B., Auchincloss H. Jr., Ferran C., Sayegh M. H. Depleting anti-CD4 monoclonal antibody cures new-onset diabetes, prevents recurrent autoimmune diabetes, and delays allograft rejection in nonobese diabetic mice. *Transplantation,* 77, 990–997 (2004).

[34] Calderon B., Suri A., Unanue E. R. In CD4+ T-cell-induced diabetes, macrophages are the final effector cells that mediate islet beta-cell killing: studies from an acute model. *Am. J. Pathol.*, 169, 2137–2147 (2006).

[35] Sakaguchi S., Vignali D. A., Rudensky A. Y., Niec R. E., Waldmann H. The plasticity and stability of regulatory T cells. *Nat. Rev. Immunol.*, 13, 461–467 (2013).

[36] Min Li, Lu-Jun Song, Xin-Yu Qin. Advances in the cellular immunological pathogenesis of type 1 diabetes. *J. Cell. Mol. Med.*, 5, 749-758 (2014).

[37] Kawamoto K., Pahuja A., Nettles A., Honkanen-Scott M., Bansal-Pakala P. Downregulation of TGF-beta RII in T effector cells leads to increased resistance to TGFbeta- mediated suppression of autoimmune responses in type I diabetes. *Autoimmunity*, 45: 310–9 (2012).

[38] Jana S., Campbell H., Woodliff J., Waukau J., Jailwala P., Ghorai J., Ghosh S., Glisic S. The type of responder T-cell has a significant impact in a human in vitro suppression assay. *PLoS ONE*, 5: e15154 (2010).

[39] Tonkin D. R., Haskins K. Regulatory T cells enter the pancreas during suppression of type 1 diabetes and inhibit effector T cells and macrophages in a TGF-beta-dependent manner. *Eur. J. Immunol.*, 39: 1313–22 (2009).

[40] Tarbell K. V., Yamazaki S., Olson K., Toy P., Steinman R. M. CD25(+) CD4(+) T cells, expanded with dendritic cells presenting a single autoantigenic peptide, suppress autoimmune diabetes. *J. Exp. Med.*, 199: 1467–77 (2004).

[41] Li L., Nishio J., van Maurik A., Mathis D., Benoist C. Differential response of regulatory and conventional CD4(+) lymphocytes to CD3 engagement: clues to a possible mechanism of anti-CD3 action? *J. Immunol.*, 191: 3694–704 (2013).

[42] Shameli A., Yamanouchi J., Tsai S., Yang Y., Clemente-Casares X., Moore A., Serra P., Santamaria P. IL-2 promotes the function of memory-like autoregulatory CD8+ T cells but suppresses their development via FoxP3+ Treg cells. *Eur. J. Immunol.*, 43: 394–403 (2013).

[43] Kornete M., Mason E. S., Piccirillo C. A. Immune regulation in T1D and T2D: prospective role of Foxp3+ treg cells in disease pathogenesis and treatment. *Front. Endocrinol.*, 4: 76 (2013).

[44] Marek-Trzonkowska N., Myśliwec M., Siebert J., Trzonkowski P. Clinical application of regulatory T cells in type 1 diabetes. *Pediatr. Diabetes,* 14: 322–32 (2013).

[45] Romagnani S. Th1/Th2 cells. *Inflamm. Bowel Dis.*, 5: 285–94 (1999).

[46] Lin M. S., Tse H. M., Delmastro M. M., Bertera S., Wong C. T., Lakomy R., He J., Sklavos M. M., Coudriet G. M., Pietropaolo M., Trucco M. M., Piganelli J. D. A multivalent vaccine for type 1 diabetes skews T cell subsets to Th2 phenotype in NOD mice. *Immunol. Res.*, 50: 213–20 (2011).

[47] Sharif S., Arreaza G. A., Zucker P., Delovitch T. L. Regulatory natural killer T cells protect against spontaneous and recurrent type 1 diabetes. *Ann. N Y Acad. Sci.*, 958: 77–88 (2002).

[48] Marwaha A. K., Crome S. Q., Panagiotopoulos C., Berg K. B., Qin H., Ouyang Q., Xu L., Priatel J. J., Levings M. K., Tan R. Cutting edge increased IL-17- secreting T cells in children with new-onset type 1 diabetes. *J. Immunol.*, 185: 3814–8 (2010).

[49] Lee I. F., Wang X., Hao J., Akhoundsadegh N., Chen L., Liu L., Langermann S., Ou D., Warnock G. L. B7-H4.Ig inhibits the development of Type 1 diabetes by regulating Th17 cells in NOD mice. *Cell Immunol.*, 282: 1–8 (2013).

[50] Rasche S., Busick R. Y., Quinn A. GAD65- Specific Cytotoxic T lymphocytes mediate beta-cell death and loss of function. *Rev. Diabet. Stud.,* 6: 43–53 (2009).

[51] Katz J., Benoist C., Mathis D. Major histocompatibility complex class-i molecules are required for the development of insulitis in nonobese diabetic mice. *Eur. J. Immunol.*, 23: 3358–60 (1993).

[52] Bulek A. M. 1., Cole D. K., Skowera A., Dolton G., Gras S., Madura F., Fuller A., Miles J. J., Gostick E., Price D. A., Drijfhout J. W., Knight R. R., Huang G. C., Lissin N., Molloy P. E., Wooldridge L., Jakobsen B. K., Rossjohn J., Peakman M., Rizkallah P. J., Sewell A. K. Structural basis for the killing of human beta cells by CD8(+) T cells in type 1 diabetes. *Nat. Immunol.*, 13: 283–1521 (2012).

[53] Coppieters K. T., Dotta F., Amirian N., Campbell P. D., Kay T. W., Atkinson M. A., Roep B. O., von Herrath M. G. Demonstration of islet-autoreactive CD8 T cells in insulitic lesions from recent onset and long-term type 1 diabetes patients. *J. Exp. Med.,* 209: 51–60 (2012).

[54] Chen X. L., Bobbala D., Rodriguez G. M., Mayhue M., Chen Y. G., Ilangumaran S., Ramanathan S. Induction of autoimmune diabetes in nonobese diabetic mice requires

interleukin- 21-dependent activation of autoreactive CD8(+) T cells. *Clin. Exp. Immunol.*, 173: 184–94 (2013).

[55] O'Neill S. K., Liu E., Cambier J. C. Change you can B(cell)eive in recent progress confirms a critical role for B cells in type 1 diabetes. *Curr. Opin. Endocrinol.*, 16: 293–8 (2009).

[56] Cox S. L., Silveira P. A. Emerging roles for B lymphocytes in Type 1 diabetes. *Expert Rev. Clin. Immunol.,* 5: 311–24 (2009).

[57] Brodie G. M., Wallberg M., Santamaria P., Wong F. S., Green E. A. B-cells promote intra-islet CD8+ cytotoxic T-cell survival to enhance type 1 diabetes. *Diabetes,* 57: 909–17 (2008).

[58] Tai N., Yasuda H., Xiang Y., Zhang L., Rodriguez-Pinto D., Yokono K., Sherwin R., Wong F. S., Nagata M., Wen L. IL-10- conditioned dendritic cells prevent autoimmune diabetes in NOD and humanized HLA-DQ8 RIP-B7.1 mice. *Clin. Immunol.*, 139: 336–49 (2011).

[59] Saxena V., Ondr J. K., Magnusen A. F., Munn D. H., Katz J. D. The countervailing actions of myeloid and plasmacytoid dendritic cells control autoimmune diabetes in the nonobese diabetic mouse. *J. Immunol.*, 179, 5041–5053 (2007).

[60] Morel P. A. Dendritic cell subsets in type 1 diabetes: friend or foe? *Front. Immunol.*, 4:415 (2013).

[61] Ganguly D., Haak S., Sisirak V., Reizis B. The role of dendritic cells in autoimmunity. *Nat. Rev. Immunol.*, 13: 566–77 (2013).

[62] Blanco P. 1., Palucka A. K., Pascual V., Banchereau J. Dendritic cells and cytokines in human inflammatory and autoimmune diseases. *Cytokine Growth Factor Rev.*, 19: 41–52 (2008).

[63] Calderon, B. and Unanue, E. R. Antigen presentation events in autoimmune diabetes. *Curr. Opin. Immunol.*, 24, 119–128 (2012).

[64] Kim H. S., Lee M. S. Role of innate immunity in triggering and tuning of autoimmune diabetes. *Curr. Mol. Med.*, 9: 30–44 (2009).

[65] Hutchings P., Rosen H., O'Reilly L., Simpson E., Gordon S., Cooke A. Transfer of diabetes in mice prevented by blockade of adhesion-promoting receptor on macrophages. *Nature,* 348, 639–642 (1990).

[66] Uno S., Imagawa A., Okita K., Sayama K., Moriwaki M., Iwahashi H., Yamagata K., Tamura S., Matsuzawa Y., Hanafusa T., Miyagawa J., Shimomura I. Macrophages and dendritic cells infiltrating islets with or without beta cells produce tumour necrosis factor-alpha in patients with recent-onset type 1 diabetes. *Diabetologia,* 50, 596–601 (2007).

[67] Diana J., Gahzarian L., Simoni Y., Lehuen A. Innate immunity in type 1 diabetes. *Discov. Med.*, 11: 513–20 (2011).

[68] Meyers A. J., Shah R. R., Gottlieb P. A., Zipris D. Altered toll-like receptor signaling pathways in human type 1 diabetes. *J. Mol. Med.,* 88: 1221–31 (2010).

[69] Mandrup-Poulsen T., Pickersgill L., Donath M. Y. Blockade of interleukin 1 in type 1 diabetes mellitus. *Nat. Rev. Endocrinol.*, 6: 158–66 (2010).

[70] Lee L. F., Xu B., Michie S. A., Beilhack G. F., Warganich T., Turley S., McDevitt H. O. The role of TNF-alpha in the pathogenesis of type 1 diabetes in the nonobese diabetic mouse: analysis of dendritic cell maturation. *Proc. Natl. Acad. Sci. USA,* 102: 15995–6000 (2005).

[71] Gur C., Porgador A., Elboim M., Gazit R., Mizrahi S., Stern-Ginossar N., Achdout H., Ghadially H., Dor Y., Nir T., Doviner V., Hershkovitz O., Mendelson M., Naparstek Y., Mandelboim O. The activating receptor NKp46 is essential for the development of type 1 diabetes. *Nat. Immunol.,* 11: 121–37 (2010).

[72] Rodacki M., Milech A., De Oliveira J. NK cells and type 1 diabetes. *Clin. Dev. Immunol.*, 13: 101–7 (2006).

[73] Chandy K. G., Charles M. A., Buckingham B., Waldeck N., Kershnar A., Gupta S. Deficiency of monoclonal-antibody (leu7) defined nk-cells in newly diagnosed insulin-dependent diabetes-mellitus. *Immunol. Lett.,* 8: 89–91 (1984).

[74] Dotta F., Censini S., van Halteren A. G., Marselli L., Masini M., Dionisi S., Mosca F., Boggi U., Muda A. O., Del Prato S., Elliott J. F., Covacci A., Rappuoli R., Roep B. O., Marchetti P. Coxsackie B4 virus infection of beta cells and natural killer cell insulitis in recent-onset type 1 diabetic patients. *Proc. Natl. Acad. Sci. U.S.A,* 104, 5115–5120 (2007).

[75] Flodström M., Maday A., Balakrishna D., Cleary M. M., Yoshimura A., Sarvetnick N. Target cell defense prevents the development of diabetes after viral infection. *Nat. Immunol.,* 3, 373–382 (2002).

[76] Winter W. E., Harris N., Schatz D. Immunological Markers in the Diagnosis and Prediction of Autoimmune Type 1a Diabetes, *Clin. Diab.,* 20(4), 183-191 (2002).

[77] Nayak R. C., Omar M. A. K., Rabizadeh A., Srikanta S., Eisenbarth G. S. “Cytoplasmic” islet cell antibodies: evidence that the target antigen is a sialoglycoconjugate. *Diabetes*, 34:617–619 (1985).

[78] Atkinson M. A., Kaufman D. L., Newman D., Tobin A. J., Maclaren N. K. Islet cell cytoplasmic autoantibody reactivity to glutamate decarboxylase in insulin-dependent diabetes. *J. Clin. Invest.*, 91:350–356 (1993).

[79] Myers M. A., Rabin D. U., Rowley M. J. Pancreatic islet cell cytoplasmic antibody in diabetes is represented by antibodies to islet cell antigen 512 and glutamic acid decarboxylase. *Diabetes*, 44:1290–1295 (1995).

[80] Atkinson M. A., Maclaren N. K., Riley W. J., Winter W. E., Fisk D. D., Spillar R. P. Are insulin autoantibodies markers for insulin dependent diabetes? *Diabetes,* 35:894–898 (1986).

[81] Pietropaolo M., Towns R., Eisenbarth G. S. Humoral Autoimmunity in Type 1 Diabetes: Prediction, Significance, and Detection of Distinct Disease Subtypes. *Cold Spring Harb. Perspect Med.*, 2(10) (2012).

[82] Vives-Pi M. 1., Somoza N., Vargas F., Armengol P., Sarri Y., Wu J. Y., Pujol-Borrell R. Expression of glutamic acid decarboxylase (GAD) in the alpha, beta and delta cells of normal and diabetic pancreas: implications for the pathogenesis of type I diabetes. *Clin. Exp. Immunol.,* 92:391–396 (1993).

[83] Solimena M., Dirkx R. Jr., Hermel J. M., Pleasic-Williams S., Shapiro J. A., Caron L., Rabin D. U.: ICA 512, an autoantigen of type I diabetes, is an intrinsic membrane protein of neurosecretory granules. *EMBO J.,* 15:2102–2114 (1996).

[84] Morran M. P., Vonberg A., Khadra A., Pietropaolo M. Immunogenetics of type 1 diabetes mellitus. *Molecular Aspects of Medicine*, 42:42–60 (2015).

In: Alterations in Glucose Homeostasis in Children
Editor: E. Otto-Buczkowska
ISBN: 978-1-63483-861-0

Chapter 2

INTENSIVE INSULIN THERAPY IN CLINICAL PRACTICE

A. Szypowska*
Medical University of Warsaw, Warsaw, Poland

ABSTRACT

Achieving tight glycemic control without increasing the incidence of hypoglycemia is challenging in type 1 diabetic children and adolescents; therefore insulin analogues associated with low incidence of hypoglycemia have been recommended. The new long-lasting insulin analogues (e.g., insulin degludec and insulin glargine 300 units/ml) offer tight glycemic control and further lower the risk of hypoglycemia.

Continuous subcutaneous insulin infusion (CSII) is the most physiologic method of insulin delivery and offers more precise prandial insulin administration and modification of basal insulin infusion in comparison with multiple daily injections (MDI). Flexibility of insulin pump therapy allows adjustment of basal insulin pattern to the patient's individual needs and facilitates maintenance of the patient's glucose profile during serious illness or physical activity.

Poor nutrition influences incidence of insulin overdose, obesity and insulin resistance; therefore, balanced meals and accurate carbohydrate/fat/protein/calorie counting is crucial to reduce postprandial glucose variability. Our algorithm calculates insulin doses to reduce postprandial hyperglycemia. This algorithm is based on fat/protein content in a meal.

Real-time continuous glucose monitoring systems (RT-CGM) transmit glucose measurements from subcutaneous tissue to a receiver. RT-CGM users receive not only data about glucose values but also information on glucose trends, if levels are rising or falling. This data is helpful in making diabetes management decisions to enable better glucose level control. Meta-analysis showed that appropriate use of RT-CGM decreased glucose variability with a marked reduction in the risk of hypoglycemia in CSII and MDI patients. However, low effectiveness of RT-CGM usage was observed in children and adolescents due to a lack of adherence to RT-CGM therapy.

* E-mail: agnieszka.szypowska@gmail.com.

The randomized study showed that automatic suspension of insulin in the insulin pump combined with RT-CGM and with a low-glucose suspend function significantly reduced the duration and severity of hypoglycemia without causing rebound hyperglycemia.

Unfortunately, despite the use of modern intensive insulin therapy many patients are unable to reach their glycemic goal. Therefore, automated closed-loop insulin delivery is a promising option to maintain glycemic control with a low risk of hypoglycemia. Closed-loop insulin delivery as well as bihormonal pancreas, which automatically deliver insulin and glucagon, showed a significantly higher proportion of time with glucose in the target range and lower rate of hypoglycemia compared to conventional therapy. Currently tested artificial pancreas does not provide complete normoglycemia as well as does not protect completely against hypoglycemia, so the question is which patients will benefit the most from automated insulin delivery.

Based on literature and personal experience, this chapter will discuss opportunities offered by current intensive insulin therapy as well as its limitations.

Keywords: Insulin analogues, exchange counting, insulin pump, continuous glucose monitoring, closed-loop

INTRODUCTION

Type 1 diabetes mellitus is an autoimmune disease causing pancreatic beta cell destruction and deficiency of exogenous insulin. The annual incidence of type 1 diabetes in children has been increasing worldwide by 3-5% per year with the most remarkable rise in incidence among the youngest age groups [1].

Current strategies for the treatment of type 1 diabetes mellitus with insulin involve the use of intensive insulin therapy to maintain near-normoglycemia in order to prevent long-term life threatening complications. This method of insulin administration can be provided with multiple daily injections or subcutaneous insulin infusion. A major barrier to achieving good glycemic control is hypoglycemia.

The burden of diabetes management requires great effort and constant vigilance. Children with diabetes and their families face tension and anxiety at all times. Currently available modern technology (such as insulin analogues, insulin pumps, continuous glucose monitoring and computer software) provide information about fluctuations in blood glucose to allow for a precise dosage of insulin and reduction of hypoglycemic episodes. However, the dose of insulin is still determined by parents and children and therefore, a high level of education is crucial in maintaining good control of diabetes. Families should be able to adjust insulin dose according to an individual child's needs.

Children with type 1 diabetes should be under the care of a diabetes team, including a doctor, nurse, dietitian and psychologist. Solving the current problems of children with diabetes and their families is imperative to achieve good diabetes control.

INSULIN ANALOGUES IN DIABETES MANAGEMENT

Achieving tight glycemic control without increasing the incidence of hypoglycemia is challenging in type 1 diabetic children and adolescents. Optimal insulin therapy should cover the needs of children in different age groups. Management of young children poses difficulties due to unpredictable food intake and exercise, frequent infections, limited ability to verbalize symptoms of hypoglycemia and difficulties in ensuring good diabetes control at kindergarten or school. Adolescence is a time of rapid physical and psychological change as well as changing lifestyle – for example, teenagers do not want to be dependent upon their parents for blood tests and insulin administration. Moreover, in this age group and younger, fear of hypoglycemia, variable eating patterns and physical activity influence any insulin regimen. Insulin analogues in comparison with human insulin improve management of type 1 diabetic patients with respect to glycemic control and avoidance of hypoglycemia [2, 3]. Three short acting insulin analogues (insulin aspart, insulin lispro and insulin glulisine), which mimic the fast postprandial insulin release and two long-lasting insulin analogues (insulin detemir and insulin glargine) with minimal peak action are currently used in the treatment of children with type 1 diabetes [4]. Some studies performed in the pediatric population showed that insulin analogues, when compared with human insulin, were associated with similar glycemic control but offered more flexibility, lower rates of hypoglycemia and a decrease in postprandial hyperglycemia [5, 6]. Other authors indicated that treatment with insulin analogues compared with human insulin resulted in significantly better HbA1c values and lower risk of severe hypoglycemia in children and adolescents with type 1 diabetes (T1D) [7].

Despite the use of insulin analogues, hypoglycemia is still a major barrier to achieving tight glycemic control [8]. The increased risk of hypoglycemia is associated with intensive insulin regimen and glycated hemoglobin (HbA1c) reduction. Well-controlled type 1 diabetic patients achieving HbA1c <7% might spend an average of 90 min per day with a glucose level ≤70 mg/dl (3.9mmol/l) [9]. Hypoglycemia is associated with brain damage and cognitive deficits in type 1 diabetic children [10]. Therefore, the new insulin analogues which offer tight glycemic control and further lower the risk of hypoglycemia are required.

Insulin degludec is a novel ultra-long-acting basal insulin analogue that does not have the threonine at B30 position as in human insulin. The addition of hexadecanedioic acid to lysine at the B29 position allows formation of multi-hexamers in subcutaneous tissues from which insulin monomers are slowly and continuously absorbed into the blood. The stable and flat action profile causes a glucose-lowering effect, which lasts beyond 42 hours [11]. A meta-analysis of 1846 adults with type 1 diabetes showed that basal-bolus treatment with insulin degludec was superior to long-acting insulin analogues detemir and glargine in reducing the risk of increased nocturnal rate of hypoglycemia [12]. Moreover, compared to other long-acting analogues, treatment with insulin degludec reduced basal and total insulin dose with no influence on HbA1c and fasting plasma glucose.

A randomized study comparing insulin degludec with insulin detemir and prandial insulin was performed in T1D children in different age groups: 85 subjects 1-5 years old, 138 subjects 6-11 years old and 127 subjects 12-17 years old [13]. The result of the study showed that treatment with insulin degludec did not influence any statistical difference in the rate of hypoglycemia nor HbA1c value when compared with insulin detemir. However, children treated with insulin degludec achieved a significant reduction in fasting plasma glucose. A reduction in fasting plasma glucose is influenced by basal insulin activity while HbA1c values are dependent much more on postprandial glucose reduction than fasting plasma glucose reduction. Moreover, in this study a 30% reduction in basal insulin dose was observed in children using insulin degludec in comparison with insulin detemir. A reduction of the amount of insulin required seems to have a positive effect on diabetes management. Furthermore, this study observed a prolonged, stable profile of insulin degludec and flexibility in the timing of injection intervals resulting in a significant decrease in hyperglycemia with ketosis compared with the detemir group.

Another new long-acting insulin analogue – insulin glargine 300 units/ml – has been developed. A current study performed on adults with type 1 diabetes confirmed that single-dose injections of insulin glargine 300 units/ml provided a flatter and prolonged profile compared with insulin glargine 100 units/ml, maintaining glycemic control for up to 36 h in euglycemic clamp settings [14].

Despite the use of insulin analogues, some type 1 diabetic patients are unsatisfactorily controlled with multiple daily injections. The meta-analysis of 11 randomized trials showed reduction of HbA1c in subjects treated with CSII when compared to MDI [15].

Rapid acting insulin analogues are used in insulin pump therapy. Van Bone et al. showed in a randomized study no significant difference among three rapid insulin analogues (insulin glulisine, insulin aspart and insulin lispro) administered by CSII with respect to glycated hemoglobin, seven-point glucose curves, severe hypoglycemia, symptomatic ketoacidosis, daily insulin dose and infusion set occlusion [16]. In this study, insulin glulisine was associated with a higher frequency of symptomatic hypoglycemia than lispro and aspart, possibly because of slight overdosing, as previous trials suggested, due to lower insulin requirements when glulisine is initiated in type 1 diabetic patients.

Continuous Subcutaneous Insulin Infusion in Different Age Groups

Over the past decade, the use of insulin pumps has increased among patients with diabetes. Continuous subcutaneous insulin infusion (CSII) is the most physiologic method of insulin delivery. CSII offers more precise insulin administration than multiple daily injections. Insulin pumps continuously deliver small doses of rapid-acting insulin analogues called “basal insulin.” Basal insulin covers glucose flow from the liver while prandial insulin is programmed before each meal and called “bolus insulin.”

CSII therapy is especially beneficial in young children who require administration of very small insulin doses. Therefore, initiating continuous subcutaneous insulin infusion right after diabetes diagnosis helps to overcome difficulties in diabetes management in this age group [17]. A common problem in the treatment of a young child is non-acceptance of multiple injections. Insulin pump therapy reduces the number of injections, improving the quality of life of the whole family and facilitating childcare in kindergarten or at school. Another common problem in the management of young diabetic children is the unpredictable variation in food intake. A child may eat all or only a part of a prepared meal. In CSII therapy, a prandial insulin dose can be taken in parts. Thereby parents can give part of the insulin before a meal and give the remainder of the dose if the whole meal is eaten. Such a dosing regimen of insulin significantly reduces the risk of hypoglycemia [18]. Clinical practice shows that many adolescents also prefer CSII to MDI therapy. Teenagers using an insulin pump from diabetes onset report high satisfaction level, flexibility and easy adaptation of insulin dosing. Furthermore, we observe low rates of pump discontinuation in all age groups, which confirms the beneficial effects of this method of insulin administration in diabetic children.

Impact of Nutrition on Insulin Therapy

Insulin pump therapy allows the use of three different types of boluses: normal, square-wave and dual-wave for different food compositions. Normal bolus delivers insulin immediately and is used to cover carbohydrates. Square-wave bolus delivers prandial insulin over a period of time and should be selected for fat/protein-containing food. Dual-wave bolus is a combination of normal bolus followed by square-wave bolus and is used for mixed meals.

Although CSII therapy offers a more precise prandial insulin administration than MDI, this method of insulin delivery has also some limitations. Therefore, the correct composition of a meal is crucial to reduce postprandial glucose variability. A meal should contain 50% of calories from protein and fat and 50% of calories from carbohydrates. Moreover, patients should avoid foods with a high glycemic index, which cause blood glucose fluctuations [19]. Foods with a high glycemic index are absorbed quickly from the gastrointestinal tract, causing rapid fluctuations in blood glucose levels. In the first hour after the ingestion of a food with a high glycemic index, there is rapidly increasing hyperglycemia and, after two hours, a sudden drop in blood sugar, which may lead to hypoglycemia.

Our own experience shows that Polish children's diet before the diagnosis of diabetes very often does not adhere to the principles of healthy nutrition. Children receive large amounts of sweets, sugary drinks, sweetened cheese and sweetened yoghurts. They eat very few or no vegetables, sometimes eat fruit, but with preference for bananas. Consequently, a change in diet may not be easily accepted by a child initially. Other groups are young children (infants, toddlers) whose diet is high in carbohydrate products, which often have a high glycemic index. Some children require insulin treatment to be started during the neonatal or infant period when they are breast-fed. In this case, it is necessary to evaluate the amount of food consumed by weighing the baby before and after meals. Treatment with continuous subcutaneous insulin infusion in these patients leads to good glycemic control as well as normal growth and development [20]. Due to the growing problem of obesity among children, attention should also be directed to the caloric content of foods as not only the quantity but also the quality of the food affects the daily prandial insulin dose. Poor nutrition influences insulin overdose, obesity and insulin resistance and it has been confirmed that insulin resistance is associated with chronic complications in type 1 diabetes [21].

Insulin overdose and insulin resistance was observed in our 16-year-old girl with type 1 diabetes who radically changed her lifestyle. Type 1 diabetes was recognized in this girl in January 2010 and she started insulin pump therapy in May 2010. In February 2012, she was diagnosed with cataracts in both eyes. Before 2013, she had very good glycemic control with the mean HbA1c around 6.5%, without any episodes of severe hypoglycemia or ketoacidosis. In 2013, she started studying in another city, living independently with others, sharing meals.

She completely changed her lifestyle, started eating food rich in carbohydrates and sugar and reduced physical activity. The new lifestyle caused an increase in insulin requirement from 0.42 U/kg to 0.86 U/kg and an increase in BMI from 22.4 kg/m2 to 24.7 kg/m2 (in an 8-month period her weight increased by 8 kg). She presented with black hyperpigmentation of the skin on her neck, diagnosed by a dermatologist as acanthosis nigricans. She also had one episode of severe hypoglycemia due to insulin overdose. During the 8-month period, her HbA1c levels remained excellent (7%, 5.9%, 6.2%). Due to the deterioration in metabolic control, her mother decided to take her home and she resumed healthy eating and physical activity. Her daily insulin dose decreased to 0.44 U/kg, skin hyperpigmentation disappeared.

This example shows that good HbA1c does not always mean good metabolic control. In particular, insulin pump therapy facilitates eating large amounts of snacks resulting in a lot of boluses. Therefore, it is important for healthcare workers to download insulin pump data and glucometer at each control visit and to discuss insulin management with the patient.

Prandial Insulin Calculation

It is well known that protein and fat consumption leads to a prolonged increase in glucose concentration. We observed that patients who do not measure carbohydrate/fat/protein intake often increase their basal insulin to cover glucose from fat and protein. Too high a dose of basal insulin causes hypoglycemia. Therefore, in our opinion, the insulin dose calculation is crucial in reducing glucose level fluctuations to achieve tight glycemic control.

Our algorithm to calculate insulin dose for a meal consisting of fat and protein is based on the fat/protein content in a meal. We defined 1 fat/protein exchange (FPE) as 100 kcal of fat and/or protein. The insulin dose is calculated using an insulin-to-FPE ratio equal to the insulin-to-carbohydrate ratio. The duration of prolonged bolus = amount of FPE + 2 hours (e.g., 3 h for 1 FPE, 4 h for 2 FPE, 5 h for 3 FPE).

We do not recommend giving insulin for less than 1 FEP. Moreover, the youngest children as well as children in the "honeymoon period" may not need insulin for FEP, especially when the number of FEP is low. An insulin dose is also not adjusted for FEP when a meal is consumed before physical activity.

Our study showed the beneficial effect of the FEP calculation on metabolic control [22]. We noted the lowest HbA1c value in the group using two and/or more dual-wave/square-wave boluses per day compared with the group administrating less than one dual-wave/square-wave bolus per day. In our study, children with an HbA1c level <7.5% had a statistically higher number of dual-wave/square-wave boluses.

The beneficial effect on the FEP calculation was also confirmed by Kordonouri et al., who investigated the efficacy of carbohydrate plus fat/protein counting compared with the conventional carbohydrate counting using normal and dual-wave bolus in sensor augmented pump therapy [23]. Authors used our algorithm for fat/protein counting. In this study, children consumed a pizza-salami meal (50% carbohydrate, 34% fat, 16% protein; corresponding to 33% of their age-adjusted daily energy requirement) at lunchtime on four different days with normal and dual-wave bolus using carbohydrate/fat/protein counting and conventional carbohydrate counting in a randomized sequence. This study confirmed that carbohydrate/fat/protein counting led to significantly lower postprandial glucose levels independent from the type of bolus. The authors noted more hypoglycemic events in children using carbohydrate/fat/protein counting, probably due to a high proportion of basal insulin. Therefore, a check of the basal insulin was recommended before start of carbohydrate/fat/protein counting (e.g., a 6-hour fasting test).

The prandial insulin dose self-calculation usually does not take into account the active insulin, which can cause postprandial hypoglycemia. This is particularly important in children who snack between meals. The use of the bolus calculator feature on the insulin pump can simplify prandial insulin dosing. When the number of exchanges eaten are entered, the bolus calculator calculates the final dose of prandial insulin to be given. The final calculation takes into account the previously determined parameters entered into the insulin pump: insulin per exchange ratio for a specific time of the day, insulin sensitivity, active insulin and current and target blood glucose. Exchanges still need to be counted when using the calculator – when exchanges are incorrectly calculated the final dose of insulin will also be incorrectly calculated. A limitation of the bolus calculator is the lack of an option to select different insulin per exchange ratios depending on the glycemic index of the product. In addition, the bolus calculator does not account for fat/protein exchanges. Therefore, you must enter the entire number of exchanges (carbohydrate + fat/protein) and then divide the bolus dose of insulin manually into normal and square-wave bolus. Our research and the literature suggest that the use of a bolus calculator feature is particularly important in reducing the incidence of hypoglycemia [24]. Moreover, some authors noted a significant reduction in daily post-prandial blood glucose values in the group using the bolus calculator [25].

BASAL INSULIN PATTERN

CSII therapy allows the modification of basal insulin infusion. In a person with normal pancreatic beta cell function, insulin is secreted to cover the influx of glucose from food, and the influx of glucose from glycogen in the liver. Consequently, insulin infusion should be as physiological as possible, e.g., it should imitate nature by giving a bolus of insulin during meals, and provide basal insulin between meals and during the night. Thanks to this physiological insulin infusion there is greater freedom to move meal times, and change both the quantity and the type of food eaten. In the event that a meal is missed, a low basal insulin infusion reduces the risk of hypoglycemia. A completely different situation arises if the programmed basal insulin infusion dose is high; it then covers not only base needs, but also to a large extent prandial needs, however skipping meals, or changing meal times, increases the risk of hypoglycemia. In this case it may also be necessary to significantly modify the basal insulin during exercise or during an infection of the gastrointestinal tract.

There are certain differences in the basal insulin pattern of younger and older children with type 1 diabetes.

Unlike older children, who require their basal insulin infusion to be approximately 30-50% of their daily dose of insulin, basal insulin infusion in an infant usually does not exceed 0-30% of their total daily dose of insulin. The following are significant differences between children under five years of age and older children when it comes to planning the basal insulin pattern. In the youngest age group, requirements of basal insulin around 3:00-8:00 in the morning decrease but often slightly increase around 21:00-24:00. With age, there is an increasing demand for insulin in the morning because of the dawn phenomenon. Typically in school-age children it is necessary to program a higher base around 3:00-9:00 in the morning [26]. The insulin pump enables the programming of several bases such as, aside from the standard base, a higher base programmed for febrile illness, or a lower one for physical effort. If there is a need to change the dose of basal insulin for a certain period of time, the patient should start increasing/decreasing the dose of insulin for about two hours before the occurrence of the high/low glucose level. Our experience shows that hourly differences between successive doses of insulin should not be large, normally the maximal difference of 0.2 unit / hour is sufficient. Parents have responsibility for manipulating the basal insulin dose, so physicians need to pay special attention that there are no insulin peaks in the base as hypoglycemia could result. For the safety of young children, the maximum basal and prandial insulin dose should be programmed by the physician.

On days when completing different activities to normal, subjects should take advantage of the option to change the basal insulin dose temporarily. The simplest and most common function used is a percentage change in the existing basal insulin dose.

Treatment with an insulin pump enables patients to stop or increase the flow of the basal insulin infusion at any time. With this ability, patients can easily correct blood glucose values e.g., during increased activity or illness. For a child, often going to the playground is linked with the need to discontinue or decrease the basal rate.

FLEXIBILITY OF INSULIN PUMP THERAPY DURING ILLNESS

A febrile illness greatly increases the need for insulin and there is a need to increase the basal insulin rate to 130% or 200% of the demand before the illness. Increasing the basal insulin to 200% seems to be a very large increase, but e.g., small children often have a basal insulin rate of 0.05unit / hour, in this case an increase of 100% basal rate is only 0.1 unit/hour. Older children usually need to increase the basal rate to 130%-150%. When recovering from a febrile illness, the insulin requirement decreases and it is necessary at this time to be very careful, because hypoglycemia, including nocturnal hypoglycemia, may occur due to the sudden drop in demand for insulin.

The flexibility of insulin pump therapy also helps to maintain a glucose profile during a serious illness. We achieved very good glycemic control using CSII therapy during parenteral nutrition and treatment of systemic infection in our patient - an 11-year-old boy. The patient has had type 1 diabetes since 2012, treated with continuous subcutaneous insulin infusion, with good glycemic control (HbA1c 5.6%). His daily insulin requirement was 0.7 unit/kg, 10% of total insulin dose was delivered as basal insulin. During annual check-ups, laboratory tests showed increased level of anti-transglutaminase antibodies and the boy was qualified for gastroscopy to confirm celiac disease. Unfortunately, the gastroscopy was complicated by large duodenal hematoma located in the intestinal wall, measuring 12x3.6x12 cm. Moreover, systemic infection caused by Staphylococcus warneri was confirmed and intensive antibiotic treatment was ordered. In the treatment total parenteral nutrition was used. The daily insulin requirement increased to 1.2 unit/kg and 90% of the daily insulin was administered as basal insulin. The number of blood glucose daily measurements increased to 12-14 fingerpicks.

During the treatment period his mean blood glucose was mostly around 130-170 mg/dl (7.4-9.4mmol/L), without high variability and without hypoglycemic episodes.

Disqualification of Insulin Pump Therapy

Therapy with an insulin pump requires a patient to be under the constant care of the diabetes team, and should not be conducted in children whose families do not work with the diabetes center, do not follow the recommendations of the diabetologist, or do not report to the follow-up visits. In the teenage group, the most common problem is lack of cooperation with the diabetes team, and non-adherence to the instructions, which results in persistent poor control of diabetes and is an indication for disqualification from treatment with an insulin pump. Disqualification of treatment may be necessary in children with eating disorders and mental illness. Some patients are not willing to connect the insulin pump because they consider wearing the equipment inconvenient and cumbersome.

The main reason for disqualification from treatment with a continuous subcutaneous insulin infusion in the youngest group of patients is not being able to educate the child's parents about insulin therapy and insulin pump use. Our experience suggests that difficulties in training parents are rare and usually occur with people with learning difficulties. Occasionally parents do not give consent to start therapy with continuous subcutaneous insulin infusion, such a decision disqualifies a child from this method of treatment. In this situation, it is worthwhile considering the cause of the parents' negative reaction to the offered therapy. It may be due to, for example, anxiety that they will not be able to cope with the insulin pump service. It is extremely important to talk with parents, and identify reasons for their resignation from pump therapy and help resolve their issues as doing so may alter the parents' decision and allow the child to benefit from treatment with continuous subcutaneous insulin infusion.

In general, the discontinuation rate of insulin pump therapy is low [27]. Wong et al. investigated the relationship between depressive symptoms measured by the Children's Depression Inventory (CDI), method of insulin delivery and glycated hemoglobin [28]. The authors analyzed data from 150 adolescents with type 1 diabetes, of whom 63% used CSII. The 2 year observation showed that baseline pump use was associated with -0.79% lower mean HbA1c. However the mean HbA1c of 9 adolescents switching to MDI during the study was 1.38% higher than that for those who did not switch the method of insulin delivery. Moreover, a 10% increase in CDI was associated with a 0.39% increase in HbA1c, independent of pump use, and a previously higher CDI score was associated with switching from pump to MDI. The authors concluded that clinicians should be aware of the associations between depressive symptoms and insulin pump discontinuation.

De Vries et al. showed that 11.3% of 530 patients with type 1 diabetes had discontinued pump use after 3 to 5 yr.; of these, 9.1% discontinued pump use at least 3 months after initiation [29]. In this group the female gender, age older than 10 years in girls and poor metabolic control at pump initiation were associated with higher risk of pump discontinuation. In both studies authors recommended intensified individual and family support for such patients.

A different issue in the qualification for the treatment with continuous subcutaneous insulin infusion is intellectual disability in children. In such situations, the decision should be made on an individual basis. Our own experience shows that connecting the insulin pump to children with intellectual disability , e.g., cerebral palsy, trisomy 21, who were under care of their parents, gave very good, long-term diabetes control. On the other hand, the above method of treatment in children with autism has failed in spite of many attempts. The child did not accept the insulin pump, removed the cannula, and was very anxious when any attempt to connect the equipment was made.

A problem with the use of continuous subcutaneous insulin infusion may occur in children with severe skin allergies. In our data, none of the children with atopic dermatitis had indications (e.g., due to the increasing severity of allergy symptoms) for disconnecting the insulin pump. Additional diseases associated with diabetes, which may significantly affect the choice of treatment, rarely occur in children with diabetes.

If other diseases are present, an individual approach to the patient is necessary. Treatment must be tailored to ensure the best therapeutic results, and the comfort of the patient and his family.

FLEXIBILITY OF INSULIN PUMP USE DURING EXERCISE

The general recommendations for exercise in children and adolescents is at least 60 minutes or more of daily physical activity with moderate to vigorous aerobic activity as well as muscle or bone strengthening exercises [30]. The use of exercise appears to be efficacious in reducing percent body fat, improves insulin sensitivity and reduces cardiovascular disease risk factors in children and adolescents [31, 32]. Moreover, studies also demonstrated that physical activity is positively associated with cognition, academic achievements, psychosocial functioning in children and may be both protective against the onset of as well as beneficial for reducing depressive symptoms [33, 34]. All these beneficial effects of physical activity on children's health play an important role in the treatment of the type 1 diabetic youth.

Our previous study showed a high prevalence of overweight/obesity and dyslipidemia in children with type 1 diabetes including those treated with an insulin pump [35]. Moreover, we have shown that in both groups (good and poor metabolic control) 1 in 5 adolescents with type 1 diabetes had depressive symptoms [36].

Unfortunately, type 1 diabetic patients are often less active than individuals without diabetes. One of the barriers to physical activity, except environmental conditions, may be glycemic variation and fear of hypoglycemia both during and after (particularly overnight) exercise [37]. Many different factors as fixed basal/bolus insulin regimens, the type, duration of exercise and its temporal relation to meals and premeal insulin doses can led to hypoglycemia. However, hyperglycemia, with or without ketosis, may also occur during physical activity. Due to a number of factors affecting glucose during and at the end of exercise there is no uniform standards for insulin dose adjustment, which would ensure normoglycemia.

Some studies highlight the flexibility of CSII use in comparison to MDI regimen during exercise. Yardley et al. compared glucose levels during 45 min aerobic exercise and 6 hours of post-exercise recovery between T1D subjects using MDI and CSII [38]. The authors showed no difference between groups in glucose levels during exercise. However, less post-exercise hyperglycemia was observed in CSII group than MDI patients. In this study two-thirds of subjects treated with MDI experienced late-onset post-exercise hyperglycemia compared with only 1 of 10 of the CSII patients.

Insulin pump therapy gives the possibility to maintain the insulin basal dose (reduce, suspend) to prevent hypoglycemia during exercise in children. The proper adjustment of the basal rate of insulin infusion during exercise is still not clear. Tsalikian et al. evaluated whether complete suspension of basal insulin infusion could effectively prevent hypoglycemia during moderate aerobic exercise [39]. The authors noted less frequently hypoglycemia when the basal insulin was discontinued than when it was continued, however hyperglycemia without cases of blood ketone levels was more frequent without basal insulin. In this study 49 children between 8 and 17 years of age were observed during two days exercise sessions; on one day basal insulin was stopped during exercise and on the second day was continued. Each exercise session consisted of four 15-min treadmill cycles in the afternoon, followed by a 45-min observation period. In this study, the last premeal bolus was given four hours before physical activity, therefore the subcutaneous depot of rapid-acting insulin analog was low. The authors suggested that it is easier to treat hypoglycemia with 15–30g carbohydrate if the basal insulin has been temporarily reduced.

On the other hand, to achieve normoglycemia during exercise active insulin must be present. If the child has regular exercise his basal insulin requirement is lowered and active insulin comes mainly from prandial insulin. Intensive training for four hours after premeal bolus with a disconnected insulin pump might lead to hyperglycemia.

Due to insufficient active insulin during exercise we observed recurrent hyperglycemia in our 13 year-old patient, trainee football in a sports club. The boy was treated with an insulin pump, his total daily insulin dose was 0.7 unit/kg, basal insulin was 20% of total daily dose and HbA1c 7.2%. On the day of football training the boy was given prandial insulin around 10 am at school. At 2 pm he had a 1.5 hour football match. The boy always checked glucose levels before training and disconnected the insulin pump. The boy began training with glucose of less than 180mg/dl (10mmol/L), but after training has always had hyperglycemia over 300mg/dl (16.7mmol/L). Normalization of glycemia began when the boy started eating lunch at school around 12 am. An additional dose of insulin given 2 hours before training increased the amount of active insulin during exercise, which resulted in normalization of blood glucose levels.

Current studies pointed out that a too long a basal rate discontinuation during exercise might lead to hyperglycemia. Jankovec et al. showed that physical activity during a 3-hour interruption of the insulin pump proved to increase blood glucose and ketone bodies. The correction bolus normalized blood glucose and ketone bodies within another 90 min [40].

A real life situation may occur in the course of the summer holiday when a child is staying all day at the beach. Most of the insulin pumps are not waterproof and must be disconnected when children bathe in the water. Young children may have a needle-phobia and may not tolerate injections. In this situation parents are asked to serve meals at regular intervals. Prandial boluses administered by an insulin pump ensure the presence of active insulin in the body, which prevents hyperglycemia. Some of the older children during holidays give insulin using multiple daily injections instead of an insulin pump. Others administer long-acting insulin analog instead of basal insulin and adjust prandial insulin using the insulin pump. The methods of insulin delivery during the holidays should be individually tailored to the needs of a child.

Children often perform exercise shortly after having a meal. Rapid-acting insulin analog has a peak of activity one hour after administration. Therefore, physical activity shortly after a meal may lead to hypoglycemia. Previous studies tried to answer the question if during exercise, that occur shortly after a mealtime, should pump users reduce a basal or prandial insulin dose.

Admon et al. investigated glucose fluctuation during unplanned exercise after breakfast full-dose insulin bolus when the basal insulin was reduced by 50% or stopped [41]. The study was performed on 10 T1D children between 10 and 19 years of age. The authors showed no difference in the number of hypoglycemia when exercising with the pump on 50% of the basal rate or pump off. In both exercise sessions 2 children developed hypoglycemia. In both groups more subjects had late hypoglycemia. The authors concluded that basal rate discontinuation did not prevent hypoglycemia when physical activity is performed short after a meal. In this situation prandial insulin reduction might be more effective.

Sonnenberg et al. showed that hypoglycemia during 60 min moderate exercise on a cycle ergometer was prevented in patients who discontinued basal rate and reduced premeal insulin bolus by 50% [42]. The authors pointed out that 25% basal rate reduction effectively prevented late hypoglycemia during post exercise hours.

Although there is no single insulin dose reduction scheme for pump users that protects the patient against hypoglycemia the use of CSII facilitates insulin dose adjustment according to individual patient needs.

Teenagers with type 1 diabetes often exercise in the gym completing small amounts of anaerobic exercise (e.g., weight lifting) can decrease blood glucose reduction caused by aerobic exercise. However, intense aerobic exercise causes hypoglycemia and requires extra doses of insulin [43].

Ramalho et al. evaluated the effect of aerobic versus resistance training on metabolic control in 13 type 1 diabetic patients in the age 13-30 year-old [44]. In both groups the total insulin dosage and waist circumference was reduced without improvement in HbA1c values. The authors concluded that resistance exercise might offer metabolic benefits with low risk of hypoglycemia.

Some authors noted that the time of day affects the effectiveness of exercise. Gomez et al. compared the influence of morning and afternoon physical activity on glycemic control and hypoglycemia [45]. The study was performed on type 1 diabetic adults treated with sensor-augmented insulin pump. Patients underwent one moderate-intensity exercise sessions in the morning and one in the afternoon. The authors noted most hypoglycemic events 15-24 hours after the exercise session. Late hypoglycemia occurred more frequently in patients exercising in the evening than in the morning.

CONTINUOUS GLUCOSE MONITORING

It is well known that in patients with type 1 diabetes the number of self-monitored blood glucose correlates with glycemic control [46]. Parents of young children perform 8-10 or more glucose measurements per day, older children usually less frequently measure blood glucose. The frequency of blood glucose monitoring is not only important, but also the timing of the test. Children are usually reluctant to measure the concentration of blood glucose at school. When blood glucose self-monitoring is performed irregularly, e.g., few blood glucose measurements at school, with more tests in the afternoon, patients do not get enough information about the actual blood glucose curve.

Continuous blood glucose monitoring (CGM) offers the measurement of interstitial glucose in an ongoing fashion. Meta-analysis showed that randomized controlled trials using blinded CGM compared with blood glucose monitoring did not show significant reduction in HbA1c in T1D patient [47]. Real-time devices for continuous glucose monitoring (RT-CGM) provide important information about glucose trends, so patients can adjust the dose of insulin before the appearance of hypo or hyperglycemia preventing glucose fluctuations. Currently used RT-CGM systems offer sounding alarms when the glucose trend projects future hyper or hypoglycemia. However, patients notice there is some limitation of RT-CGM sensor use such as low accuracy, problems with calibration or skin irritation. Moreover, the cost of RT-CGM is still very high and is not reimbursed or, if given the reimbursement, is limited [48].

Research has confirmed that excellent control from diabetes onset influence “metabolic memory” and reduce the risk of long-tem diabetes complications [49]. RT-CGM implementation from T1D diagnosis might improve glycemic control. Kordonouri et al. compared children with type 1 diabetes using sensor-augmented insulin pump (a combination of insulin pump and continuous glucose monitoring system) at the onset of diagnosis with children treated with conventional insulin pump [50]. This study included 154 children with the mean age at the onset 8.8 years. Children who were able to use RT-CGM regularly (at least one sensor per week during the first year) had lower HbA1c levels than those who did not (7.1% vs. 7.6% respectively). Nineteen of 37 (51,4%) children with regular sensor use achieved HbA1c below 7% without episodes of severe hypoglycemia. Moreover, the use of sensor-augmented insulin pump was associated with a better preservation of endogenous insulin secretion after 1 year of diabetes. Children using sensor-augmented pump therapy had lower glycemic variability, which might have protective effects on beta cell function. Unfortunately, the acceptance of RT-CGM use has decreased over time, as observed in the progressive decline in the use of sensors from the initial frequency of 2.1 sensors per week to 1.1 at 52 weeks. The number of patients who used RT-CGM regularly (48%) was too low to affect the overall HbA1c values.

Buckingham et al. found a similar association between higher C-peptide level and increased sensor glucose level in therapeutic goal and lower glucose fluctuations in the first 2 years after T1D diagnosis [51].

In the Juvenile Diabetes Research Foundation Continuous Glucose Monitoring Study Group randomized controlled trial the authors noted low compliance with RT-CGM use in T1D children and adolescents [52]. Included in the study were 322 adults and children, all with type 1 diabetes of at least 1 year before randomization. After 26 weeks of RT-CGM use the statistical decrease in HbA1c was observed only in adults, with no benefit achieved by teenagers and young adults who were 15 to 24 years of age. A small 0.37%, non statistical decline in HbA1c was noted in children between 8 to 14 years of age. The greatest use of sensors as well as consistently high sensor use was noted only in adults over 25 years of age. At least 6.0 days of sensor use per week was noted in 83% adults over 25 years of age, 50% children between 8 to 14 years of age and only 30% teenagers and young adults between 15 to 24 year-old.

The authors concluded that greater parental care could influence greater sensor use by children below 14 years of age in comparison with subjects between the ages of 15 and 24 years.

Several studies implied that the use of RT-CGM above 70% of the time influenced an increased beneficial effect of the system in all age groups [53, 54]. Children with >70% sensor usage missed fewer school days; compared with the sensor-off arm and the addition of RT-CGM to CSII resulted in better metabolic control and decreased time spend in hypoglycemia [55].

Randomized studies indicated that children might benefit from RT-CGM use. However, it seems that the RT-CGM is not friendly enough and the low compliance might be related to the inconvenience of wearing the device. Moreover, families may have problems with the interpretation of data obtained from RT-CGM and the use of this technology may be too high a burden for some patients.

The great compliance (86%) in sensor use was achieved by Battelino et al., who included 120 well controlled (HbA1c<7.5%) adolescents and adults on intensive therapy for type 1 diabetes [56]. In the study, during the six-month period median sensor wear was 6.1 days per week in both pediatric and adult group. Patients in RT-CGM group had a significant 50% reduction in time spent in hypoglycemia as well as significantly longer time spent in normoglycemia defined as glycemic value 70 to 180 mg/dl (3.9-10 mmol/L). Moreover, despite the good glycemic control at baseline the mean HbA1c was reduced at 26 weeks in adolescents (10–17 years of age) by -0.23 (7.15% vs. 6.92%). No incidence of severe hypoglycemia was noted. In this study the authors showed that highly motivated adolescents, demonstrating good metabolic control might achieve great benefits from RT-CGM use.

Mauras et al. investigated the efficacy, safety and effect of RT-CGM on glycemic control and quality of life in young children, aged 4 to 9.9 years [57]. Included in the study were 146 children with mean HbA1c 7.9%, 64% of subjects used insulin pump therapy. At 26 weeks the mean change in HbA1c was similar between groups (-0.1%). Moreover in both groups glucose values over 250 mg% occured for over 20% of the day. The amount of sensor wear declined during the follow-up, only 41% of children used RT-CGM at least 6 days/week in month 6. Among the 28 children with good compliance, using RT-CGM at least 6 days per week, a slightly greater reduction in HbA1c (-0.3%) was observed. In both groups patients experienced severe hypoglycemia, without between-group differences; there was no case of diabetic ketoacidosis. The authors noted high parental satisfaction with RT-CGM. However, the use of RT-CGM was not related to decrease in parental fear of hypoglycemia. The authors concluded that the lack in glycemic control improvement may be related to low compliance with sensor use as well as with fear of hypoglycemia. Our experience also shows that the use of RT-CGM does not improve control of diabetes in people with severe fear of hypoglycemia.

RT-CGM combined with insulin pump MiniMed® Veo™ and an automated insulin suspension algorithm can reduce the number of hypoglycemic episodes. Data from a sensor, which measures glucose values in the subcutaneous liquid, is sent to an insulin pump. When a sensor detects glucose decreasing to hypoglycemia level, a pump emits a sounding alarm. If the patient does not respond to the alarm insulin suspension function automatically stops insulin infusion for up to 2 hours, after which basal insulin delivery resumes. This technology reduces the duration and frequency of hypoglycemia.

Ly et al. investigated the occurrence of severe and moderate hypoglycemia with a sensor-augmented pump with low-glucose suspension compared with standard insulin pump therapy

[58]. The randomized study was performed during a 6-month period. Included in the study were 49 children and adults aged 4-50 years with impaired hypoglycemia awareness. The study confirmed the reduction in the rate of severe and moderate hypoglycemia in the group using sensor-augmented pump therapy with low-glucose suspension. Moreover, compared with the pump-only group, the low-glucose suspension group spent less time in the hypoglycemic range during the night. In both groups there was no change in HbA1c.

Based on the results of this study a cost-effectiveness analysis was performed. The authors concluded that in the group of hypoglycemia unaware T1D patients sensor-augmented insulin pump therapy with low-glucose-suspension might be a cost-effective alternative to standard pump therapy with self-monitoring of blood glucose [59].

Bergenstal et al. noted similar reduction in nocturnal hypoglycemia without increasing HbA1c values in patients treated for 3 months with a sensor-augmented insulin-pump with low glucose suspension [60]. The study included 247 patients aged 16 to 70 years who were randomly assigned to the group and received sensor-augmented insulin pump therapy with the threshold-suspend feature or a standard sensor-augmented insulin pump. Nocturnal hypoglycemic events were observed 31.8% less frequently in the threshold-suspend group than in the control group. Moreover, automatic insulin suspension did not result in hyperglycemia at 2 and 4 hours after pump suspension. The four severe hypoglycemic events occurred in the control group, no patients from the experimental group had severe hypoglycemia, and no patients had diabetic ketoacidosis.

Jennifer et al. showed that the 2 hours suspension of basal rate in the overnight period was safe even when the sensor falsely detected low glucose levels [61]. Suspension of basal rate in the absence of low glucose caused increased fasting morning blood glucose without statistical increase of morning blood β-hydroxybutyrate.

The new generation of insulin pump MiniMed® 530G with Enlite™ is the first diabetes management system with artificial pancreas technology. This system not only suspends basal rate when the sensor detects hypoglycemia but also automatically starts insulin delivery when glucose values rises above the preset low threshold.

In clinical practice even short term use of RT-CGM can be very helpful in patients' education and solving diabetes management problems. RT-CGM data can be downloaded by physicians during a visit to an outpatient department or data can be sent by email and discussed with patients and their family. An interesting solution to help care for children with diabetes supplies is the NightScout project (www.nightscout.info), which was created and supported by volunteers. The NightScout system allows patients to attach a smartphone with NightScout software to a Dexcome RT-CGM receiver. The software downloads the glucose data and uploads it to server. This system allows it to transmit the glucose readings through the internet to different mobile devices. The technology allows parents or caregivers to monitor the child's glucose readings from afar on their tablets, laptops or Apple Watch.

ARTIFICIAL PANCREAS

Despite modern treatment methods (insulin analogues, insulin pumps, RT-CGM), psychological support and a high level of education some diabetic children do not achieve good glycemic control. Young diabetic children can not adjust insulin without help, and require a constant care. When caregivers' compliance is low, a young child has no chance to achieve his therapeutic goal. Existing methods of treatment are also not effective in the group of adolescents. Teenagers often do not measure blood sugar, do not count exchanges and often omit boluses. Clinical experience as well as current studies emphasize that new approaches to reduce glucose fluctuations in children are needed. Even in the group of young children who are under parenteral control high glucose fluctuations are observed. Tansey et al. noted persistently high glucose levels in young type 1 diabetic children [62]. Analysis of the a minimum 48 hours of continuous glucose monitoring data during more than five out of seven visits showed suboptimal glycemic control in the group of 135 children aged between 4-10 years with HbA1c 7.9%. Fifty percent of children had glucose levels over 180 mg/dL (10.0 mmol/L) for over 12 hours per day and glucose levels over 250 mg/dL (13.9 mmol/L) for over 6 hours per day.

The closed loop system insulin delivery (artificial pancreas) has received a great interest and is currently tested as a method of treatment, which may improve glycemic control in type 1 diabetic patients and prevent hypoglycemia. The artificial pancreas consists of an insulin pump combined with RT-CGM and control algorithm, which modulates insulin delivery wirelessly [63]. Currently two different closed-loop systems are tested. One system delivers only insulin and the other, called the 'bihormonal pancreas', administers both hormones insulin and glucagon.

Kumareswaran et al. reviewed the safety and efficacy of overnight insulin delivery compared with conventional CSII in children and adults from a Cambridge cohort [64]. The authors included the analysis of two studies performed on 17 children aged between 5-18 years and two studies examined 24 adults. In the group of children closed loop was used following a standard evening meal and 40 min of moderate-intensity exercise. Overnight the closed loop insulin administration lowered glycemic fluctuation. Compared with conventional therapy, the overnight closed loop insulin delivery increased time spent in the glucose range between 3.9-8.0 mmol/L from 40% to 60% in children and from 50% to 70% in adults. Moreover, the time spent below 3.9 mmol/L was reduced during closed loop insulin delivery from 4.1% to 2.1%. This review showed that although overnight closed-loop insulin delivery improved glycemic control, for 40% of the time children were outside the glucose target and some hypoglycemic episodes were noted.

The closed-loop automated insulin delivery does not completely prevent severe hypoglycemia. Thabit et al. noted two severe hypoglycemic episodes during the home overnight closed-loop delivery [65]. Both episodes appeared in adults with history of hypoglycemia unawareness when the closed loop was interrupted due to wireless connectivity with the insulin pump.

Maintaining normoglycemia at night is very difficult for patients with diabetes. Due to glucose fluctuations and fear of hypoglycemia many parents wake up every night at 2-3 o'clock to measure their child's blood sugar levels. Thus, this has a negative impact on family functioning. Current studies showed that the use of overnight closed-loop insulin delivery increases, by 20%, the time spent in the glucose target area. This avoids about 100% nocturnal hypoglycemia in comparison to conventional sensor augmented insulin pump therapy. However, due to some limitations of the system a severe hypoglycemia episode might occur.

It is more difficult to achieve tight glycemic control during the day when glucose control is complicated by meals, illness, emotional stress or physical activity, than during the night.

Luijf et al. investigated the day and night closed-loop insulin delivery [66]. The 23-hour control using two closed-loops with different computer algorithms and conventional patient self management of CSII did not show any difference in time spend in target range between groups (glucose levels between 3.0 and 8.0 mmol/L or between 3.9 and 10.0 mmol/L after meals). In this study, performed in the hospital, 48 type 1 diabetic adults received three meals and performed exercise consisting of approximately two 15 minute of exercise at 50% VO2max.

In all groups, patients spend about 60% of time in target range, while a threefold reduction in time spent in hypoglycemia <3.9 mmol/l was noted in both closed-loop systems (2.0%, 2.1%) in comparison to the open loop system (6.4%).

Leelarathna et al. investigated efficacy and safety of 1-week home closed-loop insulin delivery [67]. This study group included 17 adults with relatively good glycemic control (HbA1c 7.6%). The authors noted significant increase in the percentage of time when glucose was between 3.9-10.0 mmol/L (75%), compared to conventional insulin pump therapy (62%). The mean glucose level was lower in the closed-loop group (8.1 mmol/L) than in the control subjects (8.8 mmol/L). The authors did not observe any difference in time spent below target while time spent above target was lower during closed loop insulin delivery.

The results of the day and night closed-loop home insulin delivery suggested that the use of this system is safe and might lower glucose fluctuations. Some authors observed a 13% increase in the time spend in glucose target, others noted a decrease in hypoglycemia episodes in adults using closed-loop system in comparison with conventional sensor-augmented insulin pump therapy.

Bihormonal closed-loop artificial pancreas, bionic pancreas automatically administers insulin and glucagon doses in response to readings from a continuous glucose monitor as directed by a computer algorithm. In this system glucagon infusion prevents hypoglycemia by obtaining a more rapid increase of blood glucose than only by slowing or switching off insulin delivery.

Russell et al. investigated the 5-day unrestricted outpatient bionic pancreas therapy performed in 20 adults (HbA1c 7.1%) and 32 teenagers (HbA1c 8.2%) [68]. Patients could eat in restaurants, drink limited amount of alcohol and had access to two gyms. The authors showed in both groups lower mean glucose level on RT-CGM during the bionic pancreas period than the control period. Among adults the mean glucose level was 7.4 mmol/L during the bionic pancreas period and 8.8 mmol/L during the control period. In this group, the percentage of time with glucose <3.9 mmol/l was lower in the bionic pancreas period (7.3%) than during the control period. Among the adolescents the mean glucose level was 7.7% during closed-loop therapy, compared to 8.7% during the control period. Although there was no difference between the percentage of time with glucose <3.9 mmol/L, the mean frequency of intervention for hypoglycemia amongst adolescents was lower when the bionic pancreas was used.

Haidar et al. compared the dual-hormone artificial pancreas, the single-hormone artificial pancreas, and the conventional insulin pump [69]. The authors noted similar efficacy of both systems. No difference was noted between the dual-hormone artificial pancreas and the single-hormone artificial pancreas in the mean proportion of time spent in the plasma glucose target range (4.0-10.0 mmol/L for 2 h postprandially and 4.0-8.0 mmol/L otherwise) over 24 h (63% vs. 62%, respectively). In this study, the authors noted a lower proportion of time spent in the target range during conventional insulin pump period (51%) in comparison with both closed-loop systems. There was reduction in hypoglycemic events during both closed-loops periods compared to conventional insulin pump therapy. The authors showed 52 hypoglycemic events with conventional insulin pump therapy, 13 with the single-hormone artificial pancreas, and 9 with the dual-hormone artificial pancreas; there was no nocturnal hypoglycemia episode during both the closed-loop insulin delivery and 13 hypoglycemic events during conventional insulin pump period.

Compared with the conventional sensor-augmented insulin pump therapy, both closed-loop systems: the dual-hormone artificial pancreas and the single-hormone artificial pancreas showed beneficial effects on hypoglycemia reduction in type 1 diabetic patients. Unfortunately, the complete elimination of hypoglycemia was not obtained. Although the closed-loop systems offer better metabolic control, patients spent only about 10% longer time in the plasma glucose target in comparison with conventional insulin pump therapy. On the basis of current available studies it is difficult to answer the question what group of patients would benefit most from the closed-loop system. Additional long-term studies to assess the use of home closed-loop systems, within different age groups with type 1 diabetic patients are needed.

CONCLUSION

Current methods of diabetes management allow patients to achieve nearly normoglycemia, which prevents the development of long-term diabetic complications. Intensive insulin therapy provides by MDI or CSII allows type 1 diabetic children achieve blood-glucose levels closed to normoglycemia. However, hypoglycemia is still an important barrier to obtaining this therapeutic goal. The new long-acting insulin analogues degludec and glargine 300 reduce occurrence of hypoglycemia and are more effective than insulin glargine 100 or detemir on basal glucose control. CSII is the most physiological method of insulin delivery, which allows better metabolic control, lower glucose fluctuations, the reduction of hypoglycemic episodes and greater flexibility of exercise in comparison with MDI. RT-CGM gives more information about glucose variability and offers improvement in metabolic control compared with self-blood glucose monitoring. However, the compliance with sensor use in some children is low. The new technology combined CSII and RT-CGM improves glycemic control in comparison with conventional CSII and SMBG. Moreover, sensor augmented insulin pump with automated insulin suspension algorithm can reduce the number of hypoglycemic episodes compared with conventional insulin pump therapy. Despite the modern technology available, the diabetes treatment still requires adherence to a strict daily regimen that involves frequent monitoring of blood glucose, food intake and insulin dosing. The responsibility for diabetes management is mainly placed in the hands of patients and their families. Unfortunately, despite the modern treatment methods some children and families do not achieve the therapeutic goal. Currently tested closed loop systems with automated insulin delivery have the potential to keep diabetes under control and relieve the burden of T1D diabetes management. However, at this time an artificial pancreas is not feasible because of limitations of available technology.

REFERENCES

[1] Jarosz-Chobot P., Polanska J., Szadkowska A., Kretowski A., Bandurska-Stankiewicz E., Ciechanowska M. et al. Rapid increase in the incidence of type 1 diabetes in Polish children from 1989 to 2004, and predictions for 2010 to 2025. *Diabetologia,* 2011; 54: 508-515.

[2] Rys P., Pankiewicz O., Łach K., Kwaskowski A., Skrzekowska-Baran I., Malecki M. T. Efficacy and safety comparison of rapid-acting insulin aspart and regular human insulin in the treatment of type 1 and type 2 diabetes mellitus: a systematic review. *Diabetes Metab.,* 2011; 37: 190-200.

[3] Pedersen-Bjergaard U., Kristensen P. L., Beck-Nielsen H., Nørgaard K., Perrild H., Christiansen J. S. et al. Effect of insulin analogues on risk of severe hypoglycaemia in patients with type 1 diabetes prone to recurrent severe hypoglycaemia (HypoAna trial): a prospective, randomised, open-label, blinded-endpoint crossover trial. *Lancet Diabetes Endocrinol.*, 2014; 2: 553-561.

[4] Galli-Tsinopoulou A., Stergidou D. Insulin analogues for type 1 diabetes in children and adolescents. *Drugs Today,* (Barc). 2012; 48: 795-809.

[5] Thalange N., Bereket A., Larsen J., Hiort L. C., Peterkova V. Insulin analogues in children with Type 1 diabetes: a 52-week randomized clinical trial. *Diabet Med.*, 2013; 30: 216-225.

[6] Thalange N., Bereket A., Larsen J., Hiort L. C., Peterkova V. Treatment with insulin detemir or NPH insulin in children aged 2–5 years with type 1 diabetes mellitus. *Pediatr. Diabetes*, 2011; 12: 632–641.

[7] Plavsić L., Mitrović K., Todorović S., Vuković R., Milenković T., Zdravković D. Glycaemic control and prevalence of hypoglycaemic events in children and adolescents with type 1 diabetes mellitus treated with insulin analogues. *Vojnosanit. Pregl.*, 2014; 71: 817-820.

[8] Johnson S. R., Cooper M. N., Davis E. A., Jones T. W. Hypoglycaemia, fear of hypoglycaemia and quality of life in children with Type 1 diabetes and their parents. *Diabet Med.,* 2013; 30: 1126-1131.

[9] Juvenile Diabetes Research Foundation Continuous Glucose Monitoring Study Group, Beck R. W., Hirsch I. B., Laffel L., Tamborlane W. V., Bode B. W., Buckingham B. et al. The effect of continuous glucose monitoring in well-controlled type 1 diabetes. *Diabetes Care*, 2009; 32: 1378-1383.

[10] Rodrigues Vilela V., de Castro Ruiz Marques A., Schamber C. R., Bazotte R. B. Hypoglycemia induced by insulin as a triggering factor of cognitive deficit in diabetic children. *Scientific World Journal*, 2014; 2014: 616534.

[11] Jonassen I., Havelund S., Hoeg-Jensen T., Steensgaard D. B., Wahlund P. O., Ribel U. Design of the novel protraction mechanism of insulin degludec, an ultra-long-acting basal insulin. *Pharm. Res.,* 2012; 29: 2104-2114.

[12] Dżygało K., Golicki D., Kowalska A., Szypowska A. The beneficial effect of insulin degludec on nocturnal hypoglycaemia and insulin dose in type 1 diabetic patients: a systematic review and meta-analysis of randomised trials. *Acta Diabetol.*, 2015; 52: 231-238.

[13] Thalange N., Deeb L., Iotova V., Kawamura T., Klingensmith G., Philotheou A. et al. Insulin degludec in combination with bolus insulin aspart is safe and effective in children and adolescents with type 1 diabetes. *Pediatr. Diabetes*, 2015; 16: 164-176.
[14] Shiramoto M., Eto T., Irie S., Fukuzaki A., Teichert L., Tillner J. et al. Single-dose new insulin glargine 300 U/ml provides prolonged, stable glycaemic control in Japanese and European people with type 1 diabetes. *Diabetes Obes. Metab.*, 2015; 17: 254-260.
[15] Monami M., Lamanna C., Marchionni N., Mannucci E. Continuous subcutaneous insulin infusion versus multiple daily insulin injections in type 1 diabetes: a meta-analysis. *Acta Diabetol.*, 2010; 47 Suppl. 1: 77-81.
[16] van Bon A. C., Bode B. W., Sert-Langeron C., DeVries J. H., Charpentier G. Insulin glulisine compared to insulin aspart and to insulin lispro administered by continuous subcutaneous insulin infusion in patients with type 1 diabetes: a randomized controlled trial. *Diabetes Technol. Ther.*, 2011; 13: 607-614.
[17] Szypowska A., Lipka M., Błazik M., Groele L., Pańkowska E. Insulin requirement in preschoolers treated with insulin pumps at the onset of type 1 diabetes mellitus. *Acta Paediatr.*, 2009; 98: 527-530.
[18] Weinzimer S. A., Swan K. L., Sikes K. A., Ahern J. H. Emerging evidence for the use of insulin pump therapy in infants, toddlers, and preschool-aged children with type 1 diabetes. *Pediatr. Diabetes*, 2006; 7: 15-19.
[19] Queiroz K. C., Novato Silva I., de Cássia Gonçalves Alfenas R. Influence of the glycemic index and glycemic load of the diet in the glycemic control of diabetic children and teenagers. *Nutr. Hosp.*, 2012; 27: 510-515.
[20] Olinder A. L., Kernell A., Smide B. Treatment with CSII in two infants with neonatal diabetes mellitus. *Pediatr. Diabetes*, 2006; 7: 284-288.
[21] Pop A., Clenciu D., Anghel M., Radu S., Socea B., Mota E. et al. Insulin resistance is associated withall chronic complications in type 1 diabetes. *J. Diabetes*, 2015 Mar. 6.
[22] Pańkowska E., Szypowska A., Lipka M., Szpotańska M., Błazik M., Groele L. Application of novel dual wave meal bolus and its impact on glycated hemoglobin A1c level in children with type 1 diabetes. *Pediatr. Diabetes,* 2009; 10: 298-303.
[23] Kordonouri O., Hartmann R., Remus K., Bläsig S., Sadeghian E., Danne T. Benefit of supplementary fat plus protein counting as compared with conventional carbohydrate counting for insulin bolus calculation in children with pump therapy. *Pediatr. Diabetes*, 2012; 13: 540-544.
[24] Ramotowska A., Szypowska A. Bolus calculator and wirelessly communicated blood glucose measurement effectively reduce hypoglycaemia in type 1 diabetic children – randomized controlled trial. *Diabetes Metab. Res. Rev.,* 2014; 30: 146-153.
[25] Ramotowska A., Golicki D., Dżygało K., Szypowska A. The effect of using the insulin pump bolus calculator compared to standard insulin dosage calculations in patients with type 1 diabetes mellitus - systematic review. *Exp. Clin. Endocrinol. Diabetes,* 2013; 121: 248-254.
[26] Szypowska A., Lipka M., Błazik M., Golicka D., Groele L., Pańkowska E. Age-dependent basal insulin patterns in children with type 1 diabetes treated with continuous subcutaneous insulin infusion. *Acta Paediatr.,* 2009; 98: 523-526.
[27] Hofer S. E., Heidtmann B., Raile K., Fröhlich-Reiterer E., Lilienthal E., Berghaeuser M. A. et al. Discontinuation of insulin pump treatment in children, adolescents, and

young adults. A multicenter analysis based on the DPV database in Germany and Austria. *Pediatr. Diabetes*, 2010; 11: 116-121.

[28] Wong J. C., Dolan L. M., Yang T. T., Hood K. K. Insulin pump use and glycemic control in adolescents with type 1 diabetes: Predictors of change in method of insulin delivery across two years. *Pediatr. Diabetes*, 2014 Nov. 10.

[29] de Vries L., Grushka Y., Lebenthal Y., Shalitin S., Phillip M. Factors associated with increased risk of insulin pump discontinuation in pediatric patients with type 1 diabetes. *Pediatr. Diabetes*, 2011; 12: 506-512.

[30] President's Council on Physical Fitness and Sports: Physical activity for children: current patterns and guidelines. *Research Digest.*, 2004; 5: 1-8.

[31] Kelley G. A., Kelley K. S. Effects of exercise in the treatment of overweight and obese children and adolescents: a systematic review of meta-analyses. *J. Obes.*, 2013; 2013: 783103.

[32] Barreira T. V., Katzmarzyk P. T., Johnson W. D., Tudor-Locke C. Walking cadence and cardiovascular risk in children and adolescents: NHANES, 2005-2006. *Am. J. Prev. Med.,* 2013; 45: e27-34.

[33] Lees C., Hopkins J. Effect of aerobic exercise on cognition, academic achievement, and psychosocial function in children: a systematic review of randomized control trials. *Prev. Chronic Dis.*, 2013; 10: E174.

[34] Brown H. E., Pearson N., Braithwaite R. E., Brown W. J., Biddle S. J. Physical activity interventions and depression in children and adolescents : a systematic review and meta-analysis. *Sports Med.,* 2013; 43: 195-206.

[35] Łuczyński W., Szypowska A., Głowińska-Olszewska B., Bossowski A. Overweight, obesity and features of metabolic syndrome in children with diabetes treated with insulin pump therapy. *Eur. J. Pediatr.*, 2011; 170: 891-898.

[36] Zduńczyk B., Sendela J., Szypowska A. High prevalence of depressive symptoms in well-controlled adolescents with type 1 diabetes treated with continuous subcutaneous insulin infusion. *Diabetes Metab. Res. Rev.,* 2014; 30: 333-338.

[37] Yardley J., Mollard R., MacIntosh A., MacMillan F., Wicklow B., Berard L. et al. Vigorous intensity exercise for glycemic control in patients with type 1 diabetes. *Can. J. Diabetes*, 2013; 37: 427-432.

[38] Yardley J. E., Iscoe K. E., Sigal R. J., Kenny G. P., Perkins B. A., Riddell M. C. Insulin pump therapy is associated with less post-exercise hyperglycemia than multiple daily injections: an observational study of physically active type 1 diabetes patients. *Diabetes Technol. Ther.,* 2013; 15: 84-88.

[39] Diabetes Research in Children Network (DirecNet) Study Group, Tsalikian E., Kollman C., Tamborlane W. B., Beck R. W., Fiallo-Scharer R., Fox L. et al. Prevention of hypoglycemia during exercise in children with type 1 diabetes by suspending basal insulin. *Diabetes Care,* 2006; 29: 2200-2204.

[40] Jankovec Z., Krcma M., Gruberova J., Komorousova J., Tomesova J., Zourek M. et al. Influence of physical activity on metabolic state within a 3-h interruption of continuous subcutaneous insulin infusion in patients with type 1 diabetes. *Diabetes Technol. Ther.*, 2011; 13: 1234-1239.

[41] Admon G., Weinstein Y., Falk B., Weintrob N., Benzaquen H., Ofan R. et al. Exercise with and without an insulin pump among children and adolescents with type 1 diabetes mellitus. *Pediatrics,* 2005; 116: e348-355.

[42] Sonnenberg G. E., Kemmer F. W., Berger M. Exercise in type 1 (insulindependent) diabetic patients treated with continuous subcutaneous insulin infusion: prevention of exercise induced hypoglycemia. *Diabetologia,* 1990; 33: 696-703.

[43] Yardley J. E., Sigal R. J., Perkins B. A., Riddell M. C., Kenny G. P. Resistance exercise in type 1 diabetes. *Can. J. Diabetes*, 2013; 37: 420-426.

[44] Ramalho A. C., de Lourdes Lima M., Nunes F., Cambuí Z., Barbosa C., Andrade A. et al. The effect of resistance versus aerobic training on metabolic control in patients with type-1 diabetes mellitus. *Diabetes Res. Clin. Pract.*, 2006; 72: 271-276.

[45] Gomez A. M., Gomez C., Aschner P., Veloza A., Muñoz O., Rubio C. et al. Effects of performing morning versus afternoon exercise on glycemic control and hypoglycemia frequency in type 1 diabetes patients on sensor-augmented insulin pump therapy. *J. Diabetes Sci. Technol.*, 2015 Jan. 1.

[46] Levin B. S., Anderson B. J., Butler D. A., Antisdel J. E., Brackett J., Laffel L. M. B. Predictors of glycaemic control and short-term adverse outcomes in youth and with type 1 diabetes. *J. Pediatr.*, 2001 139:197–203.

[47] Golicki D. T., Golicka D., Groele L., Pankowska E. Continuous Glucose Monitoring System in children with type 1 diabetes mellitus: a systematic review and meta-analysis. *Diabetologia,* 2008; 51: 233-240.

[48] Heinemann L., Franc S., Phillip M., Battelino T., Ampudia-Blasco F. J., Bolinder J. et al. Reimbursement for continuous glucose monitoring: a European view. *J. Diabetes Sci. Technol.*, 2012; 6: 1498-1502.

[49] White N. H., Sun W., Cleary P. A., Danis R. P., Davis M. D., Hainsworth D. P. et al. Prolonged effect of intensive therapy on the risk of retinopathy complications in patients with type 1 diabetes mellitus: 10 years after the Diabetes Control and Complications Trial. *Arch Ophthalmol.,* 2008; 126: 1707-1715.

[50] Kordonouri O., Pankowska E., Rami B., Kapellen T., Coutant R., Hartmann R. et al. Sensor-augmented pump therapy from the diagnosis of childhood type 1 diabetes: results of the Paediatric Onset Study (ONSET) after 12 months of treatment. *Diabetologia,* 2010; 53: 2487-2495.

[51] Buckingham B., Cheng P., Beck R. W., Kollman C., Ruedy K. J., Weinzimer S. A. et al. CGM-measured glucose values have a strong correlation with C-peptide, HbA1c and IDAAC, but do poorly in predicting C-peptide levels in the two years following onset of diabetes. *Diabetologia,* 2015 Mar. 14.

[52] Juvenile Diabetes Research Foundation Continuous Glucose Monitoring Study Group, Tamborlane W. V., Beck R. W., Bode B. W., Buckingham B., Chase H. P., Clemons R. et al. Continuous glucose monitoring and intensive treatment of type 1 diabetes. *N Engl. J. Med.,* 2008; 359: 1464-1476.

[53] Bergenstal R. M., Tamborlane W. V., Ahmann A., Buse J. B., Dailey G., Davis S. N. et al. STAR 3 Study Group. Effectiveness of sensor-augmented insulin-pump therapy in type 1 diabetes. *N Engl. J. Med.,* 2010; 363: 311-320.

[54] Juvenile Diabetes Research Foundation Continuous Glucose Monitoring Study Group. Prolonged nocturnal hypoglycemia is common during 12 months of continuous glucose monitoring in children and adults with type 1 diabetes. *Diabetes Care*, 2010; 33: 1004-1008.

[55] Hommel E., Olsen B., Battelino T., Conget I., Schütz-Fuhrmann I., Hoogma R. et al. Impact of continuous glucose monitoring on quality of life, treatment satisfaction, and

use of medical care resources: analyses from the SWITCH study. *Acta Diabetol.*, 2014; 51: 845-851.

[56] Battelino T., Conget I., Olsen B., Schütz-Fuhrmann I., Hommel E., Hoogma R. et al. The use and efficacy of continuous glucose monitoring in type 1 diabetes treated with insulin pump therapy: a randomised controlled trial. *Diabetologia,* 2012; 55: 3155-3162.

[57] Mauras N., Beck R., Xing D., Ruedy K., Buckingham B., Tansey M. et al. A randomized clinical trial to assess the efficacy and safety of real-time continuous glucose monitoring in the management of type 1 diabetes in young children aged 4 to <10 years. *Diabetes Care*, 2012; 35: 204-210.

[58] Ly T. T., Nicholas J. A., Retterath A., Lim E. M., Davis E. A., Jones T. W. Effect of sensor-augmented insulin pump therapy and automated insulin suspension vs standard insulin pump therapy on hypoglycemia in patients with type 1 diabetes: a randomized clinical trial. *JAMA,* 2013; 310: 1240-1247.

[59] Ly T. T., Brnabic A. J., Eggleston A., Kolivos A., McBride M. E., Schrover R. et al. A cost-effectiveness analysis of sensor-augmented insulin pump therapy and automated insulin suspension versus standard pump therapy for hypoglycemic unaware patients with type 1 diabetes. *Value Health*, 2014; 17: 561-569.

[60] Bergenstal R. M., Klonoff D. C., Garg S. K., Bode B. W., Meredith M., Slover R. H. et al. Threshold-based insulin-pump interruption for reduction of hypoglycemia. *N Engl. J. Med.*, 2013; 369: 224-232.

[61] Sherr J. L., Palau Collazo M., Cengiz E., Michaud C., Carria L., Steffen A. T. et al. Safety of nighttime 2-hour suspension of basal insulin in pump-treated type 1 diabetes even in the absence of low glucose. *Diabetes Care*, 2014; 37:773-779.

[62] Tansey M., Beck R., Ruedy K., Tamborlane W., Cheng P., Kollman C. et al. Persistently high glucose levels in young children with type 1 diabetes. *Pediatr. Diabetes,* 2014 Dec. 11.

[63] Elleri D., Dunger D. B., Hovorka R. Closed-loop insulin delivery for treatment of type 1 diabetes. *BMC Med.,* 2011; 9: 120.

[64] Kumareswaran K. 1., Elleri D., Allen J. M., Harris J., Xing D., Kollman C. et al. Meta-analysis of overnight closed-loop randomized studies in children and adults with type 1 diabetes: the Cambridge cohort. *J. Diabetes Sci. Technol.*, 2011; 5: 1352-1362.

[65] Thabit H., Lubina-Solomon A., Stadler M., Leelarathna L., Walkinshaw E., Pernet A. et al. Home use of closed-loop insulin delivery for overnight glucose control in adults with type 1 diabetes: a 4-week, multicentre, randomized crossover study. *Lancet Diabetes Endocrinol.,* 2014; 2: 701-709.

[66] Luijf Y. M., DeVries J. H., Zwinderman K., Leelarathna L., Nodale M., Caldwell K. et al. Day and night closed-loop control in adults with type 1 diabetes: a comparison of two closed-loop algorithms driving continuous subcutaneous insulin infusion versus patient self-management. *Diabetes Care*, 2013; 36: 3882-3887.

[67] Leelarathna L., Dellweg S., Mader J. K., Allen J. M., Benesch C., Doll W. et al. Day and night home closed-loop insulin delivery in adults with type 1 diabetes: three-center randomized crossover study. *Diabetes Care,* 2014; 37: 1931-1937.

[68] Russell S. J., El-Khatib F. H., Sinha M., Magyar K. L., McKeon K., Goergen L. G. et al. Outpatient glycemic control with a bionic pancreas in type 1 diabetes. *N Engl. J. Med.*, 2014; 371: 313-325.

[69] Haidar A., Legault L., Messier V., Mitre T. M., Leroux C., Rabasa-Lhoret R. Comparison of dual-hormone artificial pancreas, single-hormone artificial pancreas, and conventional insulin pump therapy for glycaemic control in patients with type 1 diabetes: an open-label randomised controlled crossover trial. *Lancet Diabetes Endocrinol.,* 2015; 3: 17-26.

In: Alterations in Glucose Homeostasis in Children
Editor: E. Otto-Buczkowska
ISBN: 978-1-63483-861-0

Chapter 3

Diabetes Education: Techniques, Methods and Psychological Support

***B. Zduńczyk**[*,1] **and A. Szypowska**[2]*
[1] Clinical Pediatric Hospital of Medical University of Warsaw, Poland
[2] Medical University of Warsaw, Poland

Abstract

Education in diabetes self-management (DSME) is the keystone in treatment of patients with diabetes. In the process conducted by a team of professionals (a diabetologist/endocrinologist, nurse, dietician, psychologist, diabetes educator) patients gain knowledge and practical skills, how to modify their behavior in order to keep diabetes under control in daily life. Unlike individuals with type 2 diabetes, patients with diabetes type 1 are educated from the onset of the illness. They undergo further training if they are unable to meet treatment targets. Whereas adults are fully responsible for decision making and control of their condition, management of diabetes in children is placed in the hands of their parents/caregivers. Therefore, it is not unique to educate the whole family, including parents, grandparents and siblings.

Recommendations and national standards have been developed by various diabetes associations in many countries. They provide guidance on organization principles, age appropriate content, models, methods and tools which may be used to accomplish the desired objectives. Following these guidelines, diabetes centers design educational programs to convey knowledge on diabetes and motivate patients to achieve targets in metabolic control. These programs aim to improve patients' health outcomes and quality of life and at the same time, to delay diabetes complications while keeping costs as low as possible.

Researches study what is the most effective in helping patients adhere to treatment. They consider settings (individual approach, group sessions in clinics or community settings – home, school, camps for youths with diabetes), means of communication (personal approach, internet, and telemedicine), use of tools (booklets, games, various

* E-mail: betaz@interia.pl

telephone and computer applications) and psychological techniques (empowerment, family conflict solving skills). The question is also raised when an education program is not effective and a patient should be referred to a psychologist or a psychiatrist.

This chapter deals mainly with diabetes self-management education. We discuss the current trends in education of children, adolescents and their families. We also want to stress the importance of education for other groups than patients. Medical staff have to be able to communicate information on diabetes in the language appropriate to a patient's age and intellectual abilities and understand psychological processes occurring after illness diagnosis or at different stages of a child's development. We emphasize the importance of the knowledge on diabetes in the school environment. Children and adolescents are encouraged to have as normal life as their peers; therefore it is necessary for school staff to have basic knowledge on diabetes to help a pupil with illness management. We look at education programs designed for school staff and what can be done to help students to succeed in school, to have both academic achievements and meet diabetes targets. Unfortunately, in many countries, discrimination of people with diabetes still occurs as the effect of poor diabetes awareness. At the end of the chapter we want to look closely at national and international initiatives aiming at enhancing diabetes awareness and disease prevention.

Keywords: Standards for diabetes education, initial education, continuing education, information technology, psychological intervention, self-care

INTRODUCTION

Diabetes is one of the most common chronic diseases in childhood. Diabetes type 1 is no longer associated only with children and adolescents, as it can occur at any age, while diabetes type 2, commonly perceived as a problem of the old age, is now worryingly more often diagnosed in children and adolescents [1, 2]. Although the exact number of children and adolescents with diabetes type 1 and 2 is unknown epidemiological studies show steadily increasing trends in the incidence of illness [1, 3]. It is estimated that 78,000 youths worldwide are diagnosed with diabetes type 1 annually, with the most rapid increase in incidence rate in children under the age of 5 [1, 4, 5]. As diabetes poses a great risk for health outcomes for such a growing number of patients it calls for careful consideration of treatment options to provide optimal care and prevent late complications without deterioration in quality of life.

Unlike the care of individuals with other chronic health problems, the majority of decisions concerning diabetes type 1 is made not by medical staff, but by patients themselves, and in case of children – by their parents. They have to monitor glucose level, food intake, adjust insulin dose and activity. They have to know hyperglycemia and hypoglycemia symptoms and how to react to restore normoglycemia. After diabetes diagnosis it is necessary for patients to gain basic knowledge on the mechanisms of human body function, insulin action, healthy eating rules, carbohydrate and fat counting, importance of physical activity, factors affecting achievement and maintenance of good metabolic control and the prevention of future complications. Education in diabetes self-management (DSME) is considered the keystone in treatment of patients with diabetes. The work of Joslin since the 1930s paved the way to a modern concept of teaching patients the management of illness [6]. As the recognition of the importance of education in achieving and maintaining good metabolic control grew, national standards have been developed by various diabetic associations in many countries, following the recommendations of two major diabetic associations: the American Diabetic Association and the International Society for Pediatric and Adolescent Diabetes.

STANDARDS FOR DIABETES EDUCATION

Diabetes education in self-management is understood by the International Society for Pediatric and Adolescent Diabetes as "the process of providing the person with the knowledge and skills needed to perform diabetes self-care, manage crises and to make lifestyle changes to successfully manage the disease" [7, 8]. Similarly, Standards Revision Task Force, a body summoned by the American Association of Diabetes Educators and the American Diabetes Education, views education as "the ongoing process of facilitating the knowledge, skill, and ability necessary for prediabetes and diabetes self-care... to support informed decision making, self-care behaviors, problem solving, and active collaboration with the health care team and to improve clinical outcomes, health status, and quality of life" [9]. It is clear from both definitions that the aim of education is not limited only to gaining knowledge and skills affecting health outcomes, but equally important is life satisfaction of patients with diabetes. The idea of recognizing the patient's psychological and medical needs is conveyed in further guidelines on the content and methodology of the education process. While the national standards for diabetes education set up by Standards Revision Task Force apply to the whole population of individuals with diabetes, it is still possible to derive general principles relevant to pediatric patients. The ISPAD recommendations are focused entirely on the situation of children.

It is obvious that caring for a child from the birth throughout adolescence is the main responsibility of parents or other caregivers; therefore it is necessary in the process of diabetes education to include all members of family who take control of illness management. It is also stressed that other caregivers (in nurseries or kindergarten or teachers at school) should gain the knowledge on diabetes to appropriately support a child. Education should be structured, comprehensive and based on current knowledge on disease and treatment options. It should be reviewed regularly, every 5 years is recommended. Diabetes education should be delivered by a multidisciplinary team, including a physician, nurse, dietitian, and mental health professionals – psychologist, social worker, all educated and experienced in education and treatment of individuals with type 1 diabetes. Similar principles and shared common goals in diabetes education should be followed by educators working as one team. Members of educational team should tailor the educational program to the individual needs of a particular patient and his family, considering their situation, child developmental stage, lifestyle and culture. While delivering information soon after initial diagnosis educators should bear in mind that the whole family may be in the process of adaptation to the new situation, which is life with diabetes, experiencing the wide range of emotions (anger, denial, sadness, grief, depression) and is as yet unable to absorb what is being conveyed [10]. At this time members of educational team should show sensitivity and understanding. Emotional reactions to diabetes by patients and their families, necessity to modify their lifestyle obligate educators to have training not only in teaching but also in counseling and motivation and behavior change techniques.

The most comprehensive education is carried out at the time of diagnosis when a patient and his family have to gain basic knowledge and skills on diabetes management. The following theoretical topics are recommended to be included in curriculum: diabetes diagnosis with stress on nobody's blame, insulin action, treatment goals (glucose level targets to avoid complications), diet (healthy eating, carbohydrate counting), physical activity, disease progress (hypoglycemia, hyperglycemia and DKA, late complications), psychological adaptation to diabetes diagnosis, support available (medical – emergency telephone contact, reimbursement, social – benefits from the social services or diabetic associations), integration of diabetes management into everyday life. Patients have to learn the following skills: blood glucose monitoring, urine testing, insulin injection either with the use of a pen or an insulin pump, insulin dose adjustment (with bolus calculation when using pump therapy), carbohydrate counting, glucagon administration. Continuous education is integrated into routine clinical care. Similar topics appear during further education usually when problems with a patient adherence are deducted and treatment goals are not met. The education program should be adapted to the patient's age and maturity, so curriculum for adolescents includes additional topics on: peer relationships, diabetes problem-solving in everyday life, smoking, alcohol drinking, drug use, employment, vehicle driving, sex and contraception, pregnancy and childbirth, transition to adult care. Written materials prepared in appropriate language and style, comprehensive for a person with low education to fully understand should be incorporated in the educational process. Books, booklets, leaflets should reflect the philosophy and common goals shared by the education team. Children, adolescents and some parents may find modern technology useful in the learning process so videos, computer games, websites are used alongside traditional written materials [8].

Implementation of Recommendations for Diabetes Education

National diabetes associations, established in many countries, recognizing the importance of education in diabetes treatment publish their own guidelines for diabetes self-management education usually as a part of standards for diabetes care [11-15]. Still, there is not one universal program for educating patients in diabetes self-management in existence. Each diabetes center usually develops their own model depending on the resources available, the structure of the organization, center policy, staff experience and skills etc.

Initial Education

A good example of the education program developed in one center is the education delivered in the Department of Pediatrics at Clinical Pediatric Hospital of Warsaw Medical University. Department of Pediatric is one the European center of reference for pediatric diabetes, the member of the SWEET Project - set up to improve diabetes care of children and adolescents in the European Union (www.sweet-project.eu). In this center all parents/caregivers of patients diagnosed with type 1 diabetes and children, who are able to follow the program, undergo the structured diabetes self-management education plan provided. The educational program, following the ISPAD guidelines, covers different aspects of diabetes management, multiple daily injections and an insulin pump included:

- Explanation of disease pathogenesis
- Physiology of human organism
- Action of different kinds of insulin
- Different models of insulin therapy
- Goals of insulin therapy (HbA1c and other medical parameters)
- Adjustment of insulin dose
- Diet (carbohydrate and fat/protein counting, healthy eating)
- Physical exercises and its effect on glycemia
- Other illnesses and diabetes
- Hyperglycemia - symptoms and management
- Acute complications – symptoms, prevention and management:
    - Hypoglycemia
    - Ketoacidosis
- Prolonged complications
- Insulin pump therapy

These topics are taught by a pediatric diabetologist or a physician in training to become a diabetologist or endocrinologist and a diabetological nurse/educator. Additionally, parents have meetings with a psychologist experienced in the problems of patients with diabetes type 1. These meetings focus on emotional reactions to diabetes diagnosis, illness adaptation process and psychological problems, which may arise from diabetes.

Education is interactive with the use of Power Point presentations. After each module families answer some questions to check their knowledge and understanding. Also a movie prepared by Medtronic MiniMed is used in education, especially of young children. It gives some age-appropriate explanation about disease and information about insulin pump therapy. The center uses interactive on-line education materials developed by its staff for different age-groups: young children, teenagers and special materials for pump users. These materials consist of tests on three levels of difficulty. Furthermore, all families receive free education materials written by the center educators to provide further help and support. The center's experience shows that teenagers' education is more effective when mobile applications are used.

The center's experience shows that nutrition habits and proper food counting is crucial in maintaining tight glycemic control. Therefore, the great attention is paid to teaching patients and their parents/caregivers on how to practically count a meal composition. Patients treated with continuous subcutaneous insulin infusion are educated to calculate carbohydrate exchanges and supplementary fat plus protein exchanges. The beneficial effect of this strategy of meal counting was confirmed by the center and other authors [1, 2, 17, 18].

Rules of Carbohydrate and Supplementary Protein Food Counting

Accurate counting of carbohydrate, protein and fat content is essential for the success of insulin therapy. Parents and caregivers and adolescents must be educated how to estimate the composition of the meal and should also be able to make practice examples. The education sessions include theoretical parts during which patients and his families are explained the rules of a healthy diet and type of nutrition products. It is discussed with the family which products contain mainly carbohydrates, proteins or fats, the dependency of the absorption process on the type of food and its impact on the postprandial glycemic profile. This information is given to help a patient to understand the timing of insulin administered for a protein-fat product.

The family is taught the definition of an "exchange" of carbohydrate (the amount of a product which contains 10 grams of carbohydrate, which provides the body with 40 kcal) and protein-fat exchange (the amount of a product which contains 100 kcal from protein / fat). It is also important for the patient and the family to know the caloric content of one gram of protein, fat and carbohydrate (1 g protein – 4 kcal, 1 g fat – 9 kcal, 1 g carbohydrate – 4 kcal). Diabetes centers use different methods to assess the meal composition, so it is important that the proposed method is tailored to the intellectual abilities of the parents. When intellectual impairments are observed the simpler rule is proposed:

- 1 carbohydrate exchange (CE) = 10 grams of carbohydrate = 40 kcal from carbohydrate
- 1 fat/protein exchange (FPE) = 100 kcal from fat/protein

The Calculation of Exchanges for Caloric Content

A patient using the dietary tables learns to find the calorie content of a product and information on the percentage of calories from protein, fat and carbohydrates, e.g., the caloric content per 100 g of dumplings is 240 kcal, of which 7% of calories are derived from protein, 43% from fat and 50% from carbohydrates. On the basis of this data it is easy to calculate that the protein-fat calories are 50% of 240 kcal, which gives 120 kcal. Knowing that 1 protein-fat exchange provides 100 kcal, 120 kcal are converted into exchanges, (120/100) exchanges = 1,2 FPE.

Using a similar method carbohydrate exchanges are calculated in two mathematical operations. First the calories of carbohydrates are counted, which in this example is 50% of 240 kcal = 120 kcal. Then, knowing that 1 carbohydrate exchange provides 40 kcal, 120 kcal are converted into exchanges. By dividing 120 kcal by 40, a patient gets the result - 3 CE. After the above calculations, the patient learns that 100 g of dumplings contain 3 CE and 1.2 FPE.

Calculation of Exchanges from Grams

In order to calculate carbohydrate exchanges from grams of digestible carbohydrates, the fiber content is subtracted from the carbohydrate content per 100 grams of product and the resulting value is converted into product portions. For this purpose, the ratio should be calculated. A patient can find in dietary tables that 100 g of graham roll contains 50 g of absorbable carbohydrate. For example a 30 g roll, which a child wants to eat, contains (30 g x 50 g) / 100 g = 15 g carbohydrates = 1.5 carb exchanges.

In this method fat-protein exchanges can be calculated by subtracting the energy value of carbohydrates from the total energy of the product. Knowing that 1gram of carbohydrate provides 4 calories, the energy from carbohydrates is calculated by multiplying the grams of carbohydrates by 4. For example, if a 100 g potato pancake contains 230 kcal, including 20 g of carbohydrates, the amount of fat-protein- exchanges is calculated like this: 230 kcal – (20x4) = 150 kcal = 1.5 fat/protein exchange.

Fat-protein exchanges can also be calculated by multiplying the grams of protein by 4 kcal and adding fat grams multiplied by 9 kcal.

Computer Programs

Many parents find carbohydrate and fat/protein exchange calculation complicated and they look for means of making the task easier. Computer programs are very useful for calculating exchanges. One of the programs used by Polish families is Elka-Elektroniczna kucharka. The software is available free from the website www.orcan.pl. It was developed by the father of the child diagnosed with diabetes. It allows a patient to calculate carbohydrate and protein-fat exchanges using predefined nutritional tables as well as to input own recipes. The advantage of the system is the integration of a scale with the program. When a patient places a product on the scale, he just selects the name of a product and the program makes all calculations.

A program often used by Polish families is the Medtronic Mobile Food Exchange Calculator available for mobile phones or tablets, which can be easily used with a pocket electronic scale. The program created by Medtronic MiniMed is also available for free. Using this program a child simply chooses a product from nutritional table of Polish food, enters the weight and the software calculates carbohydrate and protein-fat exchanges. This program allows a patient to add some new products and to calculate a meal content.

Programs to calculate exchanges not only allow patients to make very precise calculations quickly, but can also be used for educational games for use with a small child. Young children are happy to assist in the preparation of meals. They quickly learn through play how many exchanges each product contains. It is therefore important to inform parents of this option.

The whole education process is delivered during the hospitalization of a child after diabetes diagnosis. It lasts up to 14 days depending on the patient and his parents/caregivers' ability to assimilate all information. The 2 hour theoretical sessions take place every day, usually in the morning, with sessions later to practice skills. The goal of education delivered in this way is that families should achieve independence in decision making in the treatment of their child's diabetes and to achieve good metabolic control and flexibility of life with as few restrictions as possible.

Continuing Education

The rare example of the development of a multi-center program for 11-16 year old children is The Kids in Control Of Food (KICk-OFF), an adaptation of adult Dose Adjust for Normal Eating (DAFNE) program. It is a 5 day outpatient course based on the information-motivation-behavioral model, the approach widely used in Germany. Up to 8 participants in two age groups (11-13 and 14-16) learn what diabetes is, covering the following topics: food and diabetes, insulin management, management of hypoglycemia, sick day rules, diabetes in school and social situations. They have an opportunity to practice carbohydrate counting during the meal time. Parents have their own meetings prior the beginning of the course to give them better understanding of the course and support for children. Additional support is provided through the website developed (on the purpose of the course) where participants have access to educational material and a message forum. There is also possibility to practice carbohydrate counting there [19, 20, 21].

Barriers to Implementation of Recommendations for Diabetes Education

Existing guidelines give detailed descriptions on many aspects of education delivery, but it is little known on implementation of these recommendations in real life. The survey carried out among 14 European leading centers, members of The SWEET Project, set up to improve the standards of care of children and adolescents with diabetes across the European Union, evaluated how the guidelines were implemented in the diabetes reference centers [16]. It showed the great diversity in educational practices. While all centers had multidisciplinary teams participants raised the problem of insufficient number of health professionals in relation to the number of patients. It caused, among other problems, mixing responsibilities by team members, for example patient's social problems were often left for psychologists or other professionals to be dealt with as a social worker was not included within the diabetes team. Participants also pointed out that informal communication on patients' progress was a substitute for team meetings. Although all centers followed the ISPAD guidelines for diabetes education (as the reference) not all recommended topics were included in the program content, especially at the time of the initial education. Some centers offered more what was necessary for "survival kit" including modules recommended for continuing education. Most education programs were developed locally without any established procedure of evaluation. They used many different education models, methods and materials with still limited use of modern technology. In most centers initial education took place at hospital and while a patient and his parents/caregivers were obvious participants of the training program, it often did not involve grandparents, siblings and other caregivers not related to a patient in diabetes education [22].

EDUCATION SETTING

The ISPAD guidelines suggest that initial education should take place at hospital. Still, there are some calls for increasing diabetes self-management at home or at outpatient clinic [23, 24]. There are few studies carried out on children with diabetes type 1 but their results show that home based diabetes education is equally effective [25]. Tiberg et al. report that patients achieve similar metabolic control as their peers who have education at hospital with less hypoglycemia [26]. It reduces costs as patients use less healthcare resources. Parents whose children receive home based care are more satisfied with treatment as they have less work absences because of diabetes of their child. Promoters of the idea of home based care after diabetes diagnosis emphasize better adjustment of education to suit family needs, more support for lifestyle changes – as direct observation of the family in the home environment gives better insight into eating habits, physical activity, etc. A nurse or physician visiting a patient at home may strike a better rapport than in a formal setting.

What are disadvantages of home diabetes education? First of all, results of studies may be biased. Children who qualify for home diabetes education are usually diagnosed at the early stage of diabetes, they are not admitted to hospital with ketoacidosis so don't require urgent medical attention. Very often these children come from better-off, emotionally stable families where parents pay a lot of attention to childcare so health deterioration is quickly detected. And such parents engage in diabetes treatment, which results in good metabolic control of their child. Another problem is that big medical centers, which provide service to large population living in distance, are not able to offer such home education. It would be costly and require big staff. Smaller medical centers with local community living in proximity of the clinic have an advantage of knowing their patients before diabetes onset so relationship between clients and medical staff is closer from the start.

What are advantages of undergoing diabetes self-management education at hospital? Patients and their families are usually in emotional shock after diabetes diagnosis [27, 28]. Hospitalization takes away the burden of caring for a child when parents have to deal with their own emotions and new information on the future life with diabetes. Hospital with familiar staff becomes a safe haven as patients and parents learn at their own pace what diabetes and its treatment means. They can observe medical staff and good practice, at the same time practicing all skills related to diabetes treatment. Very often the first episode of low sugar happens during hospitalization and patients and their parents can learn to recognize symptoms and their management in practical way. When there is no pressure on discharge hospitalization allows patients and medical staff to get to know each other better. It is also a field of informal observation of parents-a child and father-mother relationships, communication styles and habits. It may give insight into potential future problems.

Longer hospitalization may be the beginning of psychological interventions to change some behaviors and prevent the escalation of problems [29]. What is most valuable is patients and their parents have a chance to meet other families with children with diabetes type 1 and to see that they are not alone in this situation and others cope so they can also manage all diabetes responsibilities. This is a basis for formation of informal group support.

One Education Program

Although most medical centers taking part in the survey complied in their curriculum to the ISPAD recommendations the SWEET Project survey showed the great diversity of education programs, techniques and methods used across the European Union [22]. Do we really have to consider it as a drawback? Is it feasible to have one program in the whole Europe or even in one country? With such huge differences in cultural background, ethnic variability in some countries, food attitude, not to mention economic situation and access to modern technologies; is it possible to develop one education program in diabetes self-management to suit all needs? Does it guarantee success in the treatment of children and adolescents with diabetes type 1? By analogy, every country has their own education curriculum for teaching children and adolescents at school and we know that academic performance of a child depends on many factors. We consider the child's traits, family and school environment as crucial in the achievement of academic success. The most boring or difficult subject may become a pupil's favorite thanks to the teacher's enthusiasm, knowledge and fair attitude. It is very likely that a child whose parents are engaged in the learning process will achieve more than even a talented student without motivation to learn. It is very well known that in psychotherapy different techniques based on different theories may be used, but what works the best is the relationship between a patient and a psychotherapist [30]. The same may be applied in the treatment of children and adolescents with diabetes type 1: the alliance between the patient and the medical team may be the most important factor affecting the future health outcome.

It is important to evaluate education programs and to ask questions what stands behind patients' adherence to treatment in one center while another struggles. It is important to look for the best ways to educate patients and improve the teaching process by introducing new methods, adding novel materials, using new technologies, even "borrowing" ideas from other places, and so on. Definitely, each patient should have an access to diabetes education with a uniform set of modules. At the same time what works for one center with his own character may not work in another place and somehow one way of implementation of recommendations in education contradicts the idea of individualization of patients' treatment.

Patients' Needs

Various diabetes associations provide guidelines for diabetes self-management education, which are followed by medical centers within their educational programs. Groups of professionals from different fields participating in caring for a child and adolescent with diabetes take part in forming education principles. They usually include the latest knowledge on diabetes treatment and their idea of best patient care. At what stage can patients themselves voice their needs and show their concept of best care?

Focus groups (parents and children mean age = 13.07) organized by Waller et al. as a part of the development of new intervention for adolescents with diabetes type 1 (adaptation of adult DAFNE course) showed that teenagers learned the most not from educational sessions itself but from the presence of others in the same predicament [31]. They found communication with medical staff difficult when too much professional jargon was used (parents had to "translate" to children what was said) or if the information was directed to parents and not directly to children. Both parents and children had to cope with information overload, especially during the initial education at the time of diagnosis. Participants suggested more practical, "fun" sessions rather than only typical, didactic lessons. The location of the course didn't matter much as long the atmosphere was comfortable. Older adolescents suggested separate sessions from their parents [31].

Chaney et al. report similar findings [32]. Adolescents aged 13-19 complained about the complexity of the current education not tailored to their needs. They felt discouraged by the program focusing on what they shouldn't do and not what they could. Emphasis on negative aspects of the disease caused anxiety and worries, but it didn't necessarily increase the compliance. Adolescents stressed the need of practical information with "group discussion, practical demonstrations and fun activities" seen as the best way of information delivery. As participants prepared for the move from home, they showed the need for independence in exclusion of parents from the course. They identified text messaging as the best means of communication with healthcare professionals.

Call for practical information is evident not only from scarce research on patients' needs but, also from clinical practice. Initial education is fear-inducing and overloaded with theory. Patients overwhelmed with anxiety often ask how to use this knowledge in practice; and it is not only about carbohydrate counting. Patients often don't see the possibility of living normal life as they usually hear what they are not allowed to do and what will happen when they don't adhere to treatment. Undoubtedly, it is a challenge to find a balance between the necessity to inform on diabetes pathogenesis, management and complications which patients should be aware of and teaching just practical skills.

With a patient's maturation the need for his independence and information grows [33]. It is worth considering when a pediatric patient is able to take more responsibility for diabetes management and how the process of separation from his parents in the medical setting should look like, so transition from pediatric care to adult care will go smoothly.

TECHNOLOGY AND DIABETES EDUCATION

As the modern world becomes more and more technologically advanced more and more new technologies burst into our lives changing the way we think, our lifestyle, culture and society. Everything is faster, more convenient and more accessible. It is impossible to imagine a day without a computer or mobile phone. The internet is used everywhere, for work, education, communication and entertainment. It is not surprising that modern technologies are present in medicine – in diabetes education and treatment of pediatric and adult patients with diabetes type 1 and 2. Blood glucose monitoring, continuous glucose monitoring and insulin pumps are everyday tools in the care of children and adolescents with diabetes type 1 [34]. Internet, phones, decision support devices and telemedicine are more and more widely used to improve diabetes self-management [35].

The internet gives the best opportunity for communication. Patients use it to find information on diabetes, contact physicians about their health problems, access their medical records, send glucose level, information on medication and diet to clinicians. Physicians provide feedback to patients, deliver diabetes specific education and psycho-education intervention, inform on best possible treatment to the problem worrying their patient [34].

Each diabetes association and medical center has its own website. The American Diabetes Association at http://www.diabetes.org, the International Society for Pediatric And Adolescent Diabetes at www.ispad.org/ or the Polish Diabetes Association at http://www.cukrzyca.info.pl just to name a few. Some are designed purely for professionals (the European Association For Study of Diabetes at http://www.easd.org) with news on new developments in research and upcoming events. Some are created with patients in mind (http://www.cukrzyca-dzialdowska.pl) with basic information sheets on diabetes, answers to most frequently asked questions and contact forms. Mostly parents, but also adolescents seek information on the internet. After diabetes diagnosis, it is often their first action to consult "dr. Google" what diabetes treatment is. As Holtslander et al. report, parents seek appropriate information on medical issues – straight answers from professionals, in easily understandable language without advertisements or misleading information promising to cure diabetes with alternative medicine [36]. They want advice on child upbringing, particularly during the difficult time of adolescence when both parents and teenagers struggle with diabetes-related responsibilities and changing roles. Parents appreciate the easy access and support from others at the time of most need. Parents and adolescents choose the internet as their favorite source of support [37]. Patients share opinions on chats, forums and blogs and these ideas may influence adherence to treatment, not necessarily to its advantage.

Diabetes educators may find themselves in the difficult position of correcting patient's knowledge based on internet information. It is hard to see patient-centered websites as structured diabetes education; but it is advisable to analyze their content, both when designing such web pages and delivering information during educational sessions. Meeting the patient's needs on internet information on one hand and relating to incorrect information as a part of an educational program on the other may help to avoid misunderstandings in the future leading to problems with adherence.

The internet is an arena of psycho-education intervention for adolescents with type 1. Programs designed for such interventions aim either at directly improving diabetes management (diabetes-related decision abilities) or indirectly – by coping skills training (concept based on cognitive behavioral therapy to enhance self-efficacy and to promote constructive behaviors) [38]. Although studies show a limited effect on glycemic control in youths, problem solving abilities greatly improve in intervention group participants [39]. Grey et al. suggest that patients may benefit most if they complete both diabetes management education and behavioral interventions [40]. Adolescents who participated in 2 program group had lower HbA1c, higher quality of life, self-efficacy, social acceptance, lower perceived stress and less family conflicts in comparison with participants who finished only one program. Such an approach may be in line with changes taking part in adolescence. As adolescents mature and struggle for independence they need new skills both in diabetes management and social relationships, parents included. In this sensitive period diabetes is one of elements youths have to cope with and which interferes with other aspects of their life. A holistic approach may be required to help teenagers to manage their life in general and diabetes in particular.

Smartphones are the most popular mode of communication among adolescents and young adults. Text messaging had been identified, by adolescents, as the most preferable tool for contacting a physician [32]. Also communication phones are used to upload the glucose level and as the self-management dedicated decision support software, such as advanced bolus calculators or algorithms, for adjustment of carbohydrate ratio and insulin dosing [34, 35]. Popularity of smartphones is an ideal opportunity for interventions mostly aimed directly at glucose control improvement (blood glucose monitoring reminders, insulin use text messaging, physical activities or healthy eating) but also as an educational tool – providing information on diabetes. Different study designs, intervention and follow-up duration make difficult to compare the effects of text messaging on diabetes management. Adolescents enjoy participation in trials, but clinical results taking different variables into account (HbA1c, self-reported adherence, social support) and, in some cases, affected by data loss due to software malfunction are ambiguous [34, 41]. There is potential for the use of these devices in diabetes education but more research should be carried out to evaluate their feasibility.

The distinct use of internet and smartphones in diabetes education are digital games designed for the purpose of diabetes information delivery. Computer and video games meet children and adolescents' needs for making education more attractive. They help them to concentrate and absorb information through play, which is especially important for younger patients with shorter attention spans. Some games are designed especially for a particular educational program in collaboration with physicians, pharmaceutical and information technology companies, some are developed by parents/grandparents of children with diabetes type 1 (for example http://grandmasandy.org) or commercial computer companies [42]. They usually use problem solving situations to teach children blood glucose monitoring, healthy diet, insulin dosing, symptoms and management of hyper- and hypoglycemia. Study results show improvement in diabetes knowledge and treatment adherence in children and adolescents participating in education sessions with the use of this type of games [43]. Educational games for patients with diabetes are still a relatively new area of information technology use, with little information on benefits for health outcome and drawbacks. Information technology development should be supervised by diabetologists and their content regularly updated as our knowledge on diabetes and its treatment change. Youngsters may find games enjoyable as the practice shows, but with so much time spent on computers already other methods of education may be worth considering to teach children practical skills necessary for good diabetes management.

School Needs

Diabetes education should answer real life needs. Recommendations for educational programs concentrate mostly on patients and the families. Still, around one-third of time young people spend at school where they are expected to manage their diabetes at the same level as at home. While adolescents usually don't require assistance smaller children are not able to take responsibility for diabetes alone. What resources are available to them and who can they rely on when in need of help when parents are not present? Although the ISPAD guidelines advise the inclusion of teachers in this area of education there is little research on diabetes type 1 self-management in the school setting [8, 22].

Clinical experience and some studies indicate that head teachers and other school staff had an inadequate understanding of diabetes type 1 and were influenced by personal experience, mostly on information of diabetes type 2 [44]. In Poland there is no clear policy what support a patient with diabetes can expect from schools. In some countries, for example the UK or Poland school staff doesn't have a legal duty to administer medicines and it is only "free will" of a teacher to assist a child with diabetes administration when the school doesn't employ a nurse on a permanent basis. Children and adolescents encounter a lot of misunderstandings concerning participation in physical education lessons with teachers using diabetes as an excuse for a pupil to 'sit out' when they have limited knowledge on diabetes management [45]. Some medical centers have specialist leaflets, booklets or DVD films on diabetes, prepared for teachers, usually handed out to them by parents. They are not followed by any form of formal education on diabetes carried out by professionals as a natural course when a child with diabetes enrolls at school or if a pupil is diagnosed with diabetes during the school year. The lack of national standards for diabetes education in the school setting established in Poland results in informal education carried out by parents or some local diabetes organizations. Still, parents express a lot of dissatisfaction with school support for their children [34]. 25% of children hide their diabetes in school, 12% of children are mocked and 17% of parents experience openly negative reactions when informing the school about their child's condition, 8% are forced to find another school - these figures call for concern as Amillategui et al. show in their survey completed by 499 parents [46]. Further, 34% of children do not take part in excursions if the trip is longer than one day. Only 34% of parents were convinced of the teachers' ability to recognize symptoms of mild hypoglycemia; when the majority of participants believed the school staff should know how to manage hypoglycemia only 55% of schools had glucagon available on their premises [36].

These figures show a big gap between diabetes type 1 treatment requirements and school staff knowledge and willingness to cooperate in diabetes management. Diabetes education in the school settings is still insufficient and should be a focus of a determined effort to improve diabetes type 1 children's safety. If we want children and adolescents with diabetes type 1 to be as active as their healthy peers and at the same time manage diabetes effectively, we have to provide better conditions to adhere to treatment. The teachers' understanding and support are necessary components in achieving success.

Summer Camps for Children and Adolescents with Diabetes

Summer camps for children and adolescents with diabetes type 1 offer a great opportunity for diabetes education. Healthcare professionals obligatorily presence at camps have the unique possibility to observe the patients' behaviors, interactions with peers, diabetes-related habits (blood glucose monitoring, keeping to diet, reactions to hypoglycemia symptoms, etc.) and also test their knowledge and skills (hyperglycemia and hypoglycemia management, insulin dosage, carbohydrate counting) [47]. These observations can be directly used for educational interventions not only age appropriate, but also exactly tailored to individual needs. Knowledge gained from experiencing particular situations can be immediately tested and implemented. Education can be delivered individually and in groups. It can happen "by chance" when an opportunity occurs or as a part of the camp program with set hours and activities. Even simple discussion may turn into education when diabetes-related issues are brought into the conversation. What education will look like depends on educators' invention and participants' engagement. Patients are available the whole day so it is easier to organize educational sessions at convenient times. Additional peer support is another factor, which may motivate to change some behaviors and improve metabolic control. It may be expected new habits should result in long-lasting changes in diabetes management. This hypothesis was not confirmed by Garcia-Perez et al. [48]. Educational intervention delivered at summer camp was not effective and didn't improve HbA1c, BMI, diabetes knowledge, anxiety, psychological adaptation or use of health care services. The only effect found was improvement in the adaptation to school environment in participants of educational sessions [38]. It is very interesting that social skills learned at the camp sustained over the time, whereas, behaviors related to diabetes management were not affected by diabetes education. It raises a question what factors influenced these results: the developmental stage when the need to belong to some group and be as others is more important than the care for one's health, family situation, the lack of support for diabetes management, hiding diabetes from healthy peers may also play a part.

Psychological Intervention

HbA1c is one of the most important criteria of compliance to treatment. When a patient achieves HbA1c within recommended range, it means he has adequate knowledge of diabetes and uses it in real life. We may assume diabetes educators and the whole medical team achieved their goal: a patient's good adherence to treatment. These patients are not perceived as difficult and their problems are often overlooked [49]. Medical staff focus mainly on children and adolescents who don't achieve good metabolic control even if they participated in numerous continuous diabetes education programs. When tested, they have knowledge on diabetes and the risk of late complications, but their adherence to treatment is problematic and often signals other emotional difficulties. In this situation these patients are offered psychological care.

Although many educational programs use psychological techniques education is understood as a reduction of gaps in knowledge, with the use of various didactic methods while psychotherapeutic intervention is based on an alliance between patient and therapist; problems are explained in terms of emotions, cognitions and behaviors. They are delivered by different specialists with different skills and theoretical background. Psychological interventions used at medical settings usually aim at improving treatment adherence and illness-related distress with less attention paid to psychosocial life with diabetes [50, 51]. Patients with diabetes and diagnosed psychiatric disorders (depression, eating disorders, conduct disorder) are usually referred to a specialist clinic where they undergo appropriate treatment as their healthy peers, although there is a call for integrating a psychological care into medical service as it may increase the use of this facility [52].

Improvement in diabetes management was shown in studies using different theoretical approaches, mostly cognitive behavioral therapy, behavioral family systems therapy, family systems theory, multi-systemic therapy, coping skills for youth with T1DM [52]. Interventions can be delivered individually, in a family group or small peer groups, usually in the hospital outpatient setting, but also at home or as a new initiative – through telemedicine [53, 54]. Studies on therapy effectiveness may have a randomized design, but in everyday practice therapeutic approach is chosen considering a patient's problem, his and family engagement in participation and a psychotherapist's skills.

Individual and family factors may affect diabetes outcome. Health beliefs (self-efficacy, diabetes perception as a burden), acceptance (illness and self as a person with diabetes), stress management (coping styles, relational stress), depressive symptoms are widely recognized factors contributing to poor diabetes management [55, 56]. Especially depression in adolescents should be addressed by clinicians because even mild symptoms may interfere with diabetes treatment and parental efforts to achieve good glycemic control in their children [57]. As parental involvement in the treatment of a child with diabetes is necessary a lot of attention is paid to family functioning and factors, which directly may affect health outcome [58]. Adolescence without diabetes is difficult enough time for youngsters in the struggle for independence and disease is an additional factor causing conflicts in families. Finding balance between parental handing over and adolescent accepting responsibilities for diabetes treatment and at the same time keeping caregivers involved is one of aims of family interventions [59, 60]. Reduction of conflicts arising from parental lack of understanding teenagers' focus on the present, criticism, intrusive and blaming behaviors and rigid control is a focus of family or multicomponent interventions [61, 62].

Multicomponent interventions seem to bring the best results as studies on the effectiveness of psychological interventions show [61]. Knowledge on diabetes management is not sufficient factor for good metabolic control. Different theoretical background, study methodological design, variety of outcomes evaluated (HbA1c, quality of life, self-efficacy, family conflicts to mention a few) and duration of follow-up treatments make comparisons difficult. Still, most research shows that psychological treatment has a modest effect on improvement of metabolic control or psychological variables (quality of life, diabetes related conflict reduction) without insufficient evidence to recommend one program [51, 54, 56]. There is a strong recommendation to include emotional, social and family processes in psychological intervention as concentration on behavior change has no such strong impact on improvement of metabolic control and as its effect is not sustained continuous psychological support may be required [62].

Self-Management or Self-Care

Research results show little effectiveness of interventions by neglecting psychosocial variables and focusing solely on improvement of diabetes control. It raises a question of shifting approach from diabetes self-management to self-care. While diabetes self-management implies a patient's responsibility for decisions and behavior required in daily management of their condition self-care is a broader concept including developmental, psychosocial and health needs. It takes a perspective of promoting health and general well-being in opposition to concentrating on management and prevention of disease [63]. Self-care is not limited to behaviors directly related to diabetes as blood glucose monitoring and insulin administration but looks at healthy lifestyle in general. It promotes healthy diet, physical activity, sufficient rest with night sleep and campaigns against smoking, drug use and excess alcohol consumption. It requires a holistic view of a patient as a person with physical and psychological needs. Such an approach requires the change of patients' perception by healthcare providers, which may be hard to achieve in practice. Both pediatric patients and their parents value relationship-based care which goes beyond medical needs and to express interest in a patient's psychosocial problems requires time and counseling skills of which medical staff is often short [64]. At the same time self-care approach gives a scope for national educational campaigns aiming at increasing diabetes awareness and its prevention.

Reduction in diabetes ketoacidosis at the diagnosis or lifestyle changes to prevent obesity or reduce the risk of metabolic syndrome in obese adolescents are examples of educational programs with self-care idea at its theoretical basis [65-68]. The best example of promoting health care behaviors is World Diabetes Day celebrated on 14th November since 1991. It was created by the International Diabetes Federation and the World Health Organization to increase global awareness of diabetes.

Different activities last the whole year with peak period in November they include: radio and television programs, sports events, free screenings for diabetes and its complications, public information meetings, poster and leaflet campaigns, diabetes workshops and exhibitions, press conferences, newspaper and magazine articles and events for children and adolescents. The more general public knows about diabetes, the more chances there are for lifestyle changes to stop the escalation of the disease and its health complications. The more knowledge and information available on diabetes the more chance there is for a safer and more satisfied life for people with diabetes type 1 and 2.

Conclusion

Education in diabetes self-management (DSME) is considered the keystone in the treatment of patients with diabetes. There is no one univocal educational program as each medical center develops its own, following either the ADA or the ISPAD recommendations for diabetes education. They employ different theories, methods and setting and are delivered by multidisciplinary teams. Each patient and his/her family undergo initial education at the time of diabetes diagnosis and have access to continuous education throughout pediatric care. Deterioration of metabolic control, especially during adolescence is the main cause of participation in the intervention program. There are strong recommendations to include emotional, social and family processes in psychoeducational intervention, as concentration on behavior change has no such strong impact on the improvement of metabolic control and as such its effect is not sustained, therefore continuous psychological support may be required. The self-care concept gives a wider perspective on a patient's treatment and may be worth considering implementing in everyday practice.

References

[1] Atkinson MA, Eisenbarth GS, Michels AW. Type 1 diabetes. *Lancet.* 2014; 383: 69–82.

[2] Copeland KC, Silverstein J, Moore KR, Prazar GE, Raymer T; The American Academy of Pediatrics. Management of newly diagnosed type 2 Diabetes Mellitus (T2DM) in children and adolescents. *Pediatrics.* 2013; 131: 364-82.

[3] Patterson CC, Gyürüs E, Rosenbauer J, Cinek O, Neu, A; EURODIAB Study Group. Trends in childhood type 1 diabetes incidence in Europe during 1989–2008: evidence of non-uniformity over time in rates of increase. *Diabetologia.* 2012; 55: 2142-7.

[4] Chiang JL, Kirkman MS, Laffel LM, Peters AL; the Type 1 Diabetes Sourcebook Authors. Type 1 diabetes through the life span: a position statement of the American Diabetes Association. *Diabetes Care.* 2014; 37: 2034-54.

[5] Patterson CC, Dahlquist GG, Gyürüs E, Green A, Soltész G; EURODIAB Study Group. Incidence trends for childhood type 1 diabetes in Europe during 1989-2003 and predicted new cases 2005-20: a multicentre prospective registration study. Lancet. 2009; 373; 2027-33.

[6] Bartlett EE. Historical glimpses of patient education in the United States. *Patient Educ Couns.* 1986; 8: 135–49.

[7] Swift PG. Diabetes education in children and adolescence. *Pediatr Diabetes.* 2009; 10 (Suppl 12):51- 7.

[8] Lange K, Swift P, Pańkowska E, Danne T. Diabetes education in children and adolescents. *Pediatr Diabetes.* 2014; 15 (Suppl. 20): 77–85.

[9] Haas L, Maryniuk M, Beck J, Cox CE, Duker P; 2012 Standards Revision Force Task. National standards for diabetes self-management and support. *Diabetes Care.* 2014; 37 (Suppl 1): S144-53.

[10] Silverstein J, Klingensmith G, Copeland K, Plotnick L, Kaufman F; American Diabetes Association. Care of children and adolescents with type 1 diabetes: a statement of the American Diabetes Association. *Diabetes Care.* 2005; 28:186-212.

[11] Canadian Diabetes Association Clinical Practice Guidelines Expert Committee. Canadian Diabetes Association 2013 Clinical Practice Guidelines for the Prevention and Management of Diabetes in Canada. *Can J Diabetes.* 2013; 37 (Suppl. 1): S1–S212.

[12] Craig ME, Twigg SM, Donaghue KC et al. for the Australian Type Diabetes Guidelines Expert Advisory Group. National Evidence-based Clinical Care Guidelines for Type 1 Diabetes in Children, Adolescents and Adults. Canberra: Australian Government Department of Health and Ageing, 2011 (available from http://www.diabetessociety.com.au/position-statements.asp).

[13] DH Diabetes Policy Team. Making Every Young Person with Diabetes Matter: Report of the Children and Young People with Diabetes Working Group. UK Department of Health, London. 2007. (available from http://www.dh.gov.uk/prod_consum_dh/groups/dh_digitalassets/@dh/@en/documents/digitalasset/dh_073675.pdf).

[14] Holterhus PM, Beyer P, Burger-Busing J et al. Diagnostik, Therapie und Verlaufskontrolle des Diabetes mellitus im Kindes- und Jugendalter. S3 Leitlinie der Deutschen Diabetes Gesellschaft. [Diagnosis, Therapy and Long Term Care of Diabetes in Childhood and Adolescence – S3 Guideline]. Mainz: Kirchheim, 2009: p. 1–p306.

[15] Zespół ds. Zaleceń Polskiego Towarzystwa Diabetologicznego. Zalecenia kliniczne dotyczące postępowania u chorych na cukrzycę 2014. Stanowisko Polskiego Towarzystwa Diabetologicznego. Diabetologia kliniczna 2014; 3 (Suppl. A): A41-6.

[16] Danne T, Lion S, Madaczy L, Veeze H, Raposo F, Rurik I; SWEET Group. Criteria for Centers of Reference for pediatric diabetes – a European perspective. *Pediatr Diabetes.* 2012; 13 (Suppl 16): 62-75.

[17] Pańkowska E, Szypowska A, Lipka M, Szpotańska M, Błazik M, Groele L. Application of novel dual wave meal bolus and its impact on glycated hemoglobin A1c level in children with type 1 diabetes. *Pediatr Diabetes.* 2009; 10: 298-303.

[18] Kordonouri O, Hartmann R, Remus K, Bläsig S, Sadeghian E, Danne T. Benefit of supplementary fat plus protein counting as compared with conventional carbohydrate counting for insulin bolus calculation in children with pump therapy. *Pediatr Diabetes.* 2012; 13: 540-544.

[19] Knowles J, Waller H, Eiser C, Heller S, Roberts J. The development of an innovative education curriculum for 11-16 yr old children with type 1 diabetes mellitus (T1DM). *Pediatr Diabetes.* 2006; 7: 322-8.

[20] Waller H, Eiser C, Knowles J, Rogers N, Wharmby S. Pilot study of a novel educational programme for 11-16 year olds with type 1 diabetes mellitus: the KICk-OFF course. *Arch Dis Child.* 2008; 93: 927-31.

[21] Price KJ, Wales J, Eiser C, Knowles J, Heller S. Does an intensive self-management structured education course improve outcomes for children and young people with type 1 diabetes? The Kids In Control OF Food (KICk-OFF) cluster-randomised controlled trial protocol. *BMJ Open.* 2013; 24: 3: e002429.

[22] Martin D, Lange K, Sima A, Kownatka D, Skovlund S, Danne T et al. Recommendations for age-appropriate education of children and adolescents with diabetes and their parents in the European Union. *Pediatr Diabetes.* 2012; 13 Suppl 16: 20-8.

[23] Norris SL, Nichols PJ, Caspersen CJ, Glasgow RE, Engelgau MM; the Task Force on Community Preventive Services. Increasing Diabetes Self-Management Education in Community Settings. A Systematic Review. *Am J Prev Med.* 2002; 22: 39-66.

[24] Tonyushkina KN, Visintainer PF, Jasinski CF, Wadzinski TL, Allen HF. Site of initial diabetes education does not affect metabolic outcomes in children with T1DM. Pediatr Diabetes; 2014: 15: 135–141.

[25] Lowes L, Gregory JW. Management of newly diagnosed diabetes: home or hospital? *Arch Dis Child.* 2004; 89: 934–937.

[26] Tiberg I, Katarina SC, Carlsson A, Hallström I. Children diagnosed with type 1 diabetes: a randomized controlled trial comparing hospital versus home-based care. *Acta Paediatrica.* 2012; 101: 1069–1073.

[27] Landolt MA, Vollrath M, Laimbacher J, Gnehm HE, Sennhauser FH. Prospective Study of Posttraumatic Stress Disorder in Parents of Children With Newly Diagnosed Type 1 Diabetes. *J Am Acad Child Adolesc Psychiatry.* 2005; 44: 682-9.

[28] Streisand R, Mackey ER, Elliot BM, Mednick L, Slaughter IM, Turek J, Austin A. Parental anxiety and depression associated with caring for a child newly diagnosed with type 1 diabetes: opportunities for education and counseling. *Patient Educ Couns.* 2008; 73: 333-8.

[29] Hoff AL, Mullins LL, Gillaspy SR, Page MC, Van Pelt JC, Chaney JM. An Intervention to Decrease Uncertainty and Distress Among Parents of Children Newly Diagnosed With Diabetes: A Pilot Study. *Families, Systems, & Health.* 2005; 23: 329-342.

[30] Norcross JC, Wampold BE. Evidence-Based Therapy Relationships: Research Conclusions and Clinical Practices. Psychotherapy. 2011; 48: 98–102.

[31] Waller H, Eiser C, Heller S, Knowles J, Price K. Adolescents' and their parents' views on the acceptability and design of a new diabetes education programme: a focus group analysis. *Child Care Health Dev.* 2005; 3: 283-9.

[32] Chaney D, Coates V, Shevlin M, Carson D, McDougall A, Long A. Diabetes education: what do adolescents want? *J Clin Nurs.* 2012; 21: 216-23.

[33] Rankin D, Heller S, Lawton J. Understanding information and education gaps among people with type 1 diabetes: a qualitative investigation. *Patient Educ Couns.* 2011; 83: 87-91.

[34] Markowitz JT, Harrington KR, Laffel LM. Technology to Optimize Pediatric Diabetes Management and Outcomes. *Curr Diab Rep.* 2013; 13: 877–885.

[35] El-Gayar O, Timsina P, Nawar N, Eid W. A systematic review of IT for diabetes self-management: are we there yet? *Int J Med Inform.* 2013; 82: 637-52.

[36] Holtslander L, Kornder N, Letourneau N, Turner H, Paterson B. Finding straight answers: identifying the needs of parents and service providers of adolescents with type 1 diabetes to aid in the creation of an online support intervention. *J Clin Nurs.* 2012; 21: 2419-28.

[37] Paterson B, Brewer J. Needs for social support among parents of adolescents with diabetes. *J Nurs Healthc Chronic Illn.* 2009; 1: 177–185.

[38] Whittemore R, Jaser SS, Jeon S, Liberti L, Delamater A, Murphy K. An internet coping skills training program for youth with type 1 diabetes: six-month outcomes. *Nurs Res.* 2012; 61: 395-404.

[39] Mulvaney SA, Rothman RL, Wallston KA, Lybarger C, Dietrich MS. An internet-based program to improve self-management in adolescents with type 1 diabetes. *Diabetes Care.* 2010; 33: 602-4.

[40] Grey M, Whittemore R, Jeon S, Murphy K, Faulkner MS, Delamater A; TeenCope Study Group. Internet Psycho-Education Programs Improve Outcomes in Youth With Type 1 Diabetes. *Diabetes Care.* 2013; 36: 2475-82.

[41] Herbert L, Owen V, Pascarella L, Streisand R. Text message interventions for children and adolescents with type 1 diabetes: a systematic review. Diabetes Technol Ther. 2013; 15: 362-70.

[42] Kamel Boulos MN, Gammon S, Dixon MC, MacRury SM, Fergusson MJ, Miranda Rodrigues F et al. Digital Games for Type 1 and Type 2 Diabetes: Underpinning Theory With Three Illustrative Examples. Eysenbach G, ed. *JMIR Serious Games.* 2015; 3: e3.

[43] DeShazo J, Harris L, Pratt W. Effective intervention or child's play? A review of video games for diabetes education. *Diabetes Technol Ther.* 2010; 12: 815-22.

[44] Marshall M, Gidman W, Callery P. Supporting the care of children with diabetes in school: a qualitative study of nurses in the UK. *Diabet Med.* 2013; Jul; 30: 871-7.

[45] MacMillan F, Kirk A, Mutrie N, Moola F, Robertson K. Building physical activity and sedentary behavior support into care for youth with type 1 diabetes: patient, parent and diabetes professional perceptions. *Pediatr Diabetes.* 2014; doi: 10.1111/pedi.12247. [Epub ahead of print] PubMed PMID: 25482088.

[46] Amillategui B, Calle JR, Alvarez MA, Cardiel MA, Barrio R. Identifying the special needs of children with Type 1 diabetes in the school setting. An overview of parents' perceptions. *Diabet Med.* 2007; 24: 1073-9.

[47] American Diabetes Association. Diabetes management at camps for children with diabetes. *Diabetes Care.* 2012; 35 Suppl 1: S72-5.

[48] García-Pérez L, Perestelo-Pérez L, Serrano-Aguilar P, Del Mar Trujillo-Martín M. Effectiveness of a psychoeducative intervention in a summer camp for children with type 1 diabetes mellitus. *Diabetes Educ.* 2010 Mar-Apr; 36(2): 310-7.

[49] Zduńczyk B, Sendela J, Szypowska A. High prevalence of depressive symptoms in well-controlled adolescents with type 1 diabetes treated with continuous subcutaneous insulin infusion. *Diabetes Metab Res Rev.* 2014; 30: 333-8.

[50] Winkley K, Ismail K, Landau S, Eisler I. Psychological interventions to improve glycaemic control in patients with type 1 diabetes: systematic review and meta-analysis of randomised controlled trials. *BMJ.* 2006; 333(7558): 65.

[51] Barlow JH, Ellard DR. Psycho-educational interventions for children with chronic disease, parents and siblings: an overview of the research evidence base. *Child Care Health Dev.* 2004; 30: 637-45.

[52] Bitsko MJ, Bean MK, Bart S, Foster RH, Thacker L, Francis GL. Psychological treatment improves hemoglobin A1c outcomes in adolescents with type 1 diabetes mellitus. *J Clin Psychol Med Settings.* 2013; 20: 333-42.

[53] Murphy HR, Rayman G, Skinner TC. Psycho-educational interventions for children and young people with Type 1 diabetes. *Diabet Med.* 2006; 23: 935-43.

[54] Kichler JC, Kaugars AS, Marik P, Nabors L, Alemzadeh R. Effectiveness of groups for adolescents with type 1 diabetes mellitus and their parents. *Fam Syst Health.* 2013; 31: 280-93.

[55] Jaser SS, Faulkner MS, Whittemore R, Jeon S, Murphy K, Delamater A et al. Coping, self-management, and adaptation in adolescents with type 1 diabetes. *Ann Behav Med.* 2012; 43: 311-9.

[56] Naranjo D, Mulvaney S, McGrath M, Garnero T, Hood K. Predictors of self-management in pediatric type 1 diabetes: individual, family, systemic, and technologic influences. *Curr Diab Rep.* 2014; 14: 544.

[57] Wu YP, Hilliard ME, Rausch J, Dolan LM, Hood KK. Family involvement with the diabetes regimen in young people: the role of adolescent depressive symptoms. *Diabet Med.* 2013; 30: 596-602.

[58] Young MT, Lord JH, Patel NJ, Gruhn MA, Jaser SS. Good cop, bad cop: quality of parental involvement in type 1 diabetes management in youth. *Curr Diab Rep.* 2014; 14: 546.

[59] Dashiff C, Hardeman T, McLain R. Parent-adolescent communication and diabetes: an integrative review. *J Adv Nurs.* 2008; 62: 140-62.

[60] Ingerski LM, Anderson BJ, Dolan LM, Hood KK. Blood glucose monitoring and glycemic control in adolescence: contribution of diabetes-specific responsibility and family conflict. *J Adolesc Health.* 2010; 47: 191-7.

[61] Weinger K, O'Donnell KA, Ritholz MD. Adolescent views of diabetes-related parent conflict and support: a focus group analysis. *J Adolesc Health.* 2001; 29: 330-6.

[62] Hood KK, Rohan JM, Peterson CM, Drotar D. Interventions with adherence-promoting components in pediatric type 1 diabetes: meta-analysis of their impact on glycemic control. *Diabetes Care.* 2010; 33: 1658–64.

[63] Pelicand J, Fournier C, Le Rhun A, Aujoulat I. Self-care support in paediatric patients with type 1 diabetes: bridging the gap between patient education and health promotion? *A review. Health Expect.* 2015; 18: 303-11.

[64] Curtis-Tyler K, Arai L, Stephenson T, Roberts H. What makes for a 'good' or 'bad' paediatric diabetes service from the viewpoint of children, young people, carers and clinicians? A synthesis of qualitative findings. *Arch Dis Child.* 2015 Mar 19. pii: archdischild-2014-307517. [Epub ahead of print]

[65] King BR, Howard NJ, Verge CF, Jack MM, Govind N, Jameson K et al. A diabetes awareness campaign prevents diabetic ketoacidosis in children at their initial presentation with type 1 diabetes. *Pediatr Diabetes.* 2012;13: 647-51.

[66] Reinehr T. Lifestyle intervention in childhood obesity: changes and challenges. *Nat Rev Endocrinol. 2013*; 9: 607-14.

[67] Bianchini JA, da Silva DF, Nardo CC, Carolino ID, Hernandes F, Nardo N Jr. Multidisciplinary therapy reduces risk factors for metabolic syndrome in obese adolescents. *Eur J Pediatr.* 2013; 172: 215-21.

[68] Schneider M, DeBar L, Calingo A, Hall W, Hindes K, Sleigh A, et al. The effect of a communications campaign on middle school students' nutrition and physical activity: results of the HEALTHY study. *J Health Commun.* 2013; 18: 649-67.

In: Alterations in Glucose Homeostasis in Children
Editor: E. Otto-Buczkowska ISBN: 978-1-63483-861-0

Chapter 4

NOVEL EXPERIMENTAL THERAPIES OF TYPE 1 DIABETES AND FUTURE PERSPECTIVES

M. Gliwiński[1], D. Iwaszkiewicz-Grześ[1], M. Żalińska[2], M. Myśliwiec[2] and P. Trzonkowski[1,*]

[1]Department of Clinical Immunology and Transplantology, Medical University of Gdańsk, Poland

[2]Department of Pediatrics, Diabetology and Endocrinology, Medical University of Gdańsk, Poland

ABSTRACT

Type 1 diabetes (T1D) is a chronic autoimmune disease characterized by autodestruction of pancreatic β cells by autoreactive T cells. It creates a deficit of insulin and development of uncontrolled hyperglycemia. Thus, the patients are dependent on injections of insulin. Early interventions to spare remaining β cells aim at ceasing the activity of autoreactive T cells and restoring immunological self-tolerance.

Up-to-date, there are no fully approved immunotherapies of this kind. This chapter highlights past attempts and new perspectives in the treatment of T1D in children.

Keywords: Hyperglycemic, type 1 diabetes, immunotherapy, dendritic cells, Treg, cell therapy

INTRODUCTION

Insulin-dependent type 1 diabetes (T1D) is one of the most common chronic autoimmune diseases in children. It is characterized by a loss of self-tolerance which leads to selective destruction of insulin-producing pancreatic β cells. Autoreactive T cells and auto-antibodies against a variety of auto-antigens: glutamic acid decarboxylase-65 (GAD65), insulinoma-associated antigen-2 (IA-2), insulin and Zinc transporter ZnT8 can be detected. As a result,

[*] E-mail: ptrzon@gumed.edu.pl

the body capacity to produce insulin decreases and finally a complete loss of pancreatic islets occurs. It is dependent on many factors including genetic predisposition and environmental triggers.

It has been reported that T1D risk is associated with particular combinations of DR3 and DR4 HLA alleles. However, the risk for T1D can be significantly different between two closely related DR-DQ haplotypes. The comparison of these haplotype pairs gives an ability to identify critical amino acid positions in the structure of HLA receptor.

Environmental factors involving viral infections (with enteroviruses considered to be the main viral trigger), dietary factors (cow's milk proteins; gluten; deficiency of vitamin D) and growth consist another important risk factors of the disease involved in both initiation and further progression [2-6].

In many autoimmune disorders, a commonly accepted treatment is the substitution, i.e., administration of exogenous insulin in T1D or thyroid hormones in Hashimoto disease [7, 8]. The strategy based on immunosuppression to stop autoimmune process is another option. The administration of immunosuppressive drugs, from 'old' glucocorticosteroids to novel formulations and anti-inflammatory drugs is widely tested.

Immunosuppressive Drugs

Ciclosporin A

Ciclosporin (CsA) was discovered during the studies on antifungal agents. It is a small (11-amino acid) cyclic peptide with powerful immunosuppressive capacity. Due to the strong immunosuppressive effect, ciclosporin has been used to prevent transplant rejection as well as in the treatment of some autoimmune diseases. CsA has the ability to prevent activation and proliferation of T cells through the inhibition of interleukin 2 (IL-2) production. CsA, after binding with cyclophilin in the cell, gets the ability to bind and inhibit calcineurin. This process prevents dephosphorylation of the transcription factor NF-AT, resulting in the inhibition of IL-2 gene transcription. This effect was also noted for genes encoding other cytokines. Functional result of the CsA is a complete immunoparalysis.

The basic problem of this treatment in T1D is severe nephrotoxicity of CsA [9]. Because effect of the therapy can be only achieved when the drug is administered continuously, the risk/benefit ratio of the treatment dictated discontinuation of further trials with CsA in T1D. Unfortunately, in all treated patients when the therapy was discontinued (or even the dose of CsA was reduced), disease relapse occurred [10].

Glucocorticosteroids

The main task of the immune system is to defend the body against pathogens. However, excessive intense immune response may be disadvantageous to homeostasis or even lead to destruction of tissues and organs. The immunosuppressive effects of adrenocortical hormones secreted under stress conditions prevents unwanted escalation of immune responses. It is therefore one of the basic mechanisms of systemic control of immune responses.

There are many glucocorticosteroids useful as immunosuppressive agents, with the most commonly used prednisone (synthetic analogue of cortisol), prednisolone and methylprednisolone. Glucocorticosteroids can enter cells by diffusing across the plasma membrane and then binding their receptors located in the cytoplasm. This complex translocates to the nucleus and binds to specific GRE sequences (glucocorticoid response elements) present within the promoter region of many genes. An active glucocorticoid receptor can activate genes indirectly as a trigger of gene transcription for STAT5. In the cell, there are several dozen genes that contain the GRE. Binding of the active receptor with GRE sequence within the promoter may have a double effect on the process of gene transcription, stimulating some of them or repressing the others. Mechanism of action of steroid drugs is based on the inhibition of nuclear transcription factor (NF-κB), which is the master regulator of proinflammatory factors. Glucocorticosteroids can also exert other immunosuppressive activities, for example, inducing apoptosis of lymphocytes [11].

After 8 months of treatment of T1D patients with prednisone, a lower insulin requirement, higher level of endogenous insulin release and only minor side effects were reported in a group of twenty five patients. In the prednisone-treated group, insulin requirement was lower than in the control group at 12 months (0.33 ± 0.11 vs. 0.57 ± 0.06 U/kg/day), 18 months (0.34 ± 0.11 vs. 0.64 ± 0.06) and 24 months (0.38 ± 0.10 vs. 0.63 ± 0.05). With low doses of prednisone administered continuously, the treatment effectively restored endogenous insulin release in T1D patients [12]. However, long term treatment with glucocorticoids leads to increased resistance of the tissue to insulin and thus higher demand for the hormone and the development of steroid-dependent diabetes. The drugs may be also associated with many other severe adverse effects (severe infections, arterial hypertension, osteoporosis, ulcerations, bleeding from digestive tract, depression etc.), which altogether makes the therapy too risky for brittle T1D.

AUTO-ANTIGEN THERAPY

Antigen-specific modulation of the immune response aims to re-establish the immune tolerance to β cells. Many antigen-specific treatment strategies have been proved to be effective in murine models of T1D. These mainly include administration of islet auto-antigens (whole protein or peptide) through parental, oral and nasal routes. Due to the fact that this approach exerts only minimal adverse effects, targets only pathogenic autoreactive T cells and induces immune tolerance, it is of great interest in the treatment of T1D [13].

Insulin and proinsulin molecules have been identified to play a primary role in the initiation of the autoimmune process that ultimately leads to destruction of β cells and onset of clinical diabetes.

Insulin is the only protein produced exclusively by β cells. It is well-known that insulin is the most important auto-antigen in the NOD model. Many animal trials aimed at inducing tolerance of the immune system towards insulin in T1D have been conducted. Oral administration of the entire insulin protein has consistently resulted in disease prevention in spontaneous and induced animal models. For this reason, insulin administered as an auto-antigen has been widely studied in human prevention and new-onset T1D trials [13].

In 2002, Alleva et al. reported on NBI-6024 - alerted-peptide ligand based on insulin β-chain. A very promising approach unfortunately finished after a phase II clinical trial due to lack of effect in treated patients [14]. In the trial by Thrower et al. proinsulin peptide administration was well tolerated by all subjects. The study showed that intradermal injection of proinsulin was both well-tolerated and safe. There were no systemic allergic reactions or other treatment-related serious adverse events. Routine biochemical and hematological measurements did not change significantly during the treatment, either. In particular, the study showed no evidence that proinsulin peptide immunotherapy was associated with a risk of induction/reactivation of potentially damaging pro-inflammatory T cell responses [15].

In the absence of any observed induction of anti-peptide antibodies or emergence of proinsulin peptide specific CD4 T cells secreting proinflammatory IFN-γ (a hallmark of autoaggressive T cells in T1D), it seems reasonable to conclude that T1D exacerbation may be of limited concern after administration of both insulin or proinsulin through alternative routes [15].

The mechanisms through which peptide immunotherapy mediated its beneficial effects are not known with certainty, but include the induction of anergy and deletion of pathogenic T cells that target the administered sequence. Interestingly, there is also evidence for induction of $CD4^+$ T cells with regulatory properties. Peptide immunotherapy promotes dominant presentation by immature dendritic cells through the presentation of the peptides in the context of the major histocompatibility complex (MHC) class II molecules, conditions known to favor the induction of tolerogenic CD4 T cells secreting IL-10 [15].

ANTIBODIES THERAPY

Anti-CD3

The CD3 molecule is involved in signaling through the TCR receptor and therefore the administration of an anti-CD3 antibody (Ab) can cause non-specific stimulation of T cells and release of cytokines. On the other hand, lytic or blocking anti-CD3 mAb (monoclonal antibody) can paralyze T cell responses. For more than 20 years, an anti-ε-chain CD3 complex present on all T cells (muromonab-CD3, OKT3) has been used for the latter. Within minutes of the first dose of OKT3, T cells disappear from circulation. With continued treatment, after 3-5 days, T cells with CD4/CD8 molecules, but not expressing the CD3 complex, appear in the blood. The lack of the CD3 complex makes the cells non-responsive to TCR stimulation and unable to participate in the immune response.

In the 1990s, a new paradigm developed based on the finding that a brief course of treatment with anti-CD3 mAb in the NOD mouse model induced disease reversal. Immune tolerance might be induced even in animals with overt disease [16]. The use of anti-CD3 mAbs in the treatment of T1D in humans was limited due to their toxicity, the activation of T cells in vitro and the occurrence of the cytokine release syndrome (CRS). Humanized anti-CD3 antibodies with reduced binding to the low affinity IgG Fc receptors (i.e., CD16 and CD32) were therefore developed to minimize T cell activation, cytokine release and anti-drug humoral responses originally associated with murine-derived OKT3 in humans.

Three humanized anti-CD3 mAbs with reduced binding to IgG FcRs were created:

- Visilizumab (HuM291; Nuvion) – a humanized IgG2 monoclonal antibody was tested in inflammatory bowel disease. Cytokine release was significantly reduced, but not eliminated. With the dose selected for clinical testing, however, the drug failed to achieve a lasting therapeutic effect [17].
- Teplizumab (h0kT3gamma1, MGA031) was tested for the treatment of acute rejection of kidney and kidney/pancreas allografts and refractory psoriatic arthritis demonstrating efficacy and acceptable adverse events [18].
- Otelixizumab (CChAgly CD3) – a glycosylated chimeric/humanized monoclonal antibody that contains a humanized gamma1 heavy chain and rat/human chimeric delta light chain. A reduction in T cell activation similar to others was reported for this agent [19]

Unfortunately, the majority of trials with anti-CD3 antibodies in T1D stopped in phase III due to low efficacy of the treatment. These unsatisfactory results were mostly dependent on inclusion criteria for the patients. Too broad criteria resulted in the inclusion of patients with already destroyed islets and full onset diabetes. At the same time, post-trial analysis of the results revealed a group of patients who benefited from the treatment. These were mainly recent onset T1D patients in clinical remission with still relatively preserved function of pancreatic islets.

In fact, the experience with anti-CD3 trials is a very instructive lesson for all immunotherapy attempts in T1D. As this kind of interventions does not directly regenerate the function of islets and only inhibits their destruction, an early intervention is the most efficient.

Unfortunately, it means that immunotherapy is the treatment for a very limited number of patients who are diagnosed at a very early stage of the disease. It is still the minority of T1D patients as the screening for T1D is not a routine in current medicine.

CTLA4Ig (Abatacept)

In 1997, Marron et al. reported on an association between the onset of T1D and polymorphisms of CTLA4 [20]. Studies have indicated CTLA4 as a negative regulator in T cell activation. CTLA4Ig is a fusion protein consisting of CTLA4 (CD152) and the Fc fragment of human immunoglobulin. The molecule binds to B7.1 (CD80) and B7.2 (CD86) receptors, which are expressed by antigen-presenting cells (APC) and prevents their interaction with CD28 on activated T cells [21].

Full activation of naïve T cells needs two signals. The first one from a TCR receptor that recognizes an antigen presented on MHC and the second (co-stimulatory) signal from interactions of CD28 with B7.1/B7.2. In the absence of the co-stimulatory signal, T cells enter an anergic state instead of activation. Clinical trials showed that CTLA4Ig has positive effects in the treatment of rheumatoid arthritis and in the control of allograft rejection [22, 23]. It has been observed that blocking the second signal with CTLA4Ig can stop the development of T1D in NOD mice [24]. In a human trial, it has been shown that co-stimulation modulation with CTLA4Ig sustained over a two-year period slowed the loss of β cell function in T1D by 6-9 months [25].

Cell Therapy

Dendritic Cells (DCs)

Dendritic cells belong to the group of professional antigen-presenting cells. They have the ability to control immune responses. It has been shown that DCs can escalate or reduce autoimmunity, depending on the their state of maturation. Mature DCs, defined through a high expression of MHC and co-stimulatory molecules and production of IL-12, stimulate T cell responses. On the other hand, immature DCs characterized by a low level of co-stimulatory molecules are known to quench the immune response [26].

Mature DCs are definitively involved in the pathogenesis of T1D. It has been reported that adoptive transfer of DCs treated with tumor necrosis factor (TNF-α), which is the cytokine that stimulates maturation of DCs, induced diabetes in NOD mice [27].

The activation and maturation of DCs is strongly dependent on the activity of NF-κB. Strikingly, the development of T1D is also linked with NF-κB activation. In 2003, Ma et al. demonstrated that the administration of DCs with blocked activation of NF-κB prevented the development of T1D in NOD mice [26]. It is therefore logical that immature DCs might be useful in the immunotherapy against T1D. Postulated mechanisms of autoimmunity prevention with DCs include impeded co-activation through the B7 family of receptors, lowered production of proinflammatory interleukins (e.g., IL-12) and apoptosis of T cells through the Fas ligand [28]. First clinical results with DCs demonstrated that non-manipulated autologous DCs and autologous DCs directed ex vivo towards a suppressive state are well tolerated and safe. However, no influence on T1D progression could be found [29].

Mesenchymal Stromal Cells (MSCs)

Application of MSCs is a new and promising treatment in T1D. Clinical studies with the administration of these cells to twenty patients showed increase in β cell function (expressed by the concentration of C-peptide) in a group of patients treated with MSCs at 1-year follow-up. Moreover, no side effects were observed in the treated group [30].

Regulatory T Cells (Tregs)

T regulatory cells $CD4^{+}CD25^{+}$ are the population of lymphocytes characterized by the expression of transcription factor Forkhead box P3 (FoxP3), with inhibitory/regulatory effects on immune responses. Regulatory mechanisms of Tregs include mainly direct cell-to-cell contact via the CTLA4 receptor of APCs or autoreactive lymphocytes, secretion of immunosuppressive cytokines, destruction or modification of APCs and competition for trophic factors like IL-2. These cells are recognized as a promising candidate for immunosuppressive therapy in humans [31].

For the first time, Tregs were shown to prevent multiple autoimmune diseases in mice in 1995 [32]. It has been observed that Tregs can prevent the destruction of pancreatic islets and protect from T1D in several animal models [33, 34]. In 2009, Trzonkowski et al. successfully applied ex vivo expanded Tregs in patients with chronic graft versus host disease (GvHD) after bone marrow transplantation [35]. Studies on Tregs confirmed that the treatment can suppress autoreactive T cells and stop diabetes in NOD mice [36]. A similar effect was proved in the trial with Tregs in humans [37]. In 2014, Marek-Trzonkowska et al. presented results of one-year follow-up in T1D after adoptive transfer of Tregs [38]. The study showed that the therapy is safe and the treatment resulted in reduced requirements for exogenous insulin. Importantly, 2 out of 12 enrolled patients did not require insulin at all at one year post-administration.

Interestingly, Luopajärvi et al. observed that the level of Tregs was higher in children whose mothers were under insulin treatment as compared to children of unaffected mothers. These results suggest that insulin treatment induces the expansion of regulatory T cells in the fetus and therefore lower risk of T1D in children [39]. Interestingly, similar positive effects of insulin or proinsulin administered as auto-antigens were only seen when the patients responded to the vaccine with an increase in the level of Tregs. Similar association between clinical response and increasing level of Tregs was also observed in anti-CD3 trials in T1D [39].

Conclusion

Immunotherapy approaches to stop diabetes are – at best – still at the level of clinical trials and none of them is a clinically approved routine procedure. Nevertheless, recent progress in our understanding of the functioning of the immune system and pathogenesis of autoimmune T1D give a hope for efficient treatment in the forthcoming years. Modified autoantigen vaccines and cellular therapies are among the most promising interventions in the repertoire of university groups and companies working on early prevention of T1D.

References

[1] Erlich H, Valdes AM, Noble J, Carlson JA, Varney M, Concannon P et al.; Type 1 Diabetes Genetics Consortium. HLA DR-DQ haplotypes and genotypes and type 1 diabetes risk: analysis of the type 1 diabetes genetics consortium families. *Diabetes.* 2008; 57(4): 1084-92.

[2] Frisk G, Friman G, Tuvemo T, Fohlman J, Diderholm H. Coxsackie B virus IgM in children at onset of type 1 (insulin-dependent) diabetes mellitus: evidence for IgM induction by a recent or current infection. *Diabetologia*. 1992; 35(3): 249-53.

[3] Luopajärvi K, Savilahti E, Virtanen SM, Ilonen J, Knip M, Akerblom HK et al. Enhanced levels of cow's milk antibodies in infancy in children who develop type 1 diabetes later in childhood. *Pediatr Diabetes*. 2008; 9(5): 434-41.

[4] Hoorfar J, Buschard K, Dagnaes-Hansen F. Prophylactic nutritional modification of the incidence of diabetes in autoimmune non-obese diabetic (NOD) mice. *Br J Nutr*. 1993; 69(2): 597-607.

[5] Mathieu C, Waer M, Laureys J, Rutgeerts O, Bouillon R. Prevention of autoimmune diabetes in NOD mice by 1,25 dihydroxyvitamin D3. *Diabetologia*. 1994; 37(6): 552-8.

[6] Johansson C, Samuelsson U, Ludvigsson J. A high weight gain early in life is associated with an increased risk of type 1 (insulin-dependent) diabetes mellitus. *Diabetologia*. 1994; 37(1): 91-4.

[7] Atkinson MA, Eisenbarth GS. Type 1 diabetes: new perspectives on disease pathogenesis and treatment. *Lancet*. 2001; 358(9277): 221-9.

[8] Slatosky J, Shipton B, Wahba H. Thyroiditis: differential diagnosis and management. *Am Fam Physician*. 2000; 61(4): 1047-52.

[9] Parving HH, Tarnow L, Nielsen FS, Rossing P, Mandrup-Poulsen T, Osterby R et al. Cyclosporine nephrotoxicity in type 1 diabetic patients. A 7-year follow-up study. *Diabetes Care*. 1999; 22(3): 478-83.

[10] Martin S, Schernthaner G, Nerup J, Gries FA, Koivisto VA, Dupré J et al. Follow-up of ciclosporin A treatment in type 1 (insulin-dependent) diabetes mellitus: lack of long-term effects. *Diabetologia*. 1991; 34(6): 429-34.

[11] Barnes PJ. Anti-inflammatory actions of glucocorticoids: molecular mechanisms. *Clin Sci* (Lond). 1998; 94(6): 557-72.

[12] Secchi A, Pastore MR, Sergi A, Pontiroli AE, Pozza G. Prednisone administration in recent onset type I diabetes. *J Autoimmun*. 1990; 3(5): 593-600.

[13] Alleva DG, Gaur A, Jin L, Wegmann D, Gottlieb PA, Pahuja A et al. Immunological characterization and therapeutic activity of an altered-peptide ligand, NBI-6024, based on the immunodominant type 1 diabetes autoantigen insulin B-chain (9-23) peptide. *Diabetes*. 2002; 51(7): 2126-34.

[14] Walter M, Philotheou A, Bonnici F, Ziegler AG, Jimenez R, NBI-6024 Study Group. No effect of the altered peptide ligand NBI-6024 on beta-cell residual function and insulin needs in new-onset type 1 diabetes. *Diabetes Care*. 2009; 32(11): 2036-40.

[15] Thrower SL, James L, Hall W, Green KM, Arif S, Allen JS et al. Proinsulin peptide immunotherapy in type 1 diabetes: report of a first-in-man Phase I safety study. *Clin Exp Immunol*. 2009; 155(2): 156-65.

[16] Chatenoud L, Thervet E, Primo J, Bach JF. Anti-CD3 antibody induces long-term remission of overt autoimmunity in nonobese diabetic mice. *Proc Natl Acad Sci USA*. 1994; 91(1): 123-7.

[17] Sandborn WJ, Colombel JF, Frankel M, Hommes D, Lowder JN, Mayer L et al. Anti-CD3 antibody visilizumab is not effective in patients with intravenous corticosteroid-refractory ulcerative colitis. *Gut*. 2010; 59(11): 1485-92.

[18] Sherry N, Hagopian W, Ludvigsson J, Jain SM, Wahlen J, Ferry RJ Jr et al.; Protégé Trial Investigators. Teplizumab for treatment of type 1 diabetes (Protégé study): 1-year results from a randomised, placebo-controlled trial. *Lancet*. 2011; 378(9790): 487-97.

[19] Ambery P, Donner TW, Biswas N, Donaldson J, Parkin J, Dayan CM. Efficacy and safety of low-dose otelixizumab anti-CD3 monoclonal antibody in preserving C-peptide secretion in adolescent type 1 diabetes: DEFEND-2, a randomized, placebo-controlled, double-blind, multi-centre study. *Diabet Med*. 2014; 31(4): 399-402.

[20] Marron MP, Raffel LJ, Garchon HJ, Jacob CO, Serrano-Rios M, Martinez Larrad MT et al. Insulin-dependent diabetes mellitus (IDDM) is associated with CTLA4 polymorphisms in multiple ethnic groups. *Hum Mol Genet.* 1997; 6(8): 1275-82.

[21] Abrams JR, Kelley SL, Hayes E, Kikuchi T, Brown MJ, Kang S et al. Blockade of T lymphocyte costimulation with cytotoxic T lymphocyte-associated antigen 4-immunoglobulin (CTLA4Ig) reverses the cellular pathology of psoriatic plaques, including the activation of keratinocytes, dendritic cells, and endothelial cells. *J Exp Med.* 2000; 192(5): 681-94.

[22] Genant HK, Peterfy CG, Westhovens R, Becker JC, Aranda R, Vratsanos G et al. Abatacept inhibits progression of structural damage in rheumatoid arthritis: results from the long-term extension of the AIM trial. *Ann Rheum Dis.* 2008; 67(8): 1084-9.

[23] Vincenti F, Larsen C, Durrbach A, Wekerle T, Nashan B, Blancho G et al.; Belatacept Study Group. Costimulation blockade with belatacept in renal transplantation. *N Engl J Med.* 2005; 353(8): 770-81.

[24] Viglietta V, Kent SC, Orban T, Hafler DA. GAD65-reactive T cells are activated in patients with autoimmune type 1a diabetes. *J Clin Invest.* 2002; 109(7): 895-903.

[25] Orban T, Bundy B, Becker DJ, DiMeglio LA, Gitelman SE, Goland R et al.; Type 1 Diabetes TrialNet Abatacept Study Group. Co-stimulation modulation with abatacept in patients with recent-onset type 1 diabetes: a randomised, double-blind, placebo-controlled trial. *Lancet.* 2011; 378(9789): 412-9.

[26] Ma L, Qian S, Liang X, Wang L, Woodward JE, Giannoukakis N et al. Prevention of diabetes in NOD mice by administration of dendritic cells deficient in nuclear transcription factor-kappaB activity. *Diabetes.* 2003; 52(8): 1976-85.

[27] Green EA, Flavell RA. Tumor necrosis factor-alpha and the progression of diabetes in non-obese diabetic mice. *Immunol Rev.* 1999; 169: 11-22.

[28] Harnaha J, Machen J, Wright M, Lakomy R, Styche A, Trucco M et al. Interleukin-7 is a survival factor for CD4+ CD25+ T-cells and is expressed by diabetes-suppressive dendritic cells. *Diabetes.* 2006; 55(1): 158-70.

[29] Giannoukakis N, Phillips B, Finegold D, Harnaha J, Trucco M. Phase I (safety) study of autologous tolerogenic dendritic cells in type 1 diabetic patients. *Diabetes Care.* 2011; 34(9): 2026-32.

[30] Carlsson PO, Schwarcz E, Korsgren O, Le Blanc K. Preserved β-Cell Function in Type 1 Diabetes by Mesenchymal Stromal Cells. *Diabetes.* 2015; 64(2): 587-92.

[31] Battaglia M, Roncarolo MG. Immune intervention with T regulatory cells: past lessons and future perspectives for type 1 diabetes. *Semin Immunol.* 2011; 23(3): 182-94.

[32] Sakaguchi S, Sakaguchi N, Asano M, Itoh M, Toda M. Immunologic self-tolerance maintained by activated T cells expressing IL-2 receptor alpha-chains (CD25). Breakdown of a single mechanism of self-tolerance causes various autoimmune diseases. *J Immunol.* 1995; 155(3): 1151-64.

[33] Green EA, Gorelik L, McGregor CM, Tran EH, Flavell RA. $CD4^+CD25^+$ T regulatory cells control anti-islet $CD8^+$ T cells through TGF-beta-TGF-beta receptor interactions in type 1 diabetes. *Proc Natl Acad Sci USA.* 2003; 100(19): 10878-83.

[34] You S, Slehoffer G, Barriot S, Bach JF, Chatenoud L. Unique role of $CD4^+CD62L^+$ regulatory T cells in the control of autoimmune diabetes in T cell receptor transgenic mice. *Proc Natl Acad Sci USA.* 2004; 101(2): 14580-5.

[35] Trzonkowski P, Bieniaszewska M, Juścińska J, Dobyszuk A, Krzystyniak A, Marek N et al. First-in-man clinical results of the treatment of patients with graft versus host disease with human ex vivo expanded $CD4^{+}CD25^{+}CD127^{-}$ T regulatory cells. *Clin Immunol.* 2009; 133(1): 22-6.

[36] Bluestone JA, Tang Q. Therapeutic vaccination using $CD4^{+}CD25^{+}$ antigen-specific regulatory T cells. *Proc Natl Acad Sci USA.* 2004; 101(2): 14622-6.

[37] Marek-Trzonkowska N, Mysliwiec M, Dobyszuk A, Grabowska M, Techmanska I, Juscinska J et al. Administration of $CD4^{+}$CD25highCD127^{-} regulatory T cells preserves β-cell function in type 1 diabetes in children. *Diabetes Care.* 2012; 35(9): 1817-20.

[38] Marek-Trzonkowska N, Myśliwiec M, Dobyszuk A, Grabowska M, Derkowska I, Juścińska J et al. Therapy of type 1 diabetes with $CD4^{(+)}$CD25(high)CD127-regulatory T cells prolongs survival of pancreatic islets - results of one year follow-up. *Clin Immunol.* 2014; 153(1): 23-30.

[39] Luopajärvi K, Nieminen JK, Ilonen J, Akerblom HK, Knip M, Vaarala O. Expansion of $CD4^{+}CD25^{+}FOXP3^{+}$ regulatory T cells in infants of mothers with type 1 diabetes. *Pediatr Diabetes.* 2012; 13(5): 400-7.

In: Alterations in Glucose Homeostasis in Children
Editor: E. Otto-Buczkowska
ISBN: 978-1-63483-861-0

Chapter 5

HYPERGLYCEMIC CRISES – HYPEROSMOLAR HYPERGLYCEMIC SYNDROME (HHS), LACTIC ACIDOSIS (LA), DIABETIC KETOACIDOSIS (DKA)

E. Otto-Buczkowska[1,*] and M. Dryżałowski[2]
[1]Medical Specialist Centre in Gliwice, Poland
[2]Medical University of Silesia in Katowice, Poland

ABSTRACT

Hyperglycemic crises include diabetic ketoacidosis, hyperosmolar hyperglycemic syndrome, and lactic acidosis. Diabetic ketoacidosis and hyperosmolar hyperglycemic state are the most feared complications of uncontrolled diabetes seen in emergency medicine.

Ketoacidosis is a serious complication of diabetes mellitus, especially type 1, and its secondary consequences account for a large portion of diabetes-related hospitalizations in children with type 1 diabetes. Diabetic ketoacidosis occurs mainly in children with newly diagnosed type 1 diabetes mellitus and may later recur in association with illness or non-compliance with treatment. Therefore, a major goal is to prevent ketoacidosis by early diagnosis of diabetes in new patients and by appropriate treatment of intercurrent diseases in patients with known diabetes.

Cerebral edema is an uncommon but devastating complication of diabetic ketoacidosis in children. Risk factors for this complication have not been clearly defined.

Children with diabetic ketoacidosis who present with high initial serum urea nitrogen concentrations and low partial pressures of arterial carbon dioxide are at increased risk of cerebral edema.

To prevent cerebral edema in children with DKA, it is necessary that both rehydration and metabolic correction are performed slowly and carefully. Use of hypotonic fluids is not recommended. Correction of hyperglycemia is of secondary importance and insulin should be started at a low dosage.

In children with HHS, correction of intravascular hypovolemia is the most important element of treatment. Fluid replacement is the initial treatment of HHS. Insulin

* E-mail: em.buczkowski@pro.onet.pl

administration should be considered when serum glucose concentrations are no longer declining adequately with fluid administration alone.

Keywords: Hyperglycemic crisis, diabetic ketoacidosis, fluid replacement, cerebral edema

INTRODUCTION

Persistent hyperglycemia leads to serious metabolic disorders: diabetic ketoacidosis (DKA), hyperosmolar hyperglycemic syndrome (HHS) and lactic acidosis (LA) [1-12].

According to the available literature, the incidence of such disorders varies considerably. Rewers et al. performed a multicenter study in which DKA was found in 25,5% of cases of newly diagnosed type 1 diabetes – the percentage was reasonably higher in younger children (37,3%) and lower in adolescents (14,7%) [13]. In this study, the criteria for DKA were: HCO_3^- < 15 mmol/L, venous pH < 7,25, arterial/capillary pH < 7,30. In the author's own study, DKA was found in 33% of 186 children with newly diagnosed diabetes and the incidence was significantly higher among children under the age of 4 ($p < 0{,}001$).

The occurrence of DKA at the time of diagnosis of type 2 diabetes was estimated to be 9,7%. In his paper, Rosenbloom approximated the frequency of DKA to be 10-70% in newly diagnosed type 1 diabetes and 5-52% in newly diagnosed type 2 diabetes [14]. The risk of hyperosmolar hyperglycemic syndrome was calculated to be 4%.

All those states which are associated with rapid progression of hyperglycemia should be considered life-threatening emergencies.

Failures in therapy of the mentioned conditions can arise from both cerebral edema and other neurological complications such as intracranial hemorrhage. Many causes are listed, such as hypokalemia, hypoglycemia, peripheral vein thrombosis, acute pancreatitis, acute renal failure, sepsis and other, less common complications [15, 16].

Type of metabolic disorder, its severity and the patient's age must be considered when planning the therapy of acute hyperglycemia [17, 18].

HYPEROSMOLAR HYPERGLYCEMIC SYNDROME (HHS)

Diabetic ketoacidosis (DKA) and hyperosmolar hyperglycemic syndrome (HHS) are the two most acute life-threatening complications of diabetes mellitus and in most cases their treatment should be administered in an intensive care unit [19].

HHS used to be exclusive for elderly patients with type 1 or (less commonly) type 2 diabetes [20]. Recently, there have been numerous reports of HHS occurring in children and adolescents – both with type 1 and type 2 diabetes [21].

HHS is characterized by hyperglycemia, hyperosmolarity, severe dehydration, absent or mild metabolic acidosis and little or no ketone production. This often leads to multi-organ failure with most common major complications such as cerebral edema, rhabdomyolysis and thrombosis [22, 23, 24]. The mechanism of HHS is different from DKA because the syndrome occurs mainly due to insufficient hydration while the insulin deficit is usually less severe than in the case of ketoacidosis. This means that the insulin level might be insufficient to ensure proper glucose utilization in tissues but it is still high enough to prevent increased lipolysis and ketogenesis.

Hyperglycemic hyperosmolar syndrome is a potentially deadly complication of diabetes mellitus that can often be the presenting symptom of the condition in the pediatric population [25].

Diagnostic criteria and management strategies for diabetic ketoacidosis and hyperosmolar hyperglycemic crisis are provided [26].

A broad review of HHS in pediatric population was performed by Rosenbloom [27]. The author presented criteria for diagnosing and differentiating diabetic ketoacidosis and hyperglycemic hyperosmolar syndrome. Risk factors such as young age, greater severity of acidosis, degree of hypocapnia, and severity of dehydration were suggested. Intervention of mannitol infusion usually accompanied by intubation and hyperventilation was the preferred treatment, but recent evidence indicates outcome is adversely affected by aggressive hyperventilation.

The criteria for HHS include:

- Plasma glucose concentration > 33.3 mmol/L (600 mg/dL)
- Venous pH > 7.25; arterial pH > 7.30
- Serum bicarbonate > 15 mmol/L
- Small ketonuria, absent to mild ketonemia
- Effective serum osmolality > 320 mOsm/kg
- Altered consciousness (e.g., obtundation, combativeness) or seizures.

Appropriate treatment consists of adequate fluid administration and a cautious lowering of the serum glucose level. Low-dose and late insulin administration seems to be favourable.

The goal of initial fluid therapy is to expand the intra- and extravascular volume and restore normal renal perfusion. The rate of fluid replacement should be more rapid than is recommended for DKA. The initial bolus should be ≥ 20 mL/kg of isotonic saline (0.9% NaCl) and a fluid deficit of approximately 12–15% of body weight should be assumed. Additional fluid boluses should be given, if necessary, to restore peripheral perfusion.

Insulin administration should be initiated when serum glucose concentration is no longer declining at a rate of at least 50 mg/dL (3 mmol/L) per hour with fluid administration alone. In patients with more severe ketosis and acidosis, however, insulin administration should be initiated earlier. Continuous administration of 0.025–0.05 units/kg/h can be used initially, with the dosage titrated to achieve a decrease in serum glucose concentration of 3–4 mmol/L (50–75mg/dL) per hour.

HHS is always an emergency, the prognosis is serious with high mortality. Major complications include renal failure. Similarly to DKA, brain edema is also a serious threat in HHS.

Lactic Acidosis (LA)

Lactic acidosis is a type of metabolic acidosis [28, 29]. It is not strictly a complication of diabetes, although diabetics are especially prone to its occurrence. It is a condition associated with high mortality and is caused by excess buildup of lactic acid in the patient's blood – caused by its increased production or decreased exhaustion (usually a combination of both). LA must be suspected in newborns whose mothers suffer from diabetes [30].

In diabetics, LA and DKA may be present simultaneously – e.g., if hypotension occurs during DKA.

DIABETIC KETOACIDOSIS (DKA)

Introduction

Diabetic ketoacidosis is a serious medical emergency resulting from relative or absolute insulin deficiency and the unopposed action of counter-regulatory hormones, such as glucagon, cortisol and catecholamines. The hepatic metabolism of free fatty acids generates ketoanions such as beta-hydroxybutyrate and acetoacetate.

Despite the progress in diagnosis and treatment of type 1 diabetes, diabetic ketoacidosis is still a serious clinical problem. Recently, multicenter randomized trials were conducted in many countries in order to assess the incidence of ketoacidosis in juvenile patients with type 1 diabetes. Results of broad multicenter studies of the occurrence of ketoacidosis in children and adolescents were presented lately by the French authors [31].

They pointed out that DKA at the time of T1D diagnosis in children and adolescents is frequent and often severe. They also suggested that a campaign to prevent DKA at diagnosis can help reduce the frequency of DKA.

Similar studies were conducted by Spanish authors [32]. They concluded that children under 2 years of age seem to be at increased risk for severe DKA.

Austrian researchers found that the frequency of DKA in children with newly diagnosed type 1 diabetes in Austria is high [33]. Incidence of DKA in children and adolescents with type 1 diabetes mellitus was also studied in Poland [34]. Polish authors found that newly diagnosed type 1 diabetes is the main cause of DKA in children and adolescents.

Chumiecki et al. established that about 25% of newly diagnosed diabetes in children is accompanied by ketoacidosis and that the risk of acidosis is greater in younger children [35].

The Americans stress that approximately one-third of children at the onset of T1D presents with DKA and one-sixth with moderate or severe DKA [36].

Hyperglycemia and increase in acetoacetic and beta-hydroxybutyric acid levels lead to glycosuria and osmotic diuresis with water and electrolyte loss. The severity of those losses is determined by the severity of osmotic diuresis and its duration [37, 38].

Diagnosis and Overview

The biochemical criteria for the diagnosis of DKA are: [39]

- Hyperglycemia [BG > 11 mmol/L (≈ 200 mg/dL)]
- Venous pH < 7.3 or bicarbonate < 15 mmol/L
- Ketonemia and ketonuria

The severity of DKA is categorized by the degree of acidosis:

- Mild: venous pH < 7.3 or bicarbonate < 15 mmol/L
- Moderate: venous pH < 7.2, bicarbonate < 10 mmol/L
- Severe: venous pH < 7.1, bicarbonate < 5 mmol/L

DKA is most common in type 1 diabetes, but it can also occur in type 2 diabetes. Important observations of diabetic ketoacidosis and its complications were recently published by Bialo et al. [40] The authors reviewed the complications of pediatric diabetic ketoacidosis. Cerebral edema is the most common rare complication of diabetic ketoacidosis in children but DKA can be accompanied by other rare complications. Those include rhabdomyolysis, coagulopathy, pulmonary and gastrointestinal complications. The researchers also brought attention to the possibility of cognitive complications – the greatest risk factors for these changes in cerebral structure were degree of acidosis and younger age at presentation. Diabetic ketoacidosis can accompany new-onset type 1 diabetes mellitus or it can occur with established type 1 diabetes mellitus during the increased demands of an acute illness or with decreased insulin delivery due to omitted doses or insulin pump failure [41].

Multimodal neuromonitoring of both intracranial pressure and brain tissue oxygenation during episodes of clinically apparent diabetic ketoacidosis-associated cerebral edema allows for the detection and treatment of episodes of elevated intracranial pressure and/or brain tissue hypoxia that may be of clinical significance [42].

A detailed overview of the definitions, frequency, risk factors, pathophysiology, diagnostic considerations, and management recommendations for diabetic ketoacidosis in children and adolescents was presented by Rosenbloom [43].

Indications for hospitalization of patients with DKA are based on many factors. All cases of DKA associated with newly diagnosed diabetes should be admitted, regardless of the severity of acidosis. Unconditional hospitalization is also recommended for young children whose clinical state may be difficult to assess in the outpatient setting. The presence of vomiting is also an indication for admission regardless of severity of acidosis.

The treatment of DKA should be carried out in experienced centers, where neurological status and biochemistry can be monitored with sufficient frequency to prevent complications.

Patients with previously treated diabetes should be admitted based on the severity of ketoacidosis.

Mild acidosis in previously treated patients who received sufficient education and are in telephone contact with a specialist facility can usually be managed in an outpatient setting.

DKA can occur in both newly and previously diagnosed diabetes.

Retrospective studies by Hekkala et al., spanning a period of 20 years, showed that the frequency of DKA at the moment of diagnosing diabetes in children below the age of 2 is still high [44].

Bowden et al. pointed out that juvenile patients presenting with DKA at the moment of diagnosis were less likely to achieve remission which may suggest that the destruction of beta cells in this group was more aggressive [45].

Patients suffering from DKA in the course of normal therapy of diabetes usually experience this complication due to intercurrent conditions. The incidence of associated infections is estimated at 30-40%, although DKA can also be caused by negligence of proper treatment by the patient or their caregivers [46].

Patients treated with continuous subcutaneous insulin infusion (CSII) can experience DKA due to insulin pump malfunction and in such cases the progression of metabolic decompensation is usually extremely rapid.

Youth in adolescence can sometimes lead themselves to DKA either as an act of rebellion or an outcry for attention [47]. In this age group, the most common cause of DKA is the omission of insulin injections. In certain cases, brittle diabetes should be considered.

It is important to remember that DKA can occur in pregnant women and present with normoglycemia [48].

Patients treated in specialist diabetes centers usually possess enough knowledge and technical skills to prevent metabolic decompensation with minimal effort.

In most cases, diabetics and their caregivers have established ways of contacting the specialist team at any time. That is why even if DKA occurs in those patients, the severity of the condition is usually limited to „mild" or „moderate" and more serious ketoacidosis is very rare.

The situation is different in the case of newly diagnosed diabetes and that is when patients often reach the specialist clinics in severe condition. To make matters worse, neurological and orientation deficits in DKA are relatively late symptoms and patients seem „not so sick" for an extended period of time in spite of considerable metabolic instability.

Edge et al. point out that the level of confusion in children with DKA correlates with the severity of acidosis and depends on the pH instead of glycemia or serum sodium level [49].

Identyfying DKA in infants and young children can prove extremely challenging, especially if they were not previously diagnosed with diabetes [50, 51].

A precise assessment of acidosis can be difficult due to respiratory compensation. Patients with severe hypovolemia present with an increase of the anion gap (AG) and decrease of bicarbonate contentration.

The norm for AG is 12 ± 2 mmol/L. In DKA, it is typically 20–30 mmol/L and a level >35 mmol/L suggests concommitant lactic acidosis.

Water homeostasis imbalance associated with DKA is reflected by the sodium concentration in serum. Dehydration in ketoacidosis is hypermolal. In healthy individuals, serum molality is approximately 280 mOsm/L, increase over 320 mOsm/L indicates an excess of water loss in relation to electrolyte loss.

Treatment

Fluid repletion and insulin administration are mainstays of DKA treatment and serve to restore normal hemodynamic status while decreasing the metabolic acidosis. Careful monitoring of glucose concentrations, vital signs, and electrolytes is essential to prevent complications arising from the treatment of DKA [52].

The fluid requirements must be divided into 24 hours and an hourly rate of repletion must be calculated. The deficit can be calculated using a formula:

[Deficit (litres)] = [percentage of dehydration] x [body mass (kg)]

No more than 10% of the deficit should be repleted since aggresive fluid therapy can lead to cerebral edema.

Basic requirements are calculated based on age and body mass:

- 0-2 years – 80 ml/kg/24 h
- 3-5 years – 70 ml/kg/24 h
- 6-9 years – 60 ml/kg/24 h
- 10-14 years – 50 ml/kg/24 h
- > 15 years – 35 ml/kg/24 h

Electrolyte supplementation

- Potassium – supplementation should begin simultaneously with insulin therapy after confirming the presence of diuresis. If renal function is intact and the serum potassium level is < 5,5 mmol/L, potassium should be injected into the infusion fluids with a dosage relative to the serum concentration (0,2-0,6 mmol/kg/h) with an aim to keep serum potassium level between 4-5 mmol/L. Hypokalemia is rarely observed at the beginning of treatment of DKA, but in case of a reduced serum potassium level (< 3,3 mmol/L), its supplementation should begin before introducing insulin therapy to prevent life-threatening hypokalemia, arrhythmia and depression of respiratory muscles. At levels ≥ 6 mmol/L or in case of visible symptoms of hyperkalemia in ECG, potassium supplementation should be withheld and close monitoring implemented. Serum potassium level should be kept between 4-5 mmol/L.
- Phosphorous – supplementation is not usually necessary, unless serum level falls below 1,0 mmol/L. In that case, the deficit can be balanced by adding KPO_4 to the infusion fluids.
- Magnesium – the deficit is usually negligible and does not require supplementation
- Calcium – symptoms of hypocalcemia may sometimes occur, in such case calcium supplementation may be necessary
- Sodium – sodium reserves are depleted in DKA and the increase in serum osmolality causes a „dilution effect" – that is why it is necessary to calculate „true" sodium levels based on measurements in serum. If the initial sodium level is > 150 mmol/L, careful fluid therapy using a 0,45% solution of saline should be performed.

Resuscitation Fluids

Fluid infusion should be the first step in the treatment of DKA and the importance of this intervention is all the more important with the increasing severity of metabolic instability and clinical status.

For patients who are severely volume depleted but not in shock, volume expansion (resuscitation) should begin immediately with 0.9% saline to restore the peripheral circulation. In youngest children, 5% glucose is added at the beginning of treatment. The volume administered typically is 10–20 ml/kg over 1–2 h, and may need to be repeated until tissue perfusion is adequate.

In the rare patient with DKA in shock, rapidly restore circulatory volume with isotonic saline in 20 ml/kg boluses infused as quickly as possible through a large bore cannula with reassessment after each bolus. The rate of infusion is subsequently reduced not to exceed 4 litres per 1 m2 of BSA in the first 24 hours.

The composition of the infusion fluids is changed according to glycemia and serum electrolyte levels.

If renal function is adequate and serum potassium level is < 5,5 mmol/L, potassium should be added to the infusion fluid. If hypokalemia is present, potassium should be supplemented simultaneously with fluid therapy and insulin administration should be delayed until serum potassium level rises above 3,3 mmol/L.

If glycemia approaches 13,9 mmol/L (250 mg/dL), 5% glucose should be added to the infusion at a rate necessary to keep glucose level within the 150-200 mg/dL (8,3-11,1 mmol/L) range until acidosis subsides. The infusion is usually stretched over a period of 48 hours. It is especially important in young children and patients with hypernatremia, hyperglycemia and/or serum hyperosmolality. The rate of decrease in serum osmolality should not exceed 3 mOsm/kg H_20/h. Continuous monitoring of the patient to prevent hyperhydration is extremely important in order to avoid cerebral edema.

Deficit replacement after 4–6 h should be with a solution that has a tonicity ≥ 0.45% saline with added potassium [53, 54].

Treatment involves fluid resuscitation with insulin and electrolyte replacement under constant monitoring for cerebral edema. When DKA is recognized and treated immediately, the prognosis is excellent [55].

A detailed retrospective overview of DKA treatment outcomes in juvenile patients was published by White and Dickson [56]. According to the authors, outcomes in the studied centre were good. 20 out of 3712 patients hospitalized with the diagnosis of DKA suffered from cerebral edema.

Ugale et al. stressed the importance of careful assessment of the level of dehydration and adequate fluid therapy [57].

The therapy of DKA went through considerable changes over the years. Since the main pathology is associated with osmotic diuresis, fluid repletion becomes the basis of treatment.

A key rule should always be maintained that therapy of ketoacidosis should be decisive but extremely prudent [58, 59].

Wolfsdorf recently published critical reviews and presented the most pertinent new data published in the past decade, which have implications for diagnosis and management [60].

The correction of metabolic imbalance must not be too aggressive since such course of action may do „more harm than good." The repletion should be more careful with increasing severity of disorders, stretched over 24-48 hours and even 72 hours in critical cases.

Periods of treatment are determined by the severity of the patient's status and metabolic imbalance. In the most critical cases, the priority is to secure basic life functions and refill capillary bed to reestablish renal filtration. Every aspect of DKA treatment should be closely monitored and rapid decrease of serum osmolality should be avoided [61, 62]. Under life-threatening conditions, if a quick IV access cannot be secured, medication and fluids can be administered intraosseously in children under 6 years of age [63].

Neurological state should be assessed every hour (or more frequently) to ensure early detection of cerebral edema [64].

Capnography (i.e., the non-invasive monitoring of expiratory CO_2) is also used to monitor the effectiveness of therapy. Agus et al., in a group of 72 patients (aged 1-21) with DKA treated over a period of 3-38 hours, had conducted a comparative study between validated end-tidal carbon dioxide ($EtCO_2$) measured via capnography and other laboratory markers of acid-base homeostasis and proved that capnography had been useful in monitoring acidosis [65].

Bicarbonate

Chua et al. published a literature review on the efficacy and risk of bicarbonate administration in the emergency treatment of severe acidemia in diabetic ketoacidosis (DKA) [66]. They concluded that the evidence to date does not justify the administration of bicarbonate for the emergent treatment of DKA, especially in the pediatric population, in view of possible clinical harm and lack of sustained benefits.

Andrade et al., based on available literature, found that in the presence of high anion gap acidosis, especially lactic acidosis, diabetic acidosis and cardiopulmonary resuscitation, the use of sodium bicarbonate is not beneficial and has potential adverse effects, limiting its indication [67]. The administration of bicarbonate should not be considered standard therapy [68, 69].

In exceptional cases, bicarbonate therapy can be utilized in patients with shock, organ failure and extreme acidosis ($pH \leq 6{,}9$). Adequate hydration and insulin therapy leads to the correction of acidosis and persistent acidosis is usually caused by insufficient fluid resuscitation, inadequate insulin effect or sepsis.

Routine administration of bicarbonate should be contraindicated because:

- Bicarbonate can cause paradoxal CNS acidosis and promote cerebral edema [70]
- Rapid balancing of acidosis caused by bicarbonate exacerbates sodium overload, causes serum hypertonia, decrease in extracellular potassium concentration, hypokalemia and resulting cardiac arrhythmia.

If bicarbonate use is indicated ($pH \leq 6{,}9$, $HCO_3^- \leq 5$ mmol/l), it is administered in the form of diluted $NaHCO_3$, usually at the rate of 1-2 mmol/kg/h (children) or 50 mmol/h (adults). The composition should be 50 mmol HCO_3^- in 200 ml of fluids.

The correction of pH is to be expected no sooner than after 6-12 hours and the decrease in bicarbonate concentration lasts for a considerably longer period of time.

Insulin Therapy

In severe cases, insulin therapy should be started after the management of shock, implementation of fluid therapy and establishing serum potassium at ≥ 3,3 mEq/L.

Fluid infusion should precede insulin administration (0.1 U/kg/h) by 1-2 hours; an initial bolus of 10-20 mL/kg 0.9% saline is followed by 0.45% saline calculated to supply maintenance and replace 5%-10% dehydration. Potassium must be replaced early and sufficiently. Bicarbonate administration is contraindicated.

Insulin therapy in the youngest children is usually started at the rate of 0,05 U/kg/h. Nallasamy et al. presented a comparative study of low-dose (0.05 U/kg per hour) vs. standard-dose (0.1 U/kg per hour) insulin infusion [71]. Retrospective observations were also carried out by other researchers. Kapellen et al. compared the results of DKA therapy using low (0,025 U/kg/h) and conventional insulin dose (0,1 U/kg/h) [72]. The authors found that low dose insulin substitution is as safe as the recommended standard dose in respect to the occurrence of acute complications.

It is optimal to administer insulin as a continuous infusion using an electronic infusion device. The rate of infusion is continued based on glycemia. If blood glucose falls below 250 mg/dL (13,9 mmol/L), 2,5-5% glucose solution should be added to the IV. Some authors suggest the use of a "two-bag system" which is a method of infusion which uses two connected containers filled with solutions of identical electrolyte composition but varied glucose concentration. The two fluids mix with each other and the proportions can be regulated according to the requirements based on glycemia [73, 74].

Also, other experts believe that the two-bag system provided a faster rate of bicarbonate and ketone correction compared to the one-bag system.

White and Dickson provided a detailed retrospective analysis of the "3-bag" system using 2 bags of rehydration fluids (one with and one without dextrose) to allow rapid adjustment of glucose infusion rate, and a third bag with insulin. This system of rehydration was found to be beneficial.

The main rule of insulin therapy is to avoid an aggressive decline in glycemia (>100 mg/dL/h) that is why most authors advise against „boluses" [75].

After commencing insulin therapy, the plasma glucose concentration typically decreases at a rate of 2–5 mmol/L/h, depending on the timing and amount of glucose administration. To prevent an unduly rapid decrease in plasma glucose concentration and hypoglycemia, 5% glucose should be added to the IV fluid (e.g., 5% glucose added to 0.9 or 0.45% saline) when the plasma glucose falls to approximately 14–17 mmol/L (250–300 mg/dL), or sooner if the rate of fall is precipitous. It may be necessary to use 10% to prevent hypoglycemia while continuing to infuse insulin to correct the metabolic acidosis.

Insulin administration should never be stopped until metabolic instability is resolved. The signs of insulin effectiveness are not only the correction of hyperglycemia but also balancing of other metabolic parameters, such as pH and bicarbonate concentration.

Complications

Cerebral edema is one of the most serious complications of DKA. The incidence of clinically overt cerebral edema in national population studies is 0.5–0.9%. The complication is rarely seen after adolescence [76].

The cause of cerebral edema is controversial. Some have explained the pathogenesis as the result of rapid fluid administration with abrupt changes in serum osmolality. More recent investigations, however, have found that dehydration and cerebral hypoperfusion may be associated with DKA-related cerebral injury. It is currently known that life-threatening complications which used to be considered an effect of ill-conducted therapy can occur even before any treatment is provided, although it is known that factors intrinsic to DKA may be the cause of brain injury, which could be worsened during treatment [77, 78].

An extensive review of this topic was recently published by Glaser et al. [79].
The following are mentioned as iatrogenic causes of cerebral edema progression:

- Rehydration-related factors – that is why fluid resuscitation and correction of hyperglycemia should be gradual and closely monitored.
- Aggressive insulin therapy
- Use of bicarbonate to balance acidosis – it is a factor with considerable influence on the occurrence of cerebral edema. A sudden correction of extracellular acidosis caused by bicarbonate leads to paradoxal acidosis of cerebrospinal fluid.

Treatment of cerebral edema

- Initiate treatment as soon as the condition is suspected.
- Reduce the rate of fluid administration by one-third.
- Give mannitol, 0.5–1 g/kg IV over 10–15 min, and repeat if there is no initial response in 30 min. to 2 h.
- Hypertonic saline (3%), suggested dose 2.5–5 mL/kg over 10–15 min, may be used as an alternative to mannitol, especially if there is no initial response to mannitol.

CONCLUSION

Diabetic ketoacidosis and hyperosmolar hyperglycemic state are the most feared complications of uncontrolled diabetes seen in emergency medicine. The treatment of both conditions must be tailored to individual patients and relies on aggressive fluid resuscitation, insulin replacement and electrolyte management [80].

REFERENCES

[1] American Diabetes Association. Hyperglycemic crises in diabetes. *Diabetes Care* 2004; 27: 94-102.

[2] Bhowmick SK, Levens KL, Rettig KR. Hyperosmolar hyperglycemic crisis: an acute life-threatening event in children and adolescents with type 2 diabetes mellitus. *Endocr Pract.* 2005; 11: 23-29.

[3] Canarie MF, Bogue CW, Banasiak KJ, Weinzimer SA, Tamborlane WV. Decompensated hyperglycemic hyperosmolarity without significant ketoacidosis in the adolescent and young adult population. *J Pediatr Endocrinol Metab.* 2007; 20: 1115-1124.

[4] De Beer K, Michael S, Thacker M, Wynne E, Pattni C, Gomm M et al. Diabetic ketoacidosis and hyperglycaemic hyperosmolar syndrome - clinical guidelines. *Nurs Crit Care.* 2008; 13: 5-11.

[5] Gonzalez JM, Edlow AG, Silber A, Elovitz MA. Hyperosmolar hyperglycemic state of pregnancy with intrauterine fetal demise and preeclampsia. *Am J Perinatol.* 2007; 24: 541-543.

[6] Hirshberg E, Larsen G, Van Duker H. Alterations in glucose homeostasis in the pediatric intensive care unit: Hyperglycemia and glucose variability are associated with increased mortality and morbidity. *Pediatr Crit Care Med.* 2008; 9: 361-366.
[7] ISPAD Clinical Practice Consensus Guidelines 2014 Compendium. Diabetic ketoacidosis and hyperglycemic hyperosmolar state. *Pediatric Diabetes* 2014; 15 (Suppl. 20): 154–179.
[8] Jarosz-Chobot P, Otto-Buczkowska E. Zaburzenia hormonalnej regulacji homeostazy glukozy. Hiperglikemia. [in] Endokrynologia wieku rozwojowego – co nowego? Otto-Buczkowska E (ed), *Cornetis,* Wroclaw 2008: 158-182.
[9] Kershaw MJ, Newton T, Barrett TG, Berry K, Kirk J. Childhood diabetes presenting with hyperosmolar dehydration but without ketoacidosis: a report of three cases. *Diabet Med.* 2005; 22: 645-647.
[10] Miyake Y. Management of hyperglycemic crises and severe hypoglycemia in the emergency department. *Brain Nerve.* 2014; 66: 97-105.
[11] Otto-Buczkowska E. Kwasica ketonowa. [in] Cukrzyca typu 1. Otto-Buczkowska E (ed.). *Cornetis,* Wroclaw 2006: 183-203.
[12] Otto-Buczkowska E. Inne kwasice nieoddechowe u chorych na cukrzycę typu 1. [in] Cukrzyca typu 1. Otto-Buczkowska E (ed). Cornetis, Wroclaw 2006: 204-208.
[13] Rewers A, Klingensmith G, Davis C, Petitti DB, Pihoker C, Rodriguez B et al. Presence of diabetic ketoacidosis at diagnosis of diabetes mellitus in youth: the Search for Diabetes in Youth Study. *Pediatrics.* 2008; 121: e1258-1266.
[14] Rosenbloom AL. Hyperglycemic crises and their complications in children. *J Pediatr Endocrinol Metab.* 2007; 20: 5-18.
[15] Lin JJ, Lin KL, Wang HS, Wong AM, Hsia SH. Occult infarct with acute hemorrhagic stroke in juvenile diabetic ketoacidosis. *Brain Dev.* 2008; 30: 91-93.
[16] Mahmud FH, Ramsay DA, Levin SD, Singh RN, Kotylak T, Fraser DD. Coma with diffuse white matter hemorrhages in juvenile diabetic ketoacidosis. *Pediatrics.* 2007; 120: e1540-1546.
[17] Kitabchi AE, Umpierrez GE, Fisher JN, Murphy MB, Stentz FB. Thirty years of personal experience in hyperglycemic crises: diabetic ketoacidosis and hyperglycemic hyperosmolar state. *J Clin Endocrinol Metab.* 2008; 93: 1541-1552.
[18] Rosival V. Diabetic ketoacidosis: principles of management. *Br J Hosp Med* (Lond). 2007; 68,502; author reply 502. Comment on: *Br J Hosp Med* (Lond). 2007; 68: 184-189.
[19] Schneider S. Hyperglycemic crisis in patients with diabetes mellitus. Med Klin Intensivmed Notfmed. 2012; 107: 435-439.
[20] Chowdhury TA, Gouveia CF. Managing hyperglycaemic emergencies: an illustrative case and review of recent British guidelines. *Clin Med.* 2013; 13: 160-162.
[21] Wetter J, Gohlke BC, Woelfle JF. Hyperglycemic hyperosmolar state as initial manifestation of pediatric insulin-dependent diabetes. *Klin Padiatr.* 2012; 224: 32-33.
[22] Mul D, Meijer CR. Hyperglycaemic crises in children and adolescents. *Ned Tijdschr Geneeskd.* 2013; 157: A5185.
[23] Wetter J, Gohlke BC, Stutte S, Woelfle JF. Pediatric hyperglycemic hyperosmolar coma diabeticum: diagnostic evaluation and therapeutic concept. *Klin Padiatr.* 2012; 224: 26-31.

[24] Zeitler P, Haqq A, Rosenbloom A, Glaser N. Hyperglycemic hyperosmolar syndrome in children: pathophysiological considerations and suggested guidelines for treatment. *J Pediatr* 2011; 158: 9–14.

[25] Bassham B, Estrada C, Abramo T. Hyperglycemic hyperosmolar syndrome in the pediatric patient: a case report and review of the literature. *Pediatr Emerg Care*. 2012; 28: 699-702.

[26] Van Ness-Otunnu R, Hack JB. Hyperglycemic crisis. *J Emerg Med*. 2013; 45: 797-805.

[27] Rosenbloom AL. Hyperglycemic hyperosmolar state: an emerging pediatric problem. *J Pediatr*. 2010; 156: 180-184.

[28] Danhauser K, Smeitink JA, Freisinger P, Sperl W, Sabir H, Hadzik B et al. Treatment options for lactic acidosis and metabolic crisis in children with mitochondrial disease. *J Inherit Metab Dis*. 2015 Feb 17. [Epub ahead of print].

[29] Forni LG, McKinnon W, Lord GA, Treacher DF, Peron JM, Hilton PJ. Circulating anions usually associated with the Krebs cycle in patients with metabolic acidosis. *Crit Care*. 2005; 9: 591-595.

[30] Hussain K, Thornton PS, Otonkoski T, Aynsley-Green A. Severe transient neonatal hyperinsulinism associated with hyperlactataemia in non-asphyxiated infants. *J Pediatr Endocrinol Metab*. 2004; 17: 203-209.

[31] Choleau C, Maitre J, Filipovic Pierucci A, Elie C, Barat P, Bertrand AM et al. Ketoacidosis at diagnosis of type 1 diabetes in French children and adolescents. *Diabetes Metab*. 2014; 40: 137-142.

[32] Oyarzabal Irigoyen M, García Cuartero B, Barrio Castellanos R, Torres Lacruz M, Gómez Gila AL, González Casado I et al. Ketoacidosis at onset of type 1 diabetes mellitus in pediatric age in Spain and review of the literature. *Pediatr Endocrinol Rev*. 2012; 9: 669-671.

[33] Fritsch M, Schober E, Rami-Merhar B, Hofer S, Fröhlich-Reiterer E, Waldhoer T, Austrian Diabetes Incidence Study Group. Diabetic ketoacidosis at diagnosis in Austrian children: a population-based analysis, 1989-2011. *J Pediatr*. 2013; 163: 1484-1488.

[34] Pietrzak I, Mianowska B, Zmysłowska A, Fendler W, Młynarski W, Szadkowska A. Epidemiology and clinical course of diabetic ketoacidosis in children and adolescents with type 1 diabetes mellitus. *Pediatr Endocrinol Diabetes Metab*. 2013; 19: 137-142.

[35] Chumiecki M, Prokopowicz Z, Deja R, Jarosz-Chobot P. Frequency and clinical manifestation of diabetic ketoacidosis in children with newly diagnosed type 1 diabetes. *Pediatr Endocrinol Diabetes Metab*. 2013; 19: 143-147.

[36] Klingensmith GJ, Tamborlane WV, Wood J, Haller MJ, Silverstein J, Cengiz E et al. Diabetic ketoacidosis at diabetes onset: still an all too common threat in youth. *J Pediatr*. 2013; 162: 330-334.

[37] Sottosanti M, Morrison GC, Singh RN, Sharma AP, Fraser DD, Alawi K et al. Dehydration in children with diabetic ketoacidosis: a prospective study. *Arch Dis Child* 2012; 97: 96–100.

[38] Ugale J, Mata A, Meert KL, Sarnaik AP. Measured degree of dehydration in children and adolescents with type 1 diabetic ketoacidosis. *Pediatr Crit Care Med* 2012: 13: e103-e107.

[39] Dunger DB, Sperling MA, Acerini CL, Bohn DJ, Daneman D, Danne TP et al. ESPE/LWPES consensus statement on diabetic ketoacidosis in children and adolescents. *Arch Dis Child* 2004; 89: 188–194.

[40] Bialo SR, Agrawal S, Boney CM, Quintos JB. Rare complications of pediatric diabetic ketoacidosis. *World J Diabetes*. 2015; 6: 167-174.

[41] Bonadio W. Pediatric diabetic ketoacidosis: an outpatient perspective on evaluation and management. *Pediatr Emerg Med Pract.* 2013; 10: 1-13; quiz 14.

[42] O'Brien NF, Mella C. Brain tissue oxygenation-guided management of diabetic ketoacidosis induced cerebral edema. *Pediatr Crit Care Med*. 2012; 13: e383-388.

[43] Rosenbloom AL. The management of diabetic ketoacidosis in children. *Diabetes Ther.* 2010; 1: 103-120.

[44] Hekkala A, Knip M, Veijola R. Ketoacidosis at diagnosis of type 1 diabetes in children in northern Finland: temporal changes over 20 years. *Diabetes Care*. 2007; 30: 861-866.

[45] Bowden SA, Duck MM, Hoffman RP. Young children (<5 yr) and adolescents (>12 yr) with type 1 diabetes mellitus have low rate of partial remission: diabetic ketoacidosis is an important risk factor. *Pediatr Diabetes.* 2008; 9: 197-201.

[46] Wolfsdorf J, Craig ME, Daneman D, Dunger D, Edge J, Lee WR et al. Diabetic ketoacidosis. ISPAD Clinical Practice Consensus Guidelines 2006-2007. *Pediatr Diabetes. 2007*; 8: 28-43.

[47] Wolfsdorf J, Glaser N, Sperling M.; American Diabetes Association. Diabetic ketoacidosis in infants, children and adolescents: a consensus statement from the American Diabetes Association. *Diabetes Care*. 2006; 29: 1150-1159.

[48] Guo RX, Yang LZ, Li LX, Zhao XP. Diabetic ketoacidosis in pregnancy tends to occur at lower blood glucose levels: case-control study and a case report of euglycemic diabetic ketoacidosis in pregnancy. *J Obstet Gynaecol Res.* 2008; 34: 324-330.

[49] Edge JA, Roy Y, Bergomi A, Murphy NP, Ford-Adams ME, Ong KK, Dunger DB. Conscious level in children with diabetic ketoacidosis is related to severity of acidosis and not to blood glucose concentration. *Pediatr Diabetes*. 2006; 7: 11-15.

[50] Quinn M, Fleischman A, Rosner B, Nigrin DJ, Wolfsdorf JI. Characteristics at diagnosis of type 1 diabetes in children younger than 6 years. *J Pediatr.* 2006; 148: 366-371.

[51] Ting WH, Huang CY, Lo FS, Hung CM, Chan CJ, Li HJ et al. Clinical and laboratory characteristics of type 1 diabetes in children and adolescents: experience from a medical center. *Acta Paediatr Taiwan*. 2007; 48: 119-124.

[52] Donahey E, Folse S. Management of diabetic ketoacidosis. Adv Emerg Nurs J. 2012; 34: 209-215.

[53] Basnet S, Venepalli PK, Andoh J, Verhulst S, Koirala J. Effect of normal saline and half normal saline on serum electrolytes during recovery phase of diabetic ketoacidosis. *J Intensive Care Med* 2014; 29: 38–42.

[54] Koul P.B. Diabetic ketoacidosis: a current appraisal of pathophysiology and management. *Clin Pediatr (Phila).* 2009; 48: 135-144.

[55] Olivieri L, Chasm R. Diabetic ketoacidosis in the pediatric emergency department. *Emerg Med Clin North Am.* 2013; 31: 755-773.

[56] White PC, Dickson BA. Low morbidity and mortality in children with diabetic ketoacidosis treated with isotonic fluids. *J Pediatr*. 2013; 163: 761-776.

[57] Ugale J, Mata A, Meert KL, Sarnaik AP. Measured degree of dehydration in children and adolescents with type 1 diabetic ketoacidosis. *Pediatr Crit Care Med.* 2012; 13: e103-107.

[58] Fiordalisi I, Novotny WE, Holbert D, Finberg L, Harris GD.; Critical Care Management Group. An 18-yr prospective study of pediatric diabetic ketoacidosis: an approach to minimizing the risk of brain herniation during treatment. *Pediatr Diabetes.* 2007; 8: 142-149.

[59] Watts W, Edge JA. How can cerebral edema during treatment of diabetic ketoacidosis be avoided? *Pediatr Diabetes.* 2014; 15: 271-276.

[60] Wolfsdorf JI. The International Society of Pediatric and Adolescent Diabetes guidelines for management of diabetic ketoacidosis: Do the guidelines need to be modified? *Pediatr Diabetes.* 2014; 15: 277-286.

[61] Hoorn EJ, Carlotti AP, Costa LA, MacMahon B, Bohn G, Zietse R et al. Preventing a drop in effective plasma osmolality to minimize the likelihood of cerebral edema during treatment of children with diabetic ketoacidosis. *J Pediatr.* 2007; 150: 467-473.

[62] Sherry NA, Levitsky LL. Management of diabetic ketoacidosis in children and adolescents. *Pediatric Drugs* 2008; 10,: 209-215.

[63] Alawi KA, Morrison GC, Fraser DD, Al-Farsi S, Collier C, Kornecki A. Insulin infusion via an intraosseous needle in diabetic ketoacidosis. *Anaesth Intensive Care.* 2008; 36: 110-112.

[64] Rosenbloom AL. Treatment of diabetic ketoacidosis and the risk of cerebral edema. *J Pediatr.* 2008; 152: 146-147; author reply 147-149.

[65] Agus MSD, Alexander JL, Mantell PA. Continuous non-invasive end-tidal CO_2 monitoring in pediatric in patients with diabetic ketoacidosis. *Pediatric Diabetes* 2006; 7: 196-200.

[66] Chua HR, Schneider A, Bellomo R. Bicarbonate in diabetic ketoacidosis - a systematic review. *Ann Intensive Care.* 2011; 1: 23.

[67] Andrade OV, Ihara FO, Troster EJ. Metabolic acidosis in childhood: why, when and how to treat. *J Pediatr* (Rio J). 2007; 83(2 Suppl): S11-21.

[68] Kamel KS, Halperin ML. Acid-base problems in diabetic ketoacidosis. *N Engl J Med.* 2015; 372: 546-554.

[69] Sabatini S, Kurtzman NA. Bicarbonate therapy in severe metabolic acidosis. *J Am Soc Nephrol.* 2009; 20: 692-695.

[70] Otto-Buczkowska E, Jarosz Chobot P. How dangerous is cerebral oedema in diabetic ketoacidosis in the young? *Med. Intens Rat* 2004; 7: 205-211.

[71] Nallasamy K, Jayashree M, Singhi S, Bansal A. Low-dose vs standard-dose insulin in pediatric diabetic ketoacidosis: a randomized clinical trial. *JAMA Pediatr.* 2014; 168: 999-1005.

[72] Kapellen T, Vogel C, Telleis D, Siekmeyer M, Kiess W. Treatment of diabetic ketoacidosis (DKA) with 2 different regimens regarding fluid substitution and insulin dosage (0.025 vs. 0.1 units/kg/h). *Exp Clin Endocrinol Diabetes.* 2012; 120: 273-276.

[73] Poirier MP, Greer D, Satin-Smith M. A prospective study of the “two-bag system” in diabetic ketoacidosis management. *Clin Pediatr (Phila).* 2004; 43: 809-813.

[74] So TY, Grunewalder E. Evaluation of the two-bag system for fluid management in pediatric patients with diabetic ketoacidosis. *J Pediatr Pharmacol Ther.* 2009; 14: 100-105

[75] Bradley P, Tobias JD. Serum glucose changes during insulin therapy in pediatric patients with diabetic ketoacidosis. *Am J Ther.* 2007; 14: 265-268.

[76] Glaser NS, Wootton-Gorges SL, Buonocore MH, Marcin JP, Rewers A, Strain J, DiCarlo J, Neely EK, Barnes P, Kuppermann N. Frequency of sub-clinical cerebral edema in children with diabetic ketoacidosis. *Pediatr Diabetes.* 2006; 7: 75-80.

[77] Glaser NS, Ghetti S, Casper TC, Dean JM, Kuppermann N. Pediatric diabetic ketoacidosis, fluid therapy, and cerebral injury: the design of a factorial randomized controlled trial. *Pediatr Diabetes* 2013: 14: 435–446.

[78] Hom J, Sinert R. Evidence-based emergency medicine/critically appraised topic. Is fluid therapy associated with cerebral edema in children with diabetic ketoacidosis? *Ann Emerg Med.* 2008; 52: 69-75.

[79] Glaser NS, Wootton-Gorges SL, Buonocore MH, Tancredi DJ, Marcin JP, Caltagirone R et al. Subclinical cerebral edema in children with diabetic ketoacidosis randomized to 2 different rehydration protocols. *Pediatrics.* 2013; 131: e73-80.

[80] Corwell B, Knight B, Olivieri L, Willis GC. Current diagnosis and treatment of hyperglycemic emergencies. *Emerg Med Clin North Am.* 2014; 32: 437-452.

In: Alterations in Glucose Homeostasis in Children
Editor: E. Otto-Buczkowska
ISBN: 978-1-63483-861-0

Chapter 6

REMISSION IN TYPE 1 DIABETES – WHAT'S NEW?

E. Otto-Buczkowska * ***and A. Chwalba***
Medical Specialist Centre in Gliwice, Poland

ABSTRACT

Type 1 diabetes is characterized by selective and progressive autoimmune destruction of β cells of the pancreas in genetically susceptible individuals. This autoimmune goes on for years before the patient eventually develops clinical diabetes. The symptoms of diabetes appear when the mass of insulin-secreting β cells will be reduced by about 80-90%. In this state, the amount of insulin is insufficient to ensure normoglycemia. Many patients shortly after diagnosis of diabetes type 1 and initiation of insulin therapy come to partial β cell renewal and consequently to reduce the need for exogenous insulin. This phenomenon is called the remission of the disease. This period is also called honeymoon. The glucose concentration decreases so the insulin dose should be also reduced, however complete cessation of insulin is not recommended. Accuracy to determine and compare the prevalence of type 1 diabetes remission is difficult.

Higher C-peptide level at the time of onset of diabetes is associated with the appearance of clinical partial remission during the first six months of type 1 diabetes. C-peptide level may be a good predictor of the clinical partial remission during the first year of diabetes.

Keywords: Type 1 diabetes, clinical remission, insulin secretion, insulin therapy, C-peptide level

INTRODUCTION

Type 1 diabetes is an autoimmune disease in which there is a destruction of pancreatic islet β cells. In the natural course of the disease, we deal with a gradual progress during the

* E-mail: em.buczkowski@pro.onet.pl

few years of reduction in cell β mass. The symptoms of diabetes appear when the mass of insulin-secreting β cells is reduced by about 80-90%. In this state, the amount of insulin is insufficient to ensure normoglycemia.

Many patients shortly after diagnosis of diabetes type 1 and initiation of insulin therapy come to a partial β cell renewal and consequently reduce the need for exogenous insulin (Figure 1). This phenomenon is called the remission of the disease.

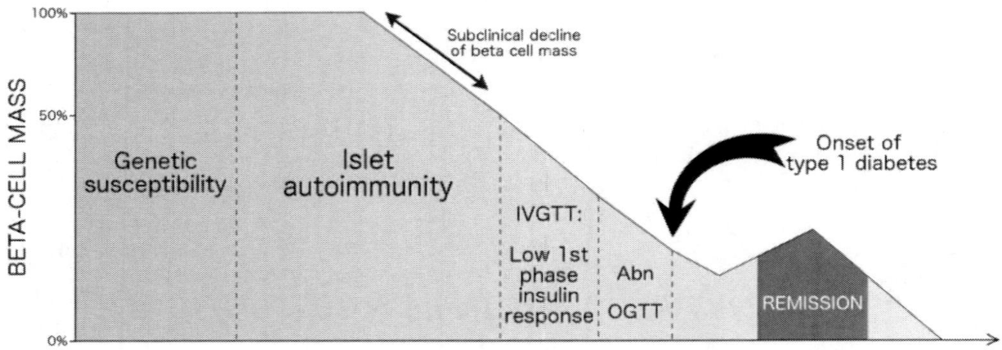

Figure 1. Progressive loss of beta cell mass in the natural history of type 1 diabetes.

The partial remission of insulin-dependent (type 1) diabetes mellitus is characterized by reduced insulin requirements while good glycemic control is maintained. Its clinical significance is the potential possibility for pharmacological intervention during this period to either slow down or arrest the ongoing destruction of the remaining β-cells.

Studies of the occurrence of remission in diabetes have a long history. In 1946, Glassberg presented a description of diabetes remission [1].

In subsequent years, there have been many reports on the occurrence and behavior of remission in juvenile diabetes mellitus and of insulin secretion during this period [2-7]. This assessment of insulin secretion also allowed the exclusion of patients with types of diabetes other than juvenile diabetes mellitus.

The first description of juvenile diabetes patients' remission in Polish literature was presented in 1971 [8].

Many studies have shown that early diagnosis and rigorous intensive insulin therapy as well as very careful metabolic disease control from the time of diagnosis determine the beginning and duration of postinitial remission. These studies have a long history [9-12].

A number of factors promote the occurrence of remission. These factors are: absence of DKA[1] at onset of clinical diabetes, higher C-peptide levels and lower HbA1c.

The severity of initial metabolic decompensation in diabetes mellitus type 1 seems to determine, at least in part, the duration of remission. The initial HbA1c concentration may serve as a useful indicator to predict the duration of the remission phase in diabetic patients [9].

The presentation with DKA at diagnosis, associated with young age, is the most important factor negatively influencing the remission rate in newly diagnosed type 1 diabetic

[1] Diabetic Ketoacidosis

patients [10]. Partial remission was less common in the ketoacidosis group than in the non-ketoacidosis group.

Diabetic ketoacidosis at diagnosis is related to a decreased capacity for β cell recovery after the clinical manifestation of IDDM in children.

Children with remission had lower blood glucose, milder hyperketonemia and ketonuria, higher pH and pCO_2 at onset than those without remission [11]. A Swedish multicenter study showed remissions in 43% of the patients with a median duration of 8 months (range 1-73) [12]. In islet antibody-positive diabetes, normal body weight was the strongest factor for entering remission, whilst a low number of islet antibodies was of importance for its duration.

Research conducted in recent years indicates that the low prevalence of remission is observed in the youngest children (< 5 years) and in adolescents (> 12 years) [13]. It is possible that the low frequency of honeymoon phase in young children reflects more aggressive β-cell destruction. In adolescents, insulin resistance contributes to a lower likelihood of having partial remission.

Other authors have similar observations [14]. They asserted that young age and severe disease in the initial period are associated with decreased residual beta-cells function what is reflected by a lower incidence of partial remission.

Many authors indicate β cell destruction and reduction of insulin secretion play an important role in the manifestation of diabetes and tissue sensitivity to the action of insulin [15-17].

In some diabetic patients in the first year there is clinical remission, characterized by a significantly reduced need for exogenous insulin at normal levels of glycemia. It is believed that its presence is related mainly to the improvement of insulin secretion. A few studies indicate that remission may also be the result of improving sensitivity for insulin.

Lower C-peptide and endogenous insulin levels in younger children may indicate a lower regeneration ability and smaller β cell numbers in this age group at the onset of T1D[2].

Patients with remission had higher C-peptide levels observed in the first 6 months of T1D than children without clinical remission [18]. C-peptide level may be a good predictor of the clinical partial remission during the first year of the disease [19].

Pecheur et al. presented the results of retrospective studies in 242 children with type 1 diabetes diagnosed on the occurrence of partial remission. They show that, at diagnosis of T1D, parameters associated with β cell mass reserve (A1C, C-peptide, and DKA) correlate with the occurrence of partial remission (PR) [20].

Martin et al. in their study show a higher spontaneous clinical remission rate than expected during the first year after diagnosis. Preserved β cell function at entry level predicts a greater chance for entering remission. More rapid loss of β cell function was seen in patients without HLA-DR3 and -DR4 [21]. Surprisingly, the beginning and the end of the remission were associated with neither major changes in C-peptide levels nor islet cell antibody and insulin-antibody level. A more rapid loss of stimulated C-peptide occurred in patients who lacked HLA-DR3 and -DR4.

The clinical significance of studies on mechanisms leading to the occurrence of partial remission are the potential possibility for pharmacological intervention during this period to either slow down or arrest the ongoing destruction of the remaining β cells [14].

[2] Type 1 Diabetes

For the assessment of β cell function, IDAA1c[3] may be used as an useful indicator. IDAA1c defined as actual HbA1c + (4 × insulin dose in units per kilogram per 24 h) was calculated and used as a surrogate marker of residual β-cell function. A calculated IDAA1c ≤ 9 corresponds to an estimated maximal C-peptide level above 300 pmol/L and was used to define clinical remission.

Many attempts have been made to stimulate the occurrence of remission and extend its duration. The mechanisms predisposing patients to remission (mentioned above), effective insulin treatment and immunological therapies may be successful in maintaining greater β cell function for extended periods.

It was pointed out that the intensification of insulin therapy is preferred since the onset of the first symptoms of the disease [22-25]. Mirouze et al. found that early effective treatment by means of the artificial pancreas may lead to frequent and sustained remissions of juvenile diabetes [26, 27]. Similar observations have also been made by other authors [28]. Attempts were also made with the use of immunosuppressive drugs [29, 30].

Recently, a series of reports devoted to the study of the incidence of remission in type 1 diabetes was published. Among other studies was one conducted by Kaas et al. in a group of juvenile patients in the first year after diagnosis of diabetes based on the monitoring of levels of C-peptide, proinsulin, GLP-1, glucagon, and insulin antibodies (IA) [31]. The authors concluded that during partial remission patients had significantly higher levels of proinsulin and a lower level of GLP-1 and glucagon. Proinsulin was positively associated with C-peptide. Mortensen et al. attempted to establish a new definition of partial remission in type 1 diabetes [32]. The study was a multicenter longitudinal investigation in 18 pediatric departments representing 15 countries in Europe and Japan. Their definition of partial remission was proposed, including both glycemic control and insulin dose. It reflects residual β cell function and has better stability compared with the conventional definitions. The authors consider stimulated C-peptide to be a very useful parameter of a residual β cell function measurement. The usefulness of this parameter was also pointed out by other authors [33].

Studies on the mechanisms of partial remission of type 1 diabetes are particularly relevant in light of the opening of new possibilities of treatment.

The evaluation of residual β cell function and severity of autoimmune processes in newly diagnosed patients with type 1 diabetes becomes important because it can provide opportunities for therapeutic interventions [34-36].

In recent years, research on the use of autologous Treg (regulatory T cells) infusions to prolong remission in recently diagnosed type 1 diabetes in children has been conducted in Poland [37-39].

Type 1 diabetes is a condition in which pancreatic islets are destroyed by self-reactive T cells. The process is facilitated by deficits in the number and suppressive activity of regulatory T cells. The infusion of autologous Tregs prolongs remission in recently diagnosed type 1 diabetes in children. The authors concluded that most of the patients responded to the therapy with increase in C-peptide levels and Treg administration resulted also in lower requirement for exogenous insulin.

---

[3] Insulin Dose-Adjusted A1c

Other methods of prolonging remission are also proposed: one of them is the use of sitagliptin. The use of sitagliptin in T1D patients could help to decrease daily requirement of insulin by delaying β-cell loss and improving endogenous insulin production [40].

Efforts are also made to implement other drugs. Anti-CD3 monoclonal antibodies significantly improved insulin secretion and affected the occurrence and prolongation of remission

Among other medication, alefacept (a protein complex of molecules CD58/LFA-3 and human IgG) has also been investigated, along with Dia-Pep277 (whose function is to stop or slow down the process of destruction of the β cell and improve insulin secretion). Dia-Pep277 is a molecule consisting of 24 amino acids obtained from a fragment of a heat shock protein (HSP60) which has immunomodulatory effects affecting T lymphocytes. On the basis of previous results, the study showed higher C-peptide levels in patients receiving DiaPep277 [40].

Conclusion

Recent studies indicate a major role of functional exhaustion and recovery of β cells during diabetes onset and remission. Early diagnosis followed by rapid normalization of metabolism at the onset of type 1 diabetes increase the possibility of preservation of some of the endogenous insulin production, which seems to facilitate diabetic control. The final results of clinical studies are still awaited.

References

[1] Baker L, Kaye R, Root AW. The early partial remission of juvenile diabetes mellitus. The roles of insulin and growth hormone. *J Pediatr.* 1967; 71: 825-831.

[2] Illig R, Prader A. Remission of juvenile diabetes. *Lancet.* 1968; 2: 1190.

[3] Johansen K, Orskov H. Plasma insulin during remission in juvenile diabetes mellitus. *Br Med J.* 1969; 1: 676-678.

[4] Ludwiczak H, Symonides-Lawecka A, Rogala H, Lawecki J, Fedorowicz B. Spontaneous remissions in the early stage of diabetes in children. *Pediatr Pol.* 1972; 48: 1335-1340.

[5] Otto-Buczkowska E. Remissions of diabetes in children. *Pol Tyg Lek.* 1971; 26: 1540-1541.

[6] Kamado K, Fujita S, Izumi K, Hoshi M. Remission phase in childhood diabetes – an investigation of summer campers in Japan. *Tohoku J Exp Med.* 1983; 141 Suppl: 191-198.

[7] Ludvigsson J, Heding LG, Larsson Y, Leander E. C-peptide in juvenile diabetics beyond the postinitial remission period. Relation to clinical manifestations at onset of diabetes, remission and diabetic control. *Acta Paediatr Scand.* 1977; 66: 177-184.

[8] Mirouze J, Selam JL, Pham TC, Orsetti A. The external artificial pancreas: an instrument to induce remissions in severe recent juvenile diabetes. Comparison with a preprogrammed insulin infusion system. *Horm Metab Res Suppl.* 1979; 8: 141-145.

[9] Vetter U, Heinze E, Beischer W, Kohne E, Kleihauer E, Teller WM. Haemoglobin A1c: a predictor for the duration of the remission phase in juvenile insulin-dependent diabetic patients. *Acta Paediatr Scand.* 1980; 69: 481-483.

[10] Böber E, Dündar B, Büyükgebiz A. Partial remission phase and metabolic control in type 1 diabetes mellitus in children and adolescents. *J Pediatr Endocrinol Metab.* 2001; 14: 435-441.

[11] Knip M, Säkkinen A, Huttunen NP, Käär ML, Länkelä S, Mustonen A et al. Postinitial remission in diabetic children – an analysis of 178 cases. *Acta Paediatr Scand.* 1982; 71: 901-908.

[12] Schölin A, Törn C, Nyström L, Berne C, Arnqvist H, Blohmé G et al. Normal weight promotes remission and low number of islet antibodies prolong the duration of remission in Type 1 diabetes. *Diabet Med.* 2004; 21: 447-455.

[13] Bowden SA, Duck MM, Hoffman RP. Young children (< 5 yr) and adolescents (>12 yr) with type 1 diabetes mellitus have low rate of partial remission: diabetic ketoacidosis is an important risk factor. *Pediatr Diabetes.* 2008; 9: 197-201.

[14] Abdul-Rasoul M, Habib H, Al-Khouly M. 'The honeymoon phase' in children with type 1 diabetes mellitus: frequency, duration, and influential factors. *Pediatr Diabetes.* 2006; 7: 101-107.

[15] Szadkowska A, Pietrzak I, Zmysłowska A, Wyka K, Bodalski J. Insulin resistance in newly diagnosed type 1 diabetic children and adolescents. *Med Wieku Rozwoj.* 2003; 7: 181-191.

[16] Szadkowska A, Pietrzak I, Markuszewski L, Bodalski J. The role of insulin secretion and insulin resistance in the clinical course of newly diagnosed diabetes mellitus in children. *Med Wieku Rozwoj.* 2005; 9: 223-234.

[17] Szadkowska A, Pietrzak I, Zmysłowska A, Bodalska-Lipińska J, Bodalski J. Insulin sensitivity – importance in clinical remission of type 1 diabetes in children and adolescents. *Przegląd Pediatryczny* 2006; 36: 205-209.

[18] Zmysłowska A, Młynarski W, Szadkowska A, Bodalski J. Prediction of clinical remission using the C-peptide level in type 1 diabetes in children. Pediatr Endocrinol Diabetes Metab. 2007; 13: 71-74.

[19] Petriczko E, Bartnik M, Cebula J, Marcisz-Grzanka K, Horodnicka-Józwa A, Walczak M. C-peptide and endogenous insulin levels in children with newly diagnosed diabetes mellitus type 1. *Pediatr Endocrinol* 2009; 8: 15-26.

[20] Pecheur A, Barrea T, Vandooren V, Beauloye V, Robert A, Lysy PA. Characteristics and determinants of partial remission in children with type 1 diabetes using the insulin-dose-adjusted A1C definition. *J Diabetes Res.* 2014; 2014: 851378.

[21] Martin S, Pawlowski B, Greulich B, Ziegler AG, Mandrup-Poulsen T, Mahon J. Natural course of remission in IDDM during 1st yr after diagnosis. *Diabetes Care.* 1992; 15: 66-74.

[22] The Diabetes Control and Complications Trial Research Group [No authors listed]. Effect of intensive therapy on residual beta-cell function in patients with type 1 diabetes in the diabetes control and complications trial. A randomized, controlled trial. *Ann Intern Med.* 1998; 128: 517-523.

[23] Guastamacchia E, Nardelli GM, Di Paolo S, Lattanzi V, Ciampolillo A, Lacasella R et al. Remission of insulin dependence induced by continuous subcutaneous insulin

infusion (CSII) in type I, recent onset diabetics: role of beta cell recovery and lymphocyte subsets distribution. *Diabete Metab.* 1988; 14: 138-139.

[24] Vague P, Vialettes B, Lassmann V, Moulin JP. The period of initial remission in insulin-dependent diabetes. Effect of optimized insulin therapy. *Journ Annu Diabetol Hotel Dieu.* 1984: 73-84.

[25] Vague P, Vialettes B, Lassmann V, Moulin JP. Initial remission period in insulin-dependent diabetes in the young subject. *Diabete Metab.* 1985; 11: 181-188.

[26] Mirouze J, Selam JL, Pham TC, Mendoza E, Orsetti A. Sustained insulin-induced remissions of juvenile diabetes by means of an external artificial pancreas. *Diabetologia.* 1978; 14: 223-227.

[27] Mirouze J, Selam JL, Pham TC, Orsetti A. The external artificial pancreas: an instrument to induce remissions in severe recent juvenile diabetes. Comparison with a preprogrammed insulin infusion system. *Horm Metab Res Suppl.* 1979; 8: 141-145.

[28] Vague P, Vialettes B, Lassmann V, Moulin JP, Mercier P. Sustained initial remission induced by intensive insulin treatment in type I diabetes. Possible role of the genetic background. *Acta Diabetol Lat.* 1985; 22: 295-304.

[29] Harrison LC, Colman PG, Dean B, Baxter R, Martin FI. Increase in remission rate in newly diagnosed type I diabetic subjects treated with azathioprine. *Diabetes.* 1985; 34: 1306-1308.

[30] Bougneres PF, Carel JC, Castano L, Boitard C, Gardin JP, Landais P. et al. Factors associated with early remission of type I diabetes in children treated with cyclosporine. *N Engl J Med.* 1988; 318: 663-670.

[31] Kaas A, Andersen ML, Fredheim S, Hougaard P, Buschard K, Petersen JS et al.; Hvidoere Study Group on childhood diabetes. Proinsulin, GLP-1, and glucagon are associated with partial remission in children and adolescents with newly diagnosed type 1 diabetes. *Pediatr Diabetes.* 2012; 13: 51-58.

[32] Mortensen HB, Hougaard P, Swift P, Hansen L, Holl RW, Hoey H et al.; Hvidoere Study Group on Childhood Diabetes. New definition for the partial remission period in children and adolescents with type 1 diabetes. *Diabetes Care.* 2009; 32: 1384-1390.

[33] Picardi A, Visalli N, Lauria A, Suraci C, Buzzetti R, Merola MK et al. Metabolic factors affecting residual beta cell function assessed by C-peptide secretion in patients with newly diagnosed type 1 diabetes. *Horm Metab Res.* 2006; 38: 668-672.

[34] Chmelova H, Cohrs CM, Chouinard JA, Petzold C, Kuhn M, Chen C et al. Distinct roles of beta cell mass and function during type 1 diabetes onset and remission. *Diabetes.* 2015 Jan 20.

[35] Kurozumi A, Okada Y, Arao T, Narisawa M, Torimoto K, Yamamoto S et al. Induction of thyroid remission using rituximab in a patient with type 3 autoimmune polyglandular syndrome including Graves' disease and type 1 diabetes mellitus: A case report. *Endocr J* 2015; 62: 69-75.

[36] Lima-Martínez MM, Guerra-Alcalá E, Contreras M, Nastasi J, Noble JA, Polychronakos C. One year remission of type 1 diabetes mellitus in a patient treated with sitagliptin. *Endocrinol Diabetes Metab Case Rep.* 2014; 2014: 140072.

[37] Marek-Trzonkowska N, Mysliwiec M, Dobyszuk A, Grabowska M, Techmanska I, Juscinska J et al. Administration of CD4+CD25highCD127- regulatory T cells preserves β-cell function in type 1 diabetes in children. *Diabetes Care.* 2012; 35: 1817-1820.

[38] Marek-Trzonkowska N, Myśliwec M, Siebert J, Trzonkowski P. Clinical application of regulatory T cells in type 1 diabetes. *Pediatr Diabetes.* 2013; 14: 322-332.

[39] Marek-Trzonkowska N, Myśliwiec M, Dobyszuk A, Grabowska M, Derkowska I, Juścińska J et al. Therapy of type 1 diabetes with CD4(+)CD25(high)CD127-regulatory T cells prolongs survival of pancreatic islets - results of one year follow-up. *Clin Immunol.* 2014; 153: 23-30.

[40] Niedźwiecki P, Zozulińska-Ziółkiewicz D. Clinical remission of type 1 diabetes. *Diabet. Klin.* 2013; 2: 185–190.

In: Alterations in Glucose Homeostasis in Children
Editor: E. Otto-Buczkowska
ISBN: 978-1-63483-861-0

Chapter 7

METABOLIC MEMORY – THE IMPLICATIONS FOR DIABETIC COMPLICATIONS

***E. Otto-Buczkowska*[1,*] *and M. Dryżałowski*[2]**
[1]Medical Specialist Centre in Gliwice, Poland
[2]Medical University of Silesia in Katowice, Poland

ABSTRACT

Many important biochemical mechanisms are activated in the presence of high concentrations of glucose which occur in diabetes. Large randomized studies have established that early intensive glycemic control reduces the risk of diabetic complications. This phenomenon has recently been dubbed 'metabolic memory'.

It has been suggested that early glycemia normalization can halt the hyperglycemia-induced pathological processes associated with increased oxidative stress and glycation of cellular proteins and lipids.

The phenomenon of metabolic memory suggests that early aggressive treatment and strict glycemic control could prevent chronic diabetic complications.

Keywords: Levels of glucose, metabolic memory, diabetic complications

INTRODUCTION

Diabetes is a complex, heterogeneous condition with the dysfunction of β cells at its core. Many factors (e.g., hyperglycemia/glucotoxicity, lipotoxicity, autoimmunity, inflammation, adipokines, islet amyloid, incretins and insulin resistance) influence the function of pancreatic β cells [1, 2, 3].

Experiments on animal models as long ago as the mid-1980s pointed to the phenomenon of 'metabolic memory'. Between 2000 and 2002, the results of clinical trials showed that an

* E-mail: em.buczkowski@pro.onet.pl

early intensive therapy aimed at normalizing metabolic control can counter negative processes in cells and minimize the risk of chronic diabetes complications.

There is a lot of evidence indicating that both genetic and environmental factors are involved in the pathogenesis of diabetic complications. Components associated with the pathology of diabetic complications (such as hyperglycemia, growth and inflammatory factors, oxidative stress etc.) can lead to dysregulation of epigenetic mechanisms and alter the expression of pathological genes in target cells such as endothelial, vascular smooth muscle, retinal and cardiac cells, without changes in the underlying DNA sequence. Furthermore, long-term persistence of these alterations in epigenome may be a key mechanism underlying the phenomenon of 'metabolic memory' and sustained vascular dysfunction despite attainment of glycemic control [4, 5].

Recent evidence shows that diabetes-induced epigenetic changes can affect gene expression in vascular endothelial and smooth muscle cells. Such alterations could account for the accelerated rates of chronic microvascular and macrovascular complications associated with diabetes [6, 7].

Such effects further influence inflammatory and insulin production pathways in cells and thus ensure a long-term memory where epigenetic changes are maintained even long after restoring normo-glycemic conditions by appropriate therapeutic approaches. Prolonged hyperglycemia induces chronic complications which persist and progress unimpeded via the phenomenon of metabolic memory even when glycemic control is pharmaceutically achieved [8].

Therefore, there is evidence that transient hyperglycemia causes profound transcriptional changes in vascular endothelial cells [9-13].

Chronic hyperglycemia may result in detrimental effects on insulin synthesis/secretion, cell survival and insulin sensitivity through multiple mechanisms. In the presence of hyperglycemia, prolonged exposure to increased concentrations of free fatty acids results in an accumulation of toxic metabolites in the cells ("lipotoxicity"), finally causing decreased insulin gene expression and impairment of insulin secretion.

It is well known that persistent hyperglycemia is responsible for the development of chronic complications of microangiopathic (retinopathy, nephropathy, neuropathy) as well as macroangiopathic (ischemic heart disease, cerebral and peripheral vessel disease) nature.

Under the conditions of chronic hyperglycemia, non-enzymatic protein glycation, polyol pathway and oxidative stress are activated [14].

Chronic hyperglycemia triggers many processes, e.g.:

- protein kinase C and phospholipase A2 activation leading to increased production of arachidonic acid metabolites
- increased expression of growth factors resulting in increased vasoconstriction
- production of advanced glycation end-products (AGE) which increase oxidative stress through their interaction with corresponding receptors (RAGE)
- glycation and oxidation of LDL
- increase in expression of the NF-κb nuclear transcription factor, rise of adhesion molecules' concentration, decrease of NO bioavailability, increase of IL-6, TNF-a and other cytokines

- protein kinase C activation and rise of DAG (diacyloglycerol) production, leading to a decrease of the NADPH pool and generation of NO.

The awareness of metabolic memory seems to be particularly important in children and young adults, in whom onset of the disease is usually so evident that the diagnosis is quickly established and treatment is introduced relatively fast. Therefore, these patients have a good chance of reduced risk of marked chronic complications.

A recently published study clearly showed that a mean of 6.5 years of intensive therapy aimed at near-normal glucose levels reduced the risk of development and progression of retinopathy by as much as 76% compared with conventional therapy [15].

The Epidemiology of Diabetes Interventions and Complications study (EDIC) observational follow-up showed that the risk of further progression of retinopathy 4 years after the previously mentioned study had ended was also greatly reduced in the former intensive group.

Exacerbation of Oxidative Stress

Free radicals (ROS) which are formed in the process of glucose autoxidation cause lipid and AGE oxidation, diacyloglycerol-protein kinase C pathway activation and increased thromboxane synthesis. The effect of free radicals also increases endothelium permeability and promotes the production of the secondary glycation end-products.

ROS cause lipid peroxidation in the cellular membranes and activate LDL oxidation, which in turn becomes toxic for the endothelium. Many experimental studies show that ROS can induce apoptosis.

There is substantial evidence linking hyperglycemia-induced overproduction of reactive oxygen species with increased oxidative stress in diabetic patients. In endothelial cells, the overproduction of ROS is sustained long after the normalization of glycemia.

In the case of glucose intolerance, an overproduction of the endogenous inhibitor of vascular nitric oxide synthase (ADMA: asymmetric dimethylarginine) can occur. One of the reasons for ADMA increase may also be the inhibition of an enzyme which degrades it: dimethylarginine dimethylaminohydrolase (DDAH). This enzyme is inactivated by oxidative stress.

Nitric oxide (NO) plays a crucial role in maintaining vascular homeostasis. Its decreased bioavailability, which can be caused either by insulin deficiency or by defective insulin sensitivity, leads to disturbances in structure and function of vessels. Hyperglycemia also induces the production of certain vasoconstrictive agents such as endothelin 1, which through its specific receptor causes contraction of smooth muscle cells in the vascular wall.

Hyperglycemia-induced molecular changes in the biology of microcirculatory vessel walls on one hand cause a change in their function and on the other are responsible for the 'hyperglycemic memory' phenomenon.

Even when the glycemia returns to normal, the glycated mitochondria keep producing the superoxide anion which in turn activates pathways involved in the development of chronic diabetes complications.

Hence, long-lasting hyperglycemia before the diagnosis of diabetes may already in this early period leave traces in cells of the circulatory system and target organs, contributing to the future development of complications. It is known that in order to prevent chronic complications, it is not only necessary to maintain good metabolic control, but also to implement efficient treatment from the moment of diagnosis. 'Metabolic memory' plays a role in the occurrence of vascular complications in children as well [16].

Due to increasing incidence of diabetes, especially in young age groups, early detection of risk factors and pathological signs becomes of paramount importance in order to facilitate early implementation of intervention strategies, which could reduce development and/or progression of complications [17].

The diagnostic usefulness of oxidative stress markers in children with type 1 diabetes has been reviewed by Krzystek-Korpacka et al. [18].

THE ACTIVATION OF NONENZYMATIC PROTEIN GLYCATION

The nonenzymatic protein glycation pathway is an important mechanism through which hyperglycemia damages endothelial cells. Endothelial dysfunction may play a significant role in the development of diabetic angiopathy.

In the process of diabetic angiopathy development, disturbances of the proteoglycans' metabolism play a significant role. The early stages of glycation are reversible, but the process eventually results in emergence of advanced glycation end products (AGE) [19, 20].

This process leads to glycation of the basement membrane, collagen, serum proteins, DNA etc.

The glycation of mitochondrial proteins might be a part of the explanation of the 'metabolic memory' phenomenon. Advanced glycation end products (AGE) accumulate in the vessel walls from the onset of diabetes and cause damage. Endothelial function disturbances play a significant role in the early stages of vascular lesions [21].

Vascular complications affect both main parts of the vessel wall (the interna and the media), including not only functional but also structural changes.

The emergence of "metabolic memory" suggests the need for very early aggressive treatment aiming to normalize metabolic control, reduce the production of cellular reactive oxygen species and glycation in addition to normalizing glucose levels in order to prevent long-term diabetic complications [22].

THE INTENSIFICATION OF GLUCOSE METABOLISM VIA THE POLYOL PATHWAY

The increased activity of the polyol pathway occurs mainly in the retina, the eye pupil, peripheral nerves and renal glomeruli, i.e., in the regions of insulin-dependent glucose transport.

Sorbitol accumulation triggers a chain of reactions which lowers the activity of sodium-potassium ATPase and thus causes structural and functional organ disturbances.

Protein Kinase C Activation

In cells where the expression of aldose reductase is low or none at all, hyperglycemia causes increased diacyloglycerol production, which up-regulates protein kinase C activity.

The excessive activation of protein kinase C leads to functional and structural changes in cells.

Hyperglycemia can have opposite effects on protein kinase C in different cell types.

For instance, retinal pericytes (which display a high expression of aldose reductase) exhibit atrophy, which causes microaneurysms. On the other hand, endothelial cells (with no enzyme at all) proliferate, leading to occlusion of the lumen of vessels [23].

At least three randomized clinical trials have demonstrated that patients treated intensively for a period of time have a lower risk of micro- and macrovascular complications that persists during subsequent follow-up, even after their tight control had stopped. The mechanisms are not fully understood.

The excess superoxide anion produced by the mitochondria in response to hyperglycemia leads, through disturbances at the nuclear level, to the accumulation of potentially harmful substances such as advanced glycated end products, protein kinase C, and nuclear factor κB, which are believed to be directly involved in the development of vascular complications in diabetes. These adverse effects are not reversed when high blood glucose is corrected, and some may be permanent because of epigenetic changes.

Probably the best strategy is to optimize the metabolic control as early as possible [24].

Conclusion

The issues concerning metabolic memory have not been fully clarified yet. However, it has been well documented that poor metabolic control at the onset of diabetes leads to excessive generation of free oxygen radicals in the mitochondria of endothelial cells and overproduction of AGEs.

Early exposure to hyperglycemia may promote the development of complications that manifest late in the progression of the disease and persist despite improved glycemic control, indicating a memory of metabolic imbalance. Very early aggressive treatment aiming to normalize glycemic control is needed to reduce levels of cellular reactive oxygen species, prevent glycation and minimize long-term diabetic complications.

It is particularly important to be aware of the existence of 'metabolic memory' in children and young adults, in whom the onset of disease is usually so evident that the diagnosis is quickly established and treatment is introduced relatively quickly. Therefore, these patients stand a good chance of a reduction in marked chronic complications.

References

[1] Cernea S, Dobreanu M. Diabetes and beta cell function: from mechanisms to evaluation and clinical implications. *Biochem Med* (Zagreb). 2013; 23(3): 266-80.

[2] Schlotterer A, Kukudov G, Bozorgmehr F, Hutter H, Du X, Oikonomou D et al. C. elegans as model for the study of high glucose-mediated life span reduction. *Diabetes* 2009; 58: 2450–2456.

[3] Turk Z. Glycotoxines, carbonyl stress and relevance to diabetes and its complications. *Physiol Res*. 2010; 59: 147-156.

[4] Thomas MC. Glycemic exposure, glycemic control, and metabolic karma in diabetic complications. *Adv Chronic Kidney Dis*. 2014; 21: 311-317.

[5] Tonna S, El-Osta A, Cooper ME, Tikellis C. Metabolic memory and diabetic nephropathy: potential role for epigenetic mechanisms. *Nat Rev Nephrol.* 2010; 6: 332-341.

[6] Keating ST, El-Osta A. Epigenetic changes in diabetes. *Clin Genet.* 2013; 84: 1-10.

[7] Mitić T, Emanueli C. Diabetes-induced epigenetic signature in vascular cells. *Endocr Metab Immune Disord Drug Targets*. 2012; 12: 107-117.

[8] Intine RV, Sarras MP Jr. Metabolic memory and chronic diabetes complications: potential role for epigenetic mechanisms. *Curr Diab Rep*. 2012; 12: 551-559.

[9] Drzewoski J, Kasznicki J, Trojanowski Z. The role of "metabolic memory" in the natural history of diabetes mellitus. *Pol Arch Med Wewn* 2009; 119: 493–500.

[10] Otto-Buczkowska E, Machnica L. Metabolic memory - the implications for diabetic complications. *Endokrynol Pol.* 2010; 61: 700-703.

[11] Reddy MA, Zhang E, Natarajan R. Epigenetic mechanisms in diabetic complications and metabolic memory. *Diabetologia*. 2015; 58: 443-455.

[12] Siebel AL, Fernandez AZ, El-Osta A. Glycemic memory associated epigenetic changes. *Biochem Pharmacol* 2010; 80: 1853-1859.

[13] Villeneuve LM, Natarajan R. The role of epigenetics in the pathology of diabetic complications. *Am J Physiol Renal Physiol* 2010; 299: F14–F25.

[14] Ceriello A. The emerging challenge in diabetes: the "metabolic memory." *Vascul Pharmacol*. 2012; 57: 133-138.

[15] Diabetes Control and Complications Trial (DCCT)/Epidemiology of Diabetes Interventions and Complications (EDIC) Research Group, Lachin JM, White NH, Hainsworth DP, Sun W, Cleary PA, Nathan DM. Effect of intensive diabetes therapy on the progression of diabetic retinopathy in patients with type 1 diabetes: 18 years of follow-up in the DCCT/EDIC. *Diabetes*. 2015 Feb; 64(2): 631-42.

[16] Cooper ME. Metabolic memory: implications for diabetic vascular complications. *Pediatr Diabetes* 2009; 10: 343–346.

[17] Marcovecchio ML, Tossavainen PH, Dunger DB. Prevention and treatment of microvascular disease in childhood type 1 diabetes *Br Med Bull* 2010; 94: 145-646.

[18] Krzystek-Korpacka M, Salmonowicz B, Boehm D, Berdowska I, Zielinski B, Patryn E et al. Diagnostic potential of oxidative stress markers in children and adolescents with type 1 diabetes. *Clin Biochem* 2008; 41: 48–55.

[19] Pietkiewicz J, Seweryn E, Bartyś A, Gamian A. Receptors for advanced glycation end products and their physiological and clinical significance. *Postepy Hig Med Dosw* (Online) 2008; 62: 511–523.

[20] Piwowar A. Advanced oxidation protein products. Part I. Mechanism of the formation, characteristics and property. *Pol Merkur Lekarski*. 2010; 28: 166–169.

[21] Yamagishi S, Matsui T. Smooth muscle cell pathophysiology and advanced glycation end products (AGEs). *Curr Drug Targets* 2010; 11: 875–881.

[22] Ceriello A. The emerging challenge in diabetes: the "metabolic memory." *Vascul Pharmacol*. 2012; 57: 133-138.

[23] Kowluru RA, Zhong Q, Kanwar M. Metabolic memory and diabetic retinopathy: role of inflammatory mediators in retinal pericytes. *Exp Eye Res* 2010; 90: 617–623.

[24] Aschner PJ, Ruiz AJ. Metabolic memory for vascular disease in diabetes. *Diabetes Technol Ther*. 2012; 14 Suppl 1: S68-74.

In: Alterations in Glucose Homeostasis in Children
Editor: E. Otto-Buczkowska
ISBN: 978-1-63483-861-0

Chapter 8

PERIOPERATIVE CARE OF PATIENTS WITH TYPE 1 DIABETES

***E. Otto-Buczkowska*[1,*] *and M. Dryżałowski*[2]**
[1]Medical Specialist Centre in Gliwice, Poland
[2]Medical University of Silesia in Katowice, Poland

ABSTRACT

Maintaining glycemic and metabolic control is difficult in diabetic patients who are undergoing surgery. The preoperative evaluation of all patients with diabetes should include careful screening for asymptomatic cardiac or renal disease. Frequent self-monitoring of glucose levels is important before surgery.

Acute hyperglycemia is commonly present in the perioperative period among patients undergoing surgery or with critical illness. Hyperglycemia occurs commonly among acutely ill patients due to a combination of counterregulatory and stress responses as well as insulin resistance and deficiency.

Frequent monitoring of electrolytes and acid-base status is important during the perioperative period, especially in patients with type 1 diabetes because ketoacidosis can develop at modest levels of hyperglycemia.

Surgical stress often produces hyperglycemia in the perioperative period. It is the general consensus that strict glycemic control is beneficial and should be achieved for diabetic patients in the perioperative period.

Monitoring, infusion of insulin, electrolytes and fluids should be continued after surgery in the same manner as during the procedure until the patient regains consciousness and resumes oral feeding.

Keywords: Type 1 diabetes, perioperative care, minor surgery, major surgery

[*] E-mail: em.buczkowski@pro.onet.pl

INTRODUCTION

Every surgical procedure and associated anesthesia involve a risk of more or less life-threatening complications. Impairment of glucose metabolism is one example of such [1, 2].

In diabetic patients undergoing surgery, it is helpful to divide procedures into two categories:

- minor surgery or procedures that require a brief general anesthesia (GA) or heavy sedation, usually of less than 1h duration, and that should not have a major impact on glycemic control
- major surgery that requires prolonged GA and is associated with greater risk of metabolic decompensation

Extensive data have shown that acute hyperglycemia is commonly present in the perioperative period among patients undergoing surgery or with critical illness. It is generally believed that intensive insulin therapy (IIT – target blood glucose of 4.4 mmol/L) reduces in-hospital mortality. However, recent large trials have questioned the efficacy and safety of IIT and raised concerns about increased mortality rates due to hypoglycemia. However, it appears that beside the risk associated with IIT, the influence of other factors must be considered – such as the effect of fluid management, anesthetic agents and anesthesia techniques on glucose homeostasis [3-7]. Perioperative hyperglycemia is a common phenomenon affecting patients both with and without a known prior history of diabetes [8].

Optimal glycemic control and proper diagnosis of chronic diabetes complications prior to surgery greatly influence the outcomes of therapy [9].

The major risk factors for diabetics undergoing surgery are the associated end-organ diseases: cardiovascular disease, autonomic neuropathy, limited joint mobility syndrome, and immune deficiency [10]. It is crucial to determine diabetes comorbidities preoperatively in order to avoid perioperative renal and cardiovascular complications [11]. Identification of chronic complications is necessary in type 2 diabetes [12]. It is also very important in type 1 diabetes, even in juvenile patients [13-18].

MANAGEMENT OF YOUNG DIABETIC PATIENTS REQUIRING SURGERY

Preoperative preparation of a pediatric patient is a task for a well-coordinated team. Whenever possible, surgery on children and adolescents with diabetes should be performed in centers with appropriate personnel and facilities to care for children with diabetes [19-24].

The objectives for patient management are to achieve the best possible metabolic control and avoid hypo- and hyperglycemia both during and after surgery. Regulating hyperglycemia can reduce morbidity, mortality, and postoperative infections [25]. It is unclear if tight glucose control (TGC) with intensive insulin therapy (IIT) can improve outcomes in critically ill children admitted to the intensive care unit (ICU). Large multicenter studies of TGC with IIT using continuous glucose monitoring in critically ill children are needed to determine if this strategy can definitively improve clinical outcomes in this population without increasing hypoglycemia. Observations carried out in pediatric intensive care unit have shown that clinicians should use guidelines wisely, recognizing that no single source of knowledge is sufficient to guide clinical decisions [1, 26]. Results of a randomized trial of hyperglycemic control in pediatric intensive care have recently been published and they showed that severe hypoglycemia (blood glucose < 36 mg/dl or < 2.0 mmol/l) occurred in a higher proportion of children in the tight-glycemic-control group than in the conventional-glycemic-control group (7.3% vs. 1.5%, $p < 0.001$) [27]. In the last few years, insulin analogues have become increasingly available, which facilitates perioperative glycemic control [28]. The authors suggest a target glucose range of 5 to 10 mmol/L (90-180 mg/dL) and recommend frequent measurement of blood glucose.

Clinical Implications of Selected Chronic Complications of Diabetes

To achieve better surgical outcomes in patients with type 1 diabetes, it is especially important to properly manage chronic diabetes complications: nephropathy, diabetic neuropathy, cardiovascular disease, gastrointestinal neuropathy, limited joint mobility syndrome, increased blood coagulability and susceptibility to infections. The presence of chronic diabetes complications influences the available methods of anesthesia and management both during and after surgery.

Symptoms of diabetic nephropathy include proteinuria, hypertension and eventually renal failure in terminal stages [17, 29].

Presence of the above symptoms is an indication for extreme care for proper renal perfusion, adequate vascular bed volume and avoidance of medication which is either nephrotoxic or eliminated via the kidneys.

Dysfunctions in the cardiovascular system may be caused by cardiac autonomic neuropathy (CAN). Its symptoms include: resting tachycardia, orthostatic hypotension, absence of pulse variability in relation to changes in blood pressure and RR interval variability on deep respiration in ECG [14, 30].

Patients with autonomic neuropathy have impaired baroreceptor reflexes and thus react to anesthesia and intubation with reduced cardiac output and hypotension, as opposed to tachycardia and hypertension in healthy individuals. Neuropathy of cardiovascular system may also be present in diabetic patients at developmental age.

Diabetic cardiomyopathy may be an early complication which impairs the cardiovascular system. The complete picture of this disease is made up of several dysfunctions caused by diabetes and leading to heart failure. Diabetic cardiomyopathy starts relatively early, independently to other chronic complications [31].

Cardiovascular dysfunction promotes arrhythmia, intra- and postoperative hypotension which result in a serious threat of acute heart failure and death.

Cardiac autonomic neuropathy has been linked to arrhythmias and enhanced intraoperative or perioperative cardiovascular lability [15, 32]. Gastroparesis is usually asymptomatic. Because of that, patients with long-term diabetes should be treated with great caution by anesthesiologists, regardless of time passed since the last meal. The risk of aspiration during anesthesia should be taken into account.

Limited joint mobility syndrome (LJM), commonly manifesting as cheiroarthropathy, is more frequently seen in diabetic patients than in the general population. The atlanto-occipital joint involvement limits adequate extension of head and neck during laryngoscopy making intubation difficult [33, 34].

PERIOPERATIVE STRESS

Some studies suggest that critical states are combined with shifting of the anabolic action of GH, mediated by increase of insulin-like growth factor 1 (IGF-1), into the counterregulating catabolic action coupled with a decrease of IGF-1 level.

Surgical procedures and general anesthesia are closely related with perioperative stress. Surgical patients commonly develop hyperglycemia related to the hypermetabolic stress response, which increases glucose production and causes insulin resistance.

In diabetic and non-diabetic patients, the stress of surgery causes a hyperglycemic response characterized by increased levels of catecholamines, growth hormone, glucagon, cortisol and cytokines such as tumor necrosis factor (TNF), interleukin 1 (IL 1) et interleukin 6 (IL 6), with a concomitant depression of insulin levels. Hepatic glycogenolysis and gluconeogenesis, along with reduced insulin secretion and tissue insulin resistance, further contribute to the hyperglycemia [35].

Diabetic surgical patients will need more stringent control, starting with preoperative glucose measurements, as well as strict intraoperative BG monitoring.

During surgery, the patients' BG should be maintained in the 7.8-10 mmol/L (140-180 mg/dL) range [36]. Tissue insulin resistance may continue into the postoperative period due to factors which limit tissue perfusion (such as hypovolemia or large doses of vasopressor agents), hypocaloric feeding or infection. In diabetic patients, perioperative stress is of greater importance, since apart from typical hyperglycemia, disruption of ion distribution and osmotic diuresis, it can easily lead to ketoacidosis with all its complications if the metabolic control is not sufficient.

Hyperglycemia impairs wound healing and increases the risk of systemic infections.

It has been proven that glucose levels above 11.1 mmol/L (200 mg/dL) promote the development of infections.

Those effects are further amplified by angiopathy, resulting ischemia and tissue hypoxia.

Another important complication of hyperglycemia is increased blood coagulability and associated risk of thromboembolism [37, 38].

SCHEDULING SURGERY

Major Surgery

Major surgery requires prolonged general anesthesia (GA) and is associated with a greater risk of metabolic decompensation.

Major operations on diabetic patients should be scheduled in the morning and preferably in the beginning of the week.

Laboratory tests necessary before a planned surgical procedure include:

- complete 24-hour glycemic profile (7 tests during the day, including one at 3: 00 AM)
- complete blood count, platelet count and peripheral blood smear
- serum creatinine, electrolytes (Na^+, K^+), aminotransferase (AST, ALT)
- arterial blood gas
- urinalysis
- HbA1c level

Other diagnostic tests depend on the patient's age, overall health, presence of chronic diabetes complications and other comorbidities. Possible additional investigations include ophthalmoscopy, chest X-ray, ECG and full non-invasive cardiac diagnostic in patients with cardiovascular risk factors.

Proper care over patients with diabetes during the perioperative period is crucial for preventing acute complications such as hyperglycemia, hypoglycemia, diabetic ketoacidosis, hyperosmolar hyperglycemic state, dehydration, and electrolyte imbalance.

The time in which a patient should be admitted before a scheduled surgical procedure depends on various factors including the level of metabolic control, patient's education, type of surgery, presence of chronic complications and comorbidities. It is usually 2-6 days before a planned procedure. In the case of well-managed patients, it may be sufficient to admit them on the day before the planned procedure.

Major surgery is defined as a procedure with a presumed duration of more than 2 hours, requiring general anesthesia, linked with greater risk of metabolic decompensation and associated with the necessity of close observation of the patient in the hospital setting for at least 24 hours after the procedure.

General anesthesia requires the patient to refrain from ingesting solid foods for 6 hours and liquids (including breast milk) for 4 hours prior to surgery.

The transition from subcutaneous administration of insulin to the intravenous infusion of short- or fast-acting insulin is a preferred method of maintaining normoglycemia in the perioperative period [39, 40].

Intraoperative anesthetic care needs to focus on the preservation of hemodynamic stability, perioperative infection control and maintenance of glucose homeostasis.

There are conflicting views on the benefits and threats of very tight glycemic control in critically ill patients [41, 42]. Glycemic control has received intense scrutiny in the last decade as an important aspect of patient care. Earlier studies suggested that tight glycemic control (i.e., target level of 4.4-6.1 mmol/L or 80-110 mg/dL) improved outcomes in intensive care unit (ICU) patients. However, tight glycemic control appears to be more beneficial in patients without diabetes than in those with known diabetes. Diabetics must be managed more carefully because tight control was associated with a high incidence of hypoglycemia, which may decrease some of its potential benefits. Beneficial effects of tight glycemic control may be demonstrated when the appropriate glycemic targets are matched to the appropriate population [43].

Many factors influence the optimal care plan, including the usual home diabetes regimen, the level of diabetes control before surgery, the type of procedure, its duration and the amount of time the patient will be fasting. Of course, the age of the patient is also of great importance. An optimal plan of care is based on a thorough presurgical assessment and prescribing a regimen that will minimize acute complications [5].

According to the current published guidelines, it seems to be optimal to control BG at the level of 140-180 mg/dL in the ICU. A more strict BG control (i.e., 6.1-7.8 mmol/L or 110-140 mg/dL) may be appropriate in selected patients as long as it can be achieved without significant hypoglycemia [44].

In young patients, according to the recommendations, glycemic targets for surgery should be within in the range of 5-10 mmol/L (90-180 mg/dL).

Adequate infusion of fluids and insulin is necessary to maintain proper metabolic control [21].

Patients with type 1 diabetes require continuous substitution of exogenous insulin due to their inability to produce insulin themselves. The insulin dosage required is individual-specific and may change dramatically during the perioperative period. The patient may be endangered by metabolic decompensation including hypoglycemia and diabetic ketoacidosis. Thus, perioperative management should include frequent blood glucose measurements and frequent adjustment of the insulin and glucose administration. Intensive insulin therapy including fast-acting insulin is usually in use.

Tables 1 and 2 presents the current recommendations for the perioperative management of children and adolescents with type 1 diabetes mellitus [21, 24, 45].

Table 1. Fluid maintenance guide (fluid requirement per 24 hours)

Body weight (kg)	Fluid requirement (ml/kg/24h)	Sodium (mmol/kg)	Potassium (mmol/kg)
4-10	100-120	2-2.5	2-2.5
10-20	80-100	1.6-2	1.6-2
20-40	60-80	1.2-1.6	1.2-1.6
Adults	30-40	ok.1.5	0.8-0.9

Table 2. Glucose requirements in different age groups

Age group	Glucose requirement (g/kg/h)	5% glucose infusion rate (ml/kg/h)
Adults and adolescents	0.1	2
Children	0.1-0.2	2-4
Infants	0.2-0.3	4-6
Newborns	0.3-0.5	usually 10% glucose solution

Intraoperative Care

Surgical stress may cause hyperglycemia and increased insulin requirements. Anesthesia may cause vasodilatation and a drop in blood pressure. Therefore, blood pressure should be carefully monitored.

Blood glucose measurements should be performed every 30-60 minutes during and immediately after general anesthesia. Dextrose and insulin infusions should be adjusted (Table 1). If the blood glucose exceeds 14 mmol/L (~ 250 mg/dL), urine or blood ketones should also be measured.

If there is an unexpected acute drop in blood pressure, NS (0.9% NaCl) or Ringer's lactate must be infused rapidly.

Minor Surgery

'Single day' surgery can only be performed on diabetic patients in good overall condition, treated with intensive insulin therapy, maintaining optimal metabolic control and well-educated.

Procedures requiring local anesthesia or deep sedation are minor surgical procedures such as endoscopies, duodenal biopsy, adenotonsillectomy, grommet insertion, and simple orthopedic procedures. They are usually of less than 2h duration and should not have a major impact on glycemic control.

Minor surgery with local anesthesia usually does not require a change in previous therapy. Procedures performed when the patient is fasting may require a simplified algorithm such as:

- performing the procedure in the morning (e.g., 8: 00-9: 00 AM) and administering a meal and insulin immediately after its completion OR
- administering part of the usual insulin dose and later repeating small injections of short-acting insulin according to blood glucose levels (a safer method, especially in children)
- infusion of 5-10% dextrose if necessary [46, 47]

Emergency surgery

Indications for emergency surgery in diabetic patients must be as limited as possible (e.g., an uncontrollable life-threatening hemorrhage). Before emergency surgery, always check blood glucose, blood ß-hydroxybutyrate or urinary ketone concentration, serum electrolytes and blood gases.

If ketoacidosis is present, follow an established treatment protocol for diabetic ketoacidosis and delay surgery, if possible, until circulating volume and electrolyte deficits are corrected. If there is no ketoacidosis, start IV fluids and insulin management as for elective surgery.

It is important to remember that acute abdominal symptoms can be the effect of ketoacidosis and may subside after the correction of metabolic imbalance.

One must not forget, however, that metabolic acidosis may also be induced by an inflammatory process developing in a diabetic patient. That is why the aim should be to improve metabolic control which should enable differentiation between the aforementioned states.

The rules for monitoring and patient management during surgery are the same as in the case of major planned procedures. It should be stressed, however, that insulin requirements are often substantially increased due to considerable stress or the presence of inflammation.

Postoperative Care

Monitoring, infusion of insulin, electrolytes and fluids should be continued after surgery in the same manner as during the procedure until the patient regains consciousness and resumes oral feeding.

If there are no complications in the postoperative period, the usual dose of short- or fast-acting insulin is administered subcutaneously prior to the first meal and the intravenous infusion of insulin and glucose is stopped. Normal regimen of insulin therapy can usually be resumed shortly after the surgery. This is different in patients undergoing transplant surgery of pancreas and pancreatic insulae.

Patients who cannot tolerate enteral feeding should receive parenteral nutrition. The rules of parenteral nutrition in diabetic patients are the same as in non-diabetics as long as correct supply of insulin and balanced nutritional value are maintained.

Patients who are immobilized for extended periods of time are a special group. They are at extreme risk of thromboembolism or infection and should receive prophylactic anticoagulation therapy (both adults and children).

Conclusion

Whenever possible, surgery on children and adolescents with diabetes should be performed at centers with appropriate personnel and facilities to care for children with diabetes. Centers performing surgical procedures on children with diabetes should have written protocols for postoperative management of diabetes in wards where children are admitted.

Frequent blood glucose monitoring and IV infusion of dextrose are essential 5% dextrose is usually sufficient although 10% dextrose may be necessary when there is an increased risk of hypoglycemia. Elevated blood ketone and blood glucose concentrations require extra insulin and, possibly, additional IV fluids for correction.

Although it is clear that hyperglycemia is harmful, there is currently insufficient evidence to support the routine use of tight glycemic control (target BG 80-110 mg/dL) in the operating room or the ICU. With careful stepwise implementation of IIT protocols, maintaining BG less than 150 mg/dL and reducing BG variability may be both safe and effective.

REFERENCES

[1] Agus MS, Steil GM, Wypij D, Costello JM, Laussen PC, Langer M. et al. SPECS Study Investigators. Tight glycemic control versus standard care after pediatric cardiac surgery. *N Engl J Med*. 2012; 367: 1208-1219.

[2] Duncan AE. Hyperglycemia and perioperative glucose management. *Curr Pharm Des*. 2012; 18: 6195-6203.

[3] Kadoi Y. Anesthetic considerations in diabetic patients. Part II: intraoperative and postoperative management of patients with diabetes mellitus. *J Anesth*. 2010; 24: 748-756.

[4] Kadoi Y. Blood glucose control in the perioperative period. *Minerva Anestesiol*. 2012; 78: 574-595.

[5] Levesque CM. Perioperative care of patients with diabetes. *Crit Care Nurs Clin North Am*. 2013; 25: 21-29.

[6] Pestel G, Closhen D, Zimmermann A, Werner C, Weber MM. Aspects of perioperative care in patients with diabetes. *Anaesthesist*. 2013; 62: 9-19.

[7] Pichardo-Lowden A, Gabbay RA. Management of hyperglycemia during the perioperative period. *Curr Diab Rep*. 2012 Feb; 12: 108-118.

[8] Sheehy AM, Gabbay RA. An overview of preoperative glucose evaluation, management, and perioperative impact. *J Diabetes Sci Technol*. 2009; 3: 1261-1269.

[9] Kadoi Y. Anesthetic considerations in diabetic patients. Part I: preoperative considerations of patients with diabetes mellitus. *J Anesth*. 2010; 24: 739-747.

[10] Russo N. Perioperative glycemic control. *Anesthesiol Clin*. 2012; 30: 445-466.

[11] Rittler P, Broedl UC, Hartl W, Göke B, Jauch K. Diabetes mellitus - perioperative management. *Chirurg*. 2009; 80: 410, 412-415.

[12] Carles M, Raucoules-Aimé M. Perioperative management of diabetic patient. *Presse Med*. 2011; 40: 587-595.

[13] Blankenburg M, Kraemer N, Hirschfeld G, Krumova EK, Maier C, Hechler T. et al. Childhood diabetic neuropathy: functional impairment and non-invasive screening assessment. *Diabet Med*. 2012; 29: 1425-1432.

[14] Machnica L, Deja G, Polanska J, Jarosz-Chobot P. Blood pressure disturbances and endothelial dysfunction markers in children and adolescents with type 1 diabetes. *Atherosclerosis*. 2014; 237: 129-134.

[15] Otto-Buczkowska E, Kazibutowska Z, Sołtyk J, Machnica L. Neuropathy and type 1 diabetes mellitus. *Pediatr Endocrinol Diabetes Metab*. 2008; 14: 109-116.

[16] Otto-Buczkowska E, Jarosz-Chobot P. Limited joint mobility syndrome in patients with diabetes. *Int J Clin Pract*. 2012; 66: 332-333.

[17] Otto-Buczkowska E., Tucholski K. Kidneys function in glucose homeostasis regulation and its therapeutic implications. *Fam Med Primary Care Rev* 2013; 15: 34-37.

[18] Verrotti A, Loiacono G, Mohn A, Chiarelli F. New insights in diabetic autonomic neuropathy in children and adolescents. *Eur J Endocrinol*. 2009; 161: 811-818.

[19] American Diabetes Association's (ADA) Standards of Medical Care in Diabetes. *Diabetes Care*. 2011; 34 Suppl 1: S1-2.

[20] Górska-Ciebiada M, Ciebiada M, Barylski M, Loba J. Perioperative management of the diabetic patient. *Anaesthesiology Rescue Medicine* 2010; 4: 479-484

[21] ISPAD Consensus Guidelines 2014; 15 (Suppl. 20): 224-231.

[22] Polskie Towarzystwo Diabetologiczne. Zalecenia kliniczne dotyczące postępowania u chorych na cukrzycę, 2014. *Clin Diabetol* 2014; 3 suppl A50.

[23] Skała-Zamorowska E, Gadzicka A, Jarosz-Chobot P, Młynarski W. Surgery in diabetic children [in] Cukrzyca w populacji wieku rozwojowego – co nowego? Otto-Buczkowska E. (ed) Cornetis Wrocław 2009: 199-207.

[24] Szirer G. Perioperative care of patients with type 1 diabetes [in] Cukrzyca typu 1. Otto-Buczkowska E. (ed) Cornetis Wrocław 2006: 433-444.

[25] Schroeder SM. Perioperative management of the patient with diabetes mellitus: update and overview. *Clin Podiatr Med Surg*. 2014; 31: 1-10.

[26] Kavanagh BP. Glucose in the ICU - evidence, guidelines, and outcomes. *N Engl J Med* 2012, 367: 1259-1260.

[27] Macrae D, Grieve R, Allen E, Sadique Z, Morris K, Pappachan J et al. CHiP Investigators. A randomized trial of hyperglycemic control in pediatric intensive care. *N Engl J Med*. 2014; 370: 107-118.

[28] Killen J, Tonks K, Greenfield JR, Story DA. New insulin analogues and perioperative care of patients with type 1 diabetes. *Anaesth Intensive Care*. 2010; 38: 244-249.

[29] Bogdanović R. Diabetic nephropathy in children and adolescents. *Pediatr Nephrol*. 2008; 23: 507-525.

[30] Machnica L, Deja G, Jarosz-Chobot P. Hypertension and prehypertension in children and adolescents with diabetes type 1. *Pediatr Endocrinol Diabetes Metab*. 2008; 14: 215-219.

[31] Otto-Buczkowska E, Dworzecki T, Mazur-Dworzecka U. Cardiovascular system disease in youth patients with various forms of glucose metabolism impairment - diagnosis and treatment. *Pediatr Endocrinol Diabetes Metab*. 2008; 14: 31-35.

[32] Vinik AI, Erbas T. Diabetic autonomic neuropathy. *Handb Clin Neurol*. 2013; 117: 279-94.

[33] Hashim K, Thomas M. Sensitivity of palm print sign in prediction of difficult laryngoscopy in diabetes: A comparison with other airway indices. *Indian J Anaesth*. 2014; 58: 298-302.

[34] Otto-Buczkowska E. The impact of diabetes on musculoskeletal system [in] Cukrzyca typu 1. Otto-Buczkowska E. (ed.) Cornetis Wroclaw 2006: 369-376.

[35] Otto-Buczkowska E, Dworzecki T, Mazur-Dworzecka U, Tucholski K. Alterations in blood glucose homeostasis during septic or injury stress – hyperglycemia. *Fam Med Primary Care Rev* 2008; 10: 197-205.

[36] Raju TA, Torjman MC, Goldberg ME. Perioperative blood glucose monitoring in the general surgical population. *J Diabetes Sci Technol*. 2009; 3: 1282-1287.

[37] Randriamboavonjy V, Fleming I. Platelet function and signaling in diabetes mellitus. *Curr Vasc Pharmacol*. 2012; 10: 532-538.

[38] Vazzana N, Ranalli P, Cuccurullo C, Davì G. Diabetes mellitus and thrombosis. *Thromb Res*. 2012; 129: 371-377.

[39] Betts P, Brink SJ, Swift PG, Silink M, Wolfsdorf J, Hanas R, ISPAD. Management of children with diabetes requiring surgery. *Pediatr Diabetes*. 2007; 8: 242-247.

[40] Betts P, Brink S, Silink M, Swift PG, Wolfsdorf J, Hanas R. Management of children and adolescents with diabetes requiring surgery. *Pediatr Diabetes*. 2009; 10 Suppl 12: 169-174.

[41] Srinivasan V, Agus MS. Tight glucose control in critically ill children – a systematic review and meta-analysis. *Pediatr Diabetes*. 2014; 15: 75-83.

[42] Steil GM, Agus MS. Tight glycemic control in the ICU – is the earth flat? *Crit Care*. 2014; 18: 159.

[43] Abdelmalak BB, Lansang MC. Revisiting tight glycemic control in perioperative and critically ill patients: when one size may not fit all. *J Clin Anesth*. 2013; 25: 499-507.

[44] Minakata K, Sakata R. Perioperative control of blood glucose level in cardiac surgery. *Gen Thorac Cardiovasc Surg*. 2013; 61: 61-66.

[45] Herbst A, Kiess W. Type 1 diabetes mellitus. Perioperative management of children and adolescents. *Anaesthesist* 2007; 56: 454-460.

[46] Jälevik B, Klingberg G. Pain sensation and injection techniques in maxillary dento-alveolar surgery procedures in children – a comparison between conventional and computerized injection techniques (The Wand). *Swed Dent J*. 2014; 38: 67-75.

[47] Pasarón R, Burnweit C, Zerpa J, Malvezzi L, Knight C, Shapiro T et al. Nitrous oxide procedural sedation in non-fasting pediatric patients undergoing minor surgery: a 12-year experience with 1,058 patients. *Pediatr Surg Int*. 2015; 31: 173-780.

In: Alterations in Glucose Homeostasis in Children
Editor: E. Otto-Buczkowska

ISBN: 978-1-63483-861-0

Chapter 9

ALTERATIONS OF BLOOD GLUCOSE HOMEOSTASIS IN CRITICALLY ILL YOUNGEST PATIENTS

E. Otto-Buczkowska[1,*] ***and M. Dryżałowski***[2]
[1]Medical Specialist Centre in Gliwice, Poland
[2]Medical University of Silesia in Katowice, Poland

ABSTRACT

Alterations of blood glucose homeostasis are frequent in critically ill children. Considerable alterations of glucose metabolism are found in critical states associated with stress factors such as trauma, major surgery or sepsis. Prolonged stress is characterized by progressing metabolic dysregulation in which gluconeogenesis shows resistance to the inhibiting actions of insulin and glucose. This resistance refers not only to the insulin influence on glucose homeostasis. Stress is associated also with resistance to other insulin effects, such as lipolysis suppression. The degree of insulin resistance is positively related to the severity of the stress response.

Glucose variability, including both hyperglycemia and hypoglycemia, showed the strongest association with increased morbidity and mortality rates. Stress hyperglycemia SH commonly occurs during critical illness in children. Hyperglycemia is a significant problem for children in the ICU.

Recommendations concerning the „normoglycemic" limits of glucose concentration and indications for insulin therapy in critically ill children are still largely controversial. It is unclear if tight glucose control TGC with intensive insulin therapy IIT can improve outcomes in critically ill children. Use of tight glycemic control TGC to manage hyperglycemia remains controversial, especially given the potential risk of insulin-induced hypoglycemia.

Final guidelines concerning blood glucose level values and insulin administration in critical states have not been determined yet. In recent years, however, number of publications informed about the positive influence of normoglycemia in critical states in children.

Keywords: Blood glucose homeostasis, hyperglycemia, hypoglycemia, critically ill children

[*] E-mail: em.buczkowski@pro.onet.pl.

HYPERGLYCEMIA

Physiologically, the organism attempts to maintain glucose homeostasis, typically contained within a narrow range of 60-160 mg/dl 3,3-8,9 mmol/L, regardless of food intake and energy expense. Many mechanisms which regulate glucose metabolism are involved in the maintenance of this homeostasis.

Critically ill patients suffer from a considerable disturbance of glucose metabolism caused by stress factors such as trauma, major surgery or sepsis [1].

Stress hyperglycemia and hypoglycemia are associated with increased mortality and morbidity in critically ill patients. The first who noticed that was Claude Bernard in 1877. Other authors, when confirming that phenomenon, used terms such as: "traumatic diabetes," "diabetes of injury" or "stress diabetes" [2].

In 1942, David Cutherbertson introduced the terms of ebb and flow phases for the description of hypo- and hypermetabolism that occur as a consequence of injury.

Prolonged stress is characterized by progressing metabolic dysregulation in which gluconeogenesis shows resistance to the inhibitory effects of insulin and glucose [3].

Stress-caused impairment of carbohydrate metabolism occurs through a number of mechanisms such as the increase of glucose production. Although glycogenolysis increases the hepatic production of glucose directly after the onset of a stress factor, it is only a temporary effect because glycogen stores run out very quickly. Hyperglycemia that is typical for phase two results from an increase of gluconeogenesis.

Glucagon is the main hormonal stimulus of gluconeogenesis in the course of critical states. Studies on patients after injuries showed that liver glucose production is reduced when secretion of glucagon is inhibited by somatostatin.

Adrenergic nervous system seems to play a minor role because combination of alpha and beta blockage does not reduce gluconeogenesis. Similarly, increase in the peripheral production of gluconeogenesis precursors is not of the highest importance since it has been shown that an infusion of alanine does not stimulate gluconeogenesis.

Both TNF and IL-1 influence the stimulation of hepatic glucose production and have synergistic effects with hormones which increase gluconeogenesis.

Increase of gluconeogenesis during stress is resistant to the inhibitory effect of insulin and glucose infusion. It is probably due to prolonged stimulation by glucagon and cytokines. Metabolic dysregulation characteristic of prolonged hypermetabolic stress may lead to the multiorgan failure syndrome.

INCREASE OF GLUCOSE ABSORPTION AND UTILIZATION

Heavy posttraumatic or septic stress is combined with an increase of cellular usage of glucose. Although the precise mechanism of those changes is not known, cytokines are believed to play a crucial role in the regulation of the process. In normal conditions, 80% of glucose is utilized independently of insulin NIMGU – non-insulin-mediated glucose uptake, mainly by the central nervous system. The other 20% of glucose is used by skeletal muscles and, among that, half without insulin mediation NIMGU and half with the use of insulin IMGU – insulin-mediated glucose uptake.

Dysfunction of the β-cell may also play a role in the pathophysiology of critical illness hyperglycemia CIH [4].

Level of hyperlactatemia is a useful marker of severity of stress [5, 6, 7]. Concentration of lactates in blood plasma may be an indicator of balance between the consumption of oxygen and its metabolic usage. This concentration is usually measured in arterial blood and it ranges from <2 mmol/L in a healthy person to >4 mmol/L in a sick one. Acute metabolic acidosis is frequently encountered in critically ill patients [8, 9]. It can occur as a result of either the accumulation of endogenous acids which consume bicarbonate or the loss of bicarbonate from the gastrointestinal tract or the kidneys.

Insulin Resistance

One of the earliest descriptions of insulin resistance during stress was given in 1955 by J. M. Howard who noted that soldiers and accident victims responded to an oral glucose test similarly to diabetic people.

Investigations showed that hyperglycemia which accompanies stress should be linked to insulin resistance. However, this insulin resistance does not apply solely to the insulin influence on glucose homeostasis. Stress is also associated with resistance against other effects of insulin e.g., the suppression of lipolysis. Lipotoxicity is emerging as a significant contributor to the development of insulin resistance.

Insulin resistance is an adaptive mechanism that prioritizes utilization of energy for the immune response in the presence of infection or injury [10].

Overally, hyperglycemia is a common event in the course of posttraumatic stress or infection despite many tissues showing an increase in the cellular consumption of glucose. Peripheral insulin resistance has various consequences such as the reduction of insulin mediated glucose utilization IMGU in skeletal muscles. Hepatic insulin resistance also plays a role in the genesis of hyperglycaemia during stress [11, 12, 13].

Insulin resistance favors hyperglycemia which helps to provide for increased glucose requirements necessary to cover energetic demands of cells in critical states. Despite much research, this mechanism of insulin resistance has not been explained yet. Data suggests a heterogeneity of factors responsible for the phenomenon [14]. Fundamentally, the level of hyperglycemia is directly proportional to the severity of stress response [15, 16]. Hyperlactatemia and consumption of oxygen also increase with the severity of stress.

Other Reasons for Hyperglycemia during Stress

Parenteral glucose administration as a way of feeding is a frequent cause of hyperglycemia in patients in stress. Although glucose is a preferred source of energy in tissues that take part in an immunological response to injury and glucose uptake in such conditions is maximal, the excess of calories administrated in the form of glucose may result in significant hyperglycemia i.e., blood glucose concentration over 250 mg/dl or 13.9 mmol/l. That is why blood glucose in such patients must be monitored. Patients with latent liver cirrhosis may also show glucose intolerance "hepatogenous diabetes" caused by the limited ability to store glycogen.

Pancreatits is also a very well described cause of hyperglycemia, related to some extent to hyperglucagonemia.

Various drugs may impair glucose tolerance which results in hyperglycemia. Although corticoids are the most frequent cause of hyperglycemia, other drugs may have a similar effect.

Hypokalemia can favor hyperglycemia by worsening insulin secretion.

CONTROL OF HYPERGLYCEMIA DURING STRESS

The issue of blood glucose level that should be maintained in critical states is still under discussion. Hyperglycemia is usually defined as a glucose value > 7.8 mmol/L 140 mg/dL.

According to Farrokhi et al., hyperglycemia is present in 40% of critically ill patients and in up to 80% of patients after cardiac surgery, with ~80% of hyperglycemic ICU patients having no history of diabetes prior to admission [17]. According to these authors, a target glucose level between 7.8 and 10.0 mmol/L 140 and 180 mg/dL is recommended for the majority of ICU patients.

A similar perspective was presented by Martínez-Gangoso and Fuentes-Pumarola, who upon review of the available literature concluded that blood glucose controls should be initially performed each hour, to the point of glycemia stabilization, and then they are to be performed every two-three hours [18]. The glycemia in a critically ill patient should be maintained between 7,8-10,0 mmol/L 140-180 mg/dL, which is considered moderate control.

Krinsley et al. called attention to the role of blood glucose BG measurement frequency in the maintenance of glycemic control [19]. The Monte Carlo simulation suggests that glycemic control in critically ill patients is more optimal with a blood glucose measurement interval no longer than 1 hour, with further benefit obtained through the use of a measurement interval of 15 minutes. These findings have important implications for the development of glycemic control standards.

Until year 2001, it was believed that glucose concentration in critically ill patients should be kept within the range of 8.9-12.2 mmol/L 160-220 mg/dL because this guaranteed the sufficient supply of energy to the brain, adrenals and erythrocytes. At present, significantly lower values are recommended. Attempts are made to rigorously maintain normoglycemia at the level of 4,4-6,1 mmol/L 80-110 mg/dL [20].

A more widely-accepted view, though, is that the maintenance of glycemia between 7,8-10,0 mmol/L 140-180 mg/dL is safer [21].

The problem is complex and difficult to solve distinctly since the identification of optimal glycemia limits requires the consideration of many factors, including the patient's age [22, 23, 24].

Hyperglycemia develops commonly in the critically ill and impacts outcomes in patients with diabetes, but - even more so - in patients with stress-induced hyperglycemia.

It is believed that 30-60% of patients with stress-induced diabetes suffer from previously undiagnosed glucose intolerance or diabetes. An analysis of research results on the presence of glucose homeostasis dysfunction in critically ill patients, based on literature, was presented by Fahy et al. [25].

Many authors call attention to the frequency of glucose homeostasis dysfunction, mainly hyperglycemia, in critical states among children [26, 27]. Such dysfunction is also found in neonates, especially in very low birth weight infants [28].

It has been stressed that such dysfunction is associated with increased mortality and its onset may have a prognostic value. Márquez-González et al. expressed a view that hyperglycemia was an independent risk factor for the prediction of death, with a likelihood of death of 56.8%, when it was present in critically ill neonates [29].

The level of hyperglycemia is associated with the severity of the patient's clinical state [30]. It is relatively frequent in the course of severe illnesses and may be looked upon as the possible result of an uncoordinated insulin response to the increased glucose concentration that the body may need during periods of stress.

Many authors highlight the necessity for careful glycemic control and recommend the use of tight glucose control TGC to manage stress hyperglycemia SH [31].

Recommendations concerning the „normoglycemic" limits of glucose concentration and indications for insulin therapy in critically ill children are still largely controversial [32-36].

The multicenter randomized trial of 1369 patients in pediatric intensive care units ICU with a target blood glucose range of 4.0-7.0 mmol/L 72-126 mg/dL or conventional glycemic control, with a target level below 12.0 mmol/L 216 mg/dL showed, that tight glycemic control in critically ill children had no significant effect on major clinical outcomes, although the incidence of hypoglycemia was higher with tight glucose control than with conventional glucose control [37].

Similar conclusions, based on multicenter randomized studies, were drawn by Kalfon et al. [38]. These authors found that tight computerized glucose control with the computer decision support system CDSS did not significantly change mortality and was associated with more frequent severe hypoglycemia episodes in comparison with conventional glucose control CGC, with blood glucose targets of 4.4-6.1 and <10.0 mmol/L 79,2-109,8 and < 180 mg/dL, respectively.

Also the studies of Agust et al. on a group of children who underwent cardiac surgery and were treated in the pediatric cardiac intensive care unit PICU showed, that routine implementation of tight glycemic control in the pediatric cardiac intensive care unit is controversial due to concerns about potentially deleterious effects of insulin-induced hypoglycemia on the developing brain [39].

Therefore, both glycemic limits for the implementation of insulin therapy and the recommended dosage of insulin are still discussed and require further study [40].

Before universal implementation in pediatric intensive care, both long-term effects on outcomes and development and issues regarding optimal levels of blood glucose control need to be cleared in multicenter prospective randomised controlled trials RCTs.

Technological improvement might be helpful in optimising blood glucose control [41-44]. The effectiveness of different methods for glycemic control recording in critically ill children using the Yale Insulin Infusion Protocol YIIP and eProtocol insulin ePi was presented by Faraon-Pogaceanu et al. [45]. They found that YIIP is less effective but as safe as ePi in achieving tight glycemic control. Faustino and Apkon determined the prevalence and prognostic significance of hyperglycemia among critically ill nondiabetic children [46]. This retrospective study was based on blood glucose measurements in 942 non-diabetic patients admitted to a pediatric intensive care unit. It showed that hyperglycemia occurs frequently among critically ill non-diabetic children and is correlated with a greater in-hospital mortality rate. Srinivasan et al. reviewed 152 critically ill children [47]. Hyperglycemia was found to be a common finding. Peak blood glucose and duration of hyperglycemia were proved to be independently associated with mortality.

A special case of stress hyperglycemia was published by Mittal et al. [48]. The authors presented a case of neonatal septicemia with stress induced hyperglycemia mimicking neonatal diabetes mellitus. Extremely high blood sugar values of up to 1529 mg/dl with metabolic acidosis were a diagnostic dilemma. In such case, it is vital to promptly treat hyperglycemia with insulin infusion.

Final guidelines concerning blood glucose values and insulin administration in critical states have not been determined yet. In recent years, however, a number of publications informed about positive influence of normoglycemia in critical states in children.

Thabet et al. used continuous infusion of insulin in 30 premature infants receiving parenteral nutrition, when glycemia reached 12.1 ± 3.3 mmol/L 218mg/dL [49].

The required insulin dose range to control blood glucose was 0.07-4.2 IU/kg/h. Authors believe that continous insulin infusion is a rapid and safe way of improving glucose tolerance in infants with hyperglycemia developed during total parenteral nutrition.

Similar conclusions were reached by Kairamkonda, who used continous infusion of insulin in low birth weight infants who developed hyperglycemia, glycosuria and osmotic diuresis on day 2 of parenteral nutrition [50].

Also Ng et al. state that continous insulin infusion was relatively safe and effective in the treatment of persistent hyperglycemia in premature neonates [51].

Pham et al. adopted a policy of 'intensive' insulin therapy to achieve blood glucose levels 90 to 120 mg/dL in severely burned children [52]. The authors concluded that intensive insulin therapy should be considered for children with severe burn injuries since this therapy may lower infection rates and improve survival.

Nayak et al. performed a multicenter study of insulin therapy implementation in critically ill children in the UK and found there were significant differences in the establishment of indications for such treatment [53].

One of the causes for hyperglycemia in critically ill patients treated in intensive care units is parenteral feeding through glucose infusion. In many cases the hyperglycemia is secondary to excessive glucose supply, which largely exceeds the required uptake. The surplus of calories from glucose may also stimulate the secretion of catecholamines which in turn increases hepatic lipogenesis and leads to liver steatosis. Calculation of the correct caloric values for critically ill patients is therefore extremely important. The alimentary recommendations for youngest children must follow the accepted limits for specific age groups.

Hypoglycemia

Severe hypoglycemia, defined as blood glucose <2,2 mmol/L <40 mg/dL, is independently associated with increased mortality. Safe implementation of tight glycemic control requires appropriate monitoring to reduce the risk of this complication.

Srinivasan et Agus performed prospective randomized controlled trials of tight glucose control TGC in critically ill children admitted to the intensive care unit with intensive insulin therapy IIT [54].

They found that TGC with IIT appears to reduce the risk of acquired infection in critically ill children, however it is associated with higher incidence of hypoglycemia.

Similar results were presented by Faustino et al. [55]. They suggest defining hypoglycemia as <2,2-2,5 mmol/L <40-45 mg/dL in neonates and <3,3-3,6 mmol/L <60-65 mg/dL in children. They believe that prevention of hypoglycemia, especially in the case of intravenous insulin use, will be best accomplished by the combination of accurate measuring techniques, frequent or continuous glucose monitoring, and computerized insulin titration protocols.

Randomized controlled trials in pediatric patients who underwent cardiac surgery were demonstrated by Kandil et al. [56].

Steil et al. recommend continuous glucose monitoring for minimizing severe hypoglycemia in the course of tight glycemic control [57].

Dickerman et al. found that automated adverse-event detection using triggers derived from the electronic health record EHR is an effective method of identifying hypoglycemia in pediatric patients [58].

Lately, Leite et al. raised the issue of alterations of blood glucose homeostasis in the malnourished critically ill children [59]. Both hyperglycemia and hypoglycemia increased the risk of mortality in the malnourished patients compared with the well-nourished ones.

CONCLUSION

- Considerable alterations of glucose metabolism are found in critical states associated with stress factors such as trauma, major surgery or sepsis.
- Stress hyperglycemia and hypoglycemia are associated with increased mortality and morbidity in critically ill patients.
- Safe implementation of tight glycemic control requires appropriate monitoring to reduce the risk of complications.

REFERENCES

[1] Hartl W. H., Jauch K. W. Metabolic self-destruction in critically ill patients: origins, mechanisms and therapeutic principles. *Nutrition,* 2014; 30: 261-267.

[2] Van den Berghe G. How does blood glucose control with insulin save lives in intensive care? *J. Clin. Invest.*, 2004; 114: 1187-1195.

[3] Otto-Buczkowska E., Dworzecki T., Mazur-Dworzecka U., Tucholski K. Alterations in blood glucose homeostasis during septic or injury stress–hyperglycemia. *Fam. Med. Primary Care Rev.*, 2008; 10: 197-205.

[4] Hacıhamdioğlu B., Kendirli T., Oçal G, Sıklar Z., Savaş Erdeve S., Ince E., Berberoğlu M. Pathophysiology of critical illness hyperglycemia in children. *J. Pediatr. Endocrinol. Metab.*, 2013; 26: 715-720.

[5] Ben-Hamouda N., Haesler L., Liaudet L. Hyperlactatemia and lactic acidosis in the critically ill patient. *Rev. Med. Suisse,* 2013; 9: 2335-2340.

[6] Qi Y. Z. Prognostic values of serum procalcitonin level and pediatric critical illness score in children with sepsis. *Zhongguo Dang Dai Er Ke Za Zhi,* 2014; 16: 190-193.

[7] Zhang Z., Xu X. Lactate clearance is a useful biomarker for the prediction of all-cause mortality in critically ill patients: a systematic review and meta-analysis. *Crit. Care Med.,* 2014; 42: 2118-2125.

[8] Lim S. Metabolic acidosis. *Acta Med. Indones.,* 2007; 39: 145-150.

[9] Morris C. G., Low J. Metabolic acidosis in the critically ill: part 2. Causes and treatment. *Anaesthesia,* 2008; 63: 396-411.

[10] Dhar A., Castillo L. Insulin resistance in critical illness. *Curr. Opin. Pediatr.,* 2011; 23: 269-274.

[11] Bierbrauer J., Weber-Carstens S. Insulin resistance and protein catabolism in critically ill patients. *Anasthesiol. Intensivmed Notfallmed Schmerzther,* 2011; 46: 268-274.

[12] Chen S. S., Donmoyer C. M., Pearson D. A., Poole A., Zhang Y., Lacy D. B. et al. Impact of infection on glucose-dependent liver glucose uptake during TPN: interaction with insulin. *Am. J. Physiol. Endocrinol. Metab.*, 2004; 286: E286-295.

[13] Dhar A., Castillo L. Insulin resistance in critical illness. *Curr. Opin. Pediatr.,* 2011; 23: 269-274.

[14] Bakar M. H., Sarmidi M. R., Kai C. K., Huri H. Z., Yaakob H. Amelioration of mitochondrial dysfunction-induced insulin resistance in differentiated 3T3-L1 adipocytes via inhibition of NF-κB pathways. *Int. J. Mol. Sci.,* 2014; 15: 22227-22257.

[15] Corathers S. D., Falciglia M. The role of hyperglycemia in acute illness: supporting evidence and its limitations. *Nutrition,* 2011; 27: 276-281.

[16] Lazzeri C., Tarquini R., Giunta F., Gensini G. F. Glucose dysmetabolism and prognosis in critical illness. *Intern. Emerg. Med.*, 2009; 4: 147-156.

[17] Farrokhi F., Smiley D., Umpierrez G. E. Glycemic control in non-diabetic critically ill patients. *Best Pract. Res. Clin. Endocrinol. Metab.,* 2011; 25: 813-824.

[18] Martínez-Gangoso L., Fuentes-Pumarola C. Hyperglycemia in critically ill patients. Sample of choice, controls and values: literature review. *Enferm. Intensiva*, 2014; 25: 146-163.

[19] Krinsley J. S., Bruns D. E., Boyd J. C. The Impact of Measurement Frequency on the Domains of Glycemic Control in the Critically Ill - A Monte Carlo Simulation. *J. Diabetes Sci. Technol.*, 2015 Jan. 6. pii: 1932296814566507. [Epub ahead of print]

[20] Van Herpe T., Mesotten D., Wouters P. J., Herbots J., Voets E., Buyens J. et al. LOGIC – insulin algorithm-guided versus nurse-directed blood glucose control during critical illness: the LOGIC-1 single-center, randomized, controlled clinical trial. *Diabetes Care*, 2013; 36: 188-194.

[21] Mesotten D., Van den Berghe G. Glycemic targets and approaches to management of the patient with critical illness. *Curr. Diab. Rep.*, 2012; 12: 101-107.

[22] Van den Berghe G., Mesotten D. A trial of hyperglycemic control in pediatric intensive care. *N. Engl. J. Med.,* 2014; 370: 1354-1355.

[23] Van den Berghe G., Mesotten D. Paediatric endocrinology: Tight glycaemic control in critically ill children. *Nat. Rev. Endocrinol.*, 2014; 10: 196-197.

[24] Van Herpe T., Vanhonsebrouck K., Mesotten D., De Moor B., Van den Berghe G. Glycemic control in the pediatric intensive care unit of Leuven: two years of experience. *J. Diabetes Sci. Technol.*, 2012; 6: 15-21.

[25] Fahy B. G., Sheehy A. M., Coursin D. B. Glucose control in the intensive care unit. *Crit. Care Med.*, 2009; 37: 1769-1776.

[26] Otto-Buczkowska E., Dworzecki T., Mazur U. Alterations of blood glucose homeostasis in critically ill children – hyperglycemia. *Pediatr. Endocrinol. Diabetes Metab.*, 2007; 13: 43-46.

[27] Wintergerst K. A., Foster M. B., Sullivan J. E., Woods C. R. Association of hyperglycemia, glucocorticoids, and insulin use with morbidity and mortality in the pediatric intensive care unit. *J. Diabetes Sci. Technol.*, 2012; 6: 5-14.

[28] Bottino M., Cowett R. M., Sinclair J. C. Interventions for treatment of neonatal hyperglycemia in very low birth weight infants. *Cochrane Database Syst. Rev.*, 2011; 10:CD007453.

[29] Márquez-González H., Muñoz-Ramírez M. C., Ramírez-García M. A., Pineda-Frutos M. F., Hernández-Ramírez C., Villa-Romero A. R. et al. Hyperglycemia as a risk factor for mortality in critically ill neonates. *Rev. Med. Inst. Mex. Seguro Soc.*, 2014; 52 Suppl. 2: S104-109.

[30] Preissig C. M., Rigby M. R. Pediatric critical illness hyperglycemia: risk factors associated with development and severity of hyperglycemia in critically ill children. *J. Pediatr.*, 2009; 155: 734-739.

[31] Srinivasan V. Stress hyperglycemia in pediatric critical illness: the intensive care unit adds to the stress! *J. Diabetes Sci. Technol.*, 2012; 6: 37-47.

[32] Macrae D., Tasker R. C., Elbourne D. A trial of hyperglycemic control in pediatric intensive care. *N. Engl. J. Med.*, 2014; 370: 1355-1356.

[33] Preissig C. M., Rigby M. R. A disparity between physician attitudes and practice regarding hyperglycemia in pediatric intensive care units in the United States: a survey on actual practice habits. *Crit. Care*, 2010; 14: R11.

[34] Van den Berghe G., Mesotten D. A trial of hyperglycemic control in pediatric intensive care. *N. Engl. J. Med.*, 2014; 370: 1354-1355.

[35] Vasques T. G., Ferranti J. F., de Carvalho W. B. A trial of hyperglycemic control in pediatric intensive care. *N. Engl. J. Med.*, 2014; 370: 1355.

[36] Wise J. Tight glycaemic control in critically ill children could shorten hospital stays, finds trial. *BMJ*, 2014; 348: g137.

[37] Macrae D., Grieve R., Allen E., Sadique Z., Morris K., Pappachan J. et al. A randomized trial of hyperglycemic control in pediatric intensive care. *N. Engl. J. Med.*, 2014; 370: 107-118.

[38] Kalfon P., Giraudeau B., Ichai C, Guerrini A., Brechot N., Cinotti R. et al. Tight computerized versus conventional glucose control in the ICU: a randomized controlled trial. *Intensive Care Med.*, 2014; 40: 171-181.

[39] Agus M. S., Steil G. M., Wypij D., Costello J. M., Laussen P. C., Langer M. et al. Tight glycemic control versus standard care after pediatric cardiac surgery. *N. Engl. J. Med.*, 2012; 367: 1208-1219.

[40] Branco R. G., Xavier L., Garcia P. C., Piva J. P., Fiori H. H., Baldisserotto M. et al. Prospective operationalization and feasibility of a glycemic control protocol in critically ill children. *Pediatr. Crit. Care Med.*, 2011; 12: 265-270.

[41] Asencio Y., Lamy F., Paut O. Is it necessary to obtain a tight glycemic control in critically ill children? *Ann. Fr. Anesth. Reanim.*, 2010; 29: 570-572.

[42] Van den Heuvel I., Vlasselaers D. Clinical benefits of tight glycaemic control: focus on the paediatric patient. *Best Pract. Res. Clin. Anaesthesiol.*, 2009; 23: 441-448.

[43] Verhoeven J. J., Brand J. B., Van de Polder M. M., Joosten K. F. Management of hyperglycemia in the pediatric intensive care unit; implementation of a glucose control protocol. *Pediatr. Crit. Care Med.,* 2009; 10: 648-652.

[44] Verbruggen S. C., Landzaat L. J., Reiss I. K., van Goudoever J. B., Joosten K. F. Efficacy and safety of a tight glucose control protocol in critically ill term neonates. *Neonatology,* 2012; 101: 232-238.

[45] Faraon-Pogaceanu C., Banasiak K. J., Hirshberg E. L., Faustino E. V. Comparison of the effectiveness and safety of two insulin infusion protocols in the management of hyperglycemia in critically ill children. *Pediatr. Crit. Care Med.*, 2010; 11: 741-749.

[46] Faustino E. V., Apkon M. Persistent hyperglycemia in critically ill children. *J. Pediatr.*, 2005; 146: 30-34.

[47] Srinivasan V., Spinella P. C., Drott H. R., Roth C. L., Helfaer M. A., Nadkarni V. Association of timing, duration, and intensity of hyperglycemia with intensive care unit mortality in critically ill children. *Pediatric Critical Care Medicine*, 2004; 5: 329-336.

[48] Mittal A., Gupta R., Sharma S., Aggarwal K. C. Stress induced hyperglycemia in a term baby mimicking diabetic ketoacidosis with stroke. *J. Clin. Neonatol.*, 2013; 2: 190-192.

[49] Thabet F., Bourgeois J., Guy B., Putet G. Continuous insulin infusion in hyperglycaemic velow-birth-weight infants receiving parenteral nutrition. *Clin. Nutr.*, 2003; 22: 545-547.

[50] Kairamkonda V. Does continuous insulin infusion improve glycaemic control and nutrition in hyperglycaemic very low birth weight infants? *Arch. Dis. Child.*, 2006; 91: 76-79.

[51] Ng S. M., May J. E., Emmerson A. J. Continuous insulin infusion in hyperglycaemic extremely-low-birth-weight neonates. *Biol. Neonate,* 2005; 87: 269-272.

[52] Pham T. N., Warren A. J., Phan H. H., Molitor F., Greenhalgh D. G., Palmieri T. L. Impact of tight glycemic control in severely burned children. *J. Trauma*, 2005; 59: 1148-1154.

[53] Nayak P., Lang H., Parslow R., Davies P., Morris K.; UK Paediatric Intensive Care Society Study Group PICS SG. Hyperglycemia and insulin therapy in the critically ill child. *Pediatr. Crit. Care Med.,* 2009; 10: 303-305.

[54] Srinivasan V., Agus M. S. Tight glucose control in critically ill children – a systematic review and meta-analysis. *Pediatr. Diabetes*, 2014; 15: 75-83.

[55] Faustino E. V., Hirshberg E. L., Bogue C. W. Hypoglycemia in critically ill children. *J. Diabetes Sci. Technol.*, 2012; 6: 48-57.

[56] Kandil S. B., Miksa M., Faustino E. V. Control of serum glucose concentration in critical illness. *Curr. Opin. Pediatr.*, 2013; 25: 297-303.

[57] Steil G. M., Langer M., Jaeger K., Alexander J., Gaies M., Agus M. S. Value of continuous glucose monitoring for minimizing severe hypoglycemia during tight glycemic control. *Pediatr. Crit. Care Med.,* 2011; 12: 643-648.

[58] Dickerman M. J., Jacobs B. R., Vinodrao H., Stockwell D. C. Recognizing hypoglycemia in children through automated adverse-event detection. *Pediatrics,* 2011; 127: e1035-1041.

[59] Leite H. P., de Lima L. F., de Oliveira Iglesias S. B., Pacheco J. C., de Carvalho W. B. Malnutrition may worsen the prognosis of critically ill children with hyperglycemia and hypoglycemia. *J. Parenter. Enteral Nutr.*, 2013; 37: 335-341.

In: Alterations in Glucose Homeostasis in Children
Editor: E. Otto-Buczkowska
ISBN: 978-1-63483-861-0

Chapter 10

IMPAIRED GLUCOSE HOMEOSTASIS IN ENDOCRINOLOGICAL DISEASES

B. Wikiera, A. Zubkiewicz-Kucharska and A. Noczyńska* [*]
Wrocław Medical University, Wrocław, Poland

ABSTRACT

Glucose homeostasis is achieved by the coordinated action of counterregulatory hormones, namely insulin, glucagon and glucocorticoids, however influenced by many others.

Transcription of phosphoenolpyruvate carboxykinase (PEPCK) and glucose-6-phosphatase (G6Pase) is altered in different endocrine disorders. Moreover, many endocrine disorders lead to excessive secretion of hormones antagonistic to insulin. Reduced glucose uptake and glycogen synthesis, hepatic gluconeogenesis, increased proteolysis and lipolysis are observed, as well as insulin resistance.

As the result glucose homeostasis is frequently disturbed. Impaired glucose tolerance and frank diabetes (DM) are observed in the course of the adrenal, ovarian, pituitary and thyroid diseases.

The secondary occurrence of glucose metabolism disorders in various hormonal diseases is a recurrent observation. The increased cardiovascular morbidity and mortality associated may be a consequence of increased insulin resistance that normally accompanies hormone excess. Adequate diagnosis and treatment for glucose imbalance is mandatory in these patients.

Keywords: Diabetes, impaired glucose tolerance, insulin resistance, cortisol, aldosterone, PCOS, growth hormone, thyroid hormones, prolactin

[*] E-mail: anna.noczynska@umed.wroc.pl.

INTRODUCTION

Glucose homeostasis is achieved by the coordinated action of counterregulatory hormones, namely insulin, glucagon and glucocorticoids; however it may be influenced by many others. Concentration of circulating glucose is maintained within a narrow range (irrespective of fasting or postprandial), by a delicate balance between intestinal absorption, glucose production by the liver, and glucose utilization by the peripheral tissues. Hormone response units (HRUs) present in the two genes that catalyze gluconeogenesis: phosphoenolpyruvate carboxykinase (PEPCK) and glucose-6-phosphatase (G6Pase), integrate signals from various signaling pathways triggered by hormones. Transcription of these genes is altered in different endocrine disorders.

Many endocrine disorders lead to excessive secretion of hormones antagonistic to insulin. As a result, glucose homeostasis is frequently disturbed. Impaired glucose tolerance and frank diabetes (DM) are observed in the course of adrenal, ovarian, pituitary and thyroid diseases.

ADRENAL GLANDS

In the adrenal glands, hypercortisolism, hyperaldosteronism and elevated levels of epinephrine, increase glucose levels. Diabetes mellitus is one of the most common complications of Cushing syndrome (CS). Glucocorticoids (GC) induce insulin resistance of target tissues and impair insulin secretion from the pancreas. Reduced glucose uptake and glycogen synthesis, hepatic gluconeogenesis, increased proteolysis and lipolysis, are observed in CS. GC also evoked impairment of adipokine secretion, elevated levels of leptin, decreased levels of adiponectin and alteration of TNF and IL6 were described. Accordingly, deterioration of insulin secretion followed a long time period of insulin resistance. The duration of CS and the degree of hypercortisolism, play a role in the development of DM. Genetic factors are also involved and GC receptor polymorphisms N363S, BCL1, ER22/EK impair GC sensitivity and are associated with DM.

Hypercortisolism

Hypercortisolism is defined as the elevation of cortisol level in the blood. Glucocorticoids action is displayed in Table 1. Prolonged exposure to high glucocorticoids (GC) levels can be exogenous or endogenous in origin and leads to a clinical picture of Cushing syndrome (CS). GC are widely used in the treatment of different diseases, because of their strong immunosuppressive and anti-inflammatory activity. Therapeutic doses of GC are multiple replacement doses and often result in the development of iatrogenic CS.

Although exogenous CS is the most common condition, endogenous CS also develops during childhood and adolescence [1-3]. It constitutes only 10% of CS cases in the general population, and can be ACTH dependent or ACTH independent. ACTH dependent CS accounts for 85% of cases in children older than 7 years. In the study published by Batista et al., there was no female gender predominance among children with ACTH dependent CS, in contrast to female predisposition in adults [1]. In younger children, endogenous CS is usually

ACTH independent and is caused by adrenocortical adenoma, adrenocortical carcinoma, bilateral micronodular adrenal disease, macronodular adrenal hyperplasia, McCune-Albight syndrome or MEN Type I. In the older population, 80% of endogenous CS is the result of small-sized, pituitary adenoma, secreting ACTH (Cushing disease). ACTH or CRH produced by other tissues, leads to a similar clinical picture of the disease, but is very rarely diagnosed in youth. Ectopic ACTH secretion was found in neuroblastoma, pheochromocytoma, insulinoma, bronchial carcinoma, thymoma [4]. ACTH concentration in ectopic secretion is several doses higher than in a pituitary ACTH secreting tumor. Nevertheless, clinical signs of hypercortisolemia are scant because of rapid onset of the disease and general impoverishment in the course of neoplasm. High concentration of ACTH stimulates deoxycorticosterone secretion, and leads to alkalosis and hypokalemia.

Adrenal tumors more frequently are found in girls than in boys. Benign adrenal neoplasms secrete cortisol; adrenal cancer also secrets androgens and progressive virilisation frequently develops. Adrenal tumors can coexist with genetic syndromes: Beckwith-Wiedemann and Li-Fraumeni.

ACTH independent, micronodular, adrenal disease also called primary pigmented adrenocortical disease (PPAD), is characterized by small bilateral adrenal nodules secreting excessive amounts of cortisol and androgens. Adrenal glands are not enlarged because ACTH is suppressed by cortisol. PPAD more frequently occurs in girls than in boys.

Table 1. Glucocorticoids activity

Site of the activity	Result of the activity
CNS	ADH decrease, CRH increase, deterioration of taste, smell, hearing
Pituitary gland	Retardation of the synthesis and the secretion of ACTH
Cardio-vascular system	Maintenance of sensitivity to epinephrine and norepinephrine, sensitization to vasoconstrictive factors, maintenance of microcirculation
Digestive system	Increase of the secretion of gastric acid, decrease of proliferation of the mucosa cells
Liver	Gluconeogenesis increase
Lungs	Increase of maturation and synthesis of surfactant in fetal period
Kidneys	Increase in glomerular filtration rate (GFR)
Bones	Increase of bone resorption, decrease of bone formation
Muscle	Augmentation of protein catabolism, decrease of glucose oxidation decrease of insulin sensitivity, reduction of protein synthesis
Immune system	decrease of thymus and lymph nodes volume, decrease of lymphocytes, eosinophils, and basophiles, decrease of cellular immunity
Contiguous tissue	Decrease of collagen synthesis and fibroblast activity

It can be a component of Carney complex a condition characterized by pigmented spots on the skin and mucosae, concomitant tumors: schwannoma, cardiac myxoma, pituitary adenoma, testicular Leydig-cells or Sertolli-cells tumors, medullary carcinoma of the thyroid gland. The condition also consists of an element of multiple endocrine neoplasia (MEN). Carney complex appears sporadically or as familial autosomal dominant disease. Mutations of the regulatory subunit of protein kinase encoded on the long arm of chromosome 17 and mutations of phosphodiesterase gene were confirmed.

Adrenal nodular hyperplasia can occur in McCune-Albright syndrome, caused by postmeiotic activating mutation of GNAS1 gene. It is characterized by a triad of signs: pigmented skin spots with irregular borders, fibrous dysplasia of bones, and hyper-function of different endocrine glands.

Macronodular hyperplasia is the result of stimulation of anomalous adrenal receptors through catecholamines, estrogens, LH or other peptides [5].

Clinical manifestations of CS results from elevated secretion of GC [6]. Almost 100% of children and adolescents affected by CS are found to be obese. Obesity is characterized by central fat accumulation with round face, buffalo hump, supraclavicular fat pads. In younger children, fat distribution is more generalized. Skin is thin with purple striae and facial plethora. Growth velocity is usually diminished and pubertal development is disturbed. Other signs of CS include: proximal muscle weakness, hypertension, osteopenia, bruising, emotional lability, and depression.

Diabetes mellitus is a known and frequent complication of CS. The prevalence of DM ranges from 20 to 50% [7-10]. In 70% of patients with CS glucose intolerance is confirmed [11]. The development of DM depends rather on the time of duration of the excessive GC secretion than the level of cortisol in the blood. The constant improvement in hormonal and imaging diagnostics of CS accelerates establishing diagnosis and treatment what effects in shortening of the time of GC exposal. In the recent study of Witek et al. the rate of DM is lower and relates to 16,7% of patients, and prediabetes was diagnosed in only 33,3% of patients [12].

GC influence glucose homeostasis in many ways [7, 9, 10]. GC induce insulin resistance of target tissues and impaired insulin secretion from the pancreas [13]. Reduced glucose uptake and glycogen synthesis, hepatic gluconeogenesis, increased proteolysis and lipolysis are observed in CS [7, 13].

GC stimulate gluconeogenesis in the liver mainly through the augmented activities of phosphoenolpyruvate carboxykinase and glucose-6-phosphatase. Insulin resistance can be mediated by the expression of hepatic peroxisome-proliferator-activated receptor γ (PPAR-γ).

GC evoke also impairment of adipocytokines secretion and elevated levels of leptin, decreased levels of adiponectin and alteration of TNF and IL6 were described [3, 14, 15]. Accordingly, the deterioration of insulin secretion and β-cell apoptosis following a long period of insulin resistance [13, 14]. The duration of CS and the degree of hypercortisolism play a role in DM development.

Genetic factors are also involved in the development of DM in patients with CS. Higher prevalence of family history of DM was observed and patients were predisposed to deterioration of insulin secretion [3].

GC receptor polymorphisms N363S, BCL1 impair GC sensitivity and are associated with DM [16, 17]. The presence of polymorphism ER22/EK leads to GC resistance and better metabolic parameters [18]. Similarly the presence of polymorphism A3669G is also associated with lower prevalence of DM [16].

The diagnosis of Cushing syndrome is based on clinical signs and documentation of hypercortisolism: elevated concentration of free cortisol in 24 hour urine collection, especially when fourfold greater than upper normal limits suggest CS, low-dose dexamethasone suppression test with 1 mg dexamethasone administrated orally at 10 a.m. with estimation of cortisol serum level at 8 p.m. (normal value below 1.8 ug/dl) [19]. CS is also characterized by derangement of diurnal rhythm of cortisol without evening nadir. Recently measurement of salivary cortisol at bedtime or midnight is recognized as an easy and noninvasive method for diagnosis of CS [19, 20]. Unfortunately, applying these methods does not help to differentiate between ACTH dependent and ACTH independent CS.

Plasma ACTH is usually suppressed below 10 pg/ml in the morning in patients with ACTH independent CS, whereas levels of ACTH above 20 pg/ml characterize ACTH dependent CS. The highest ACTH levels accompany ectopic source of ACTH.

High-dose dexamethasone suppression test is a useful tool in differential diagnosis. The patient is given 2 mg dexamethasone every six hours for two days. Cortisol level decrease at least by 50% after the test in most patients with pituitary adenoma secreting ACTH, whereas patients with ectopic ACTH secretion or with ACTH independent CS do not respond [21, 22]. Patients with pituitary ACTH secreting adenoma respond also to CRH infusion (1mg/kg) by increase in ACTH above 50% and cortisol above 20%. Patients with anorexia nervosa, mental disorders or under stressful conditions have unreliable results of adrenal tests.

Hormonal evaluation should be followed by imaging techniques. Gadolinium enhanced magnetic resonance imaging (MRI) of the pituitary gland can visualize pituitary adenoma, but this method is not useful in case of small microadenoma. Bilateral inferior petrosal sinus sampling (BIPSS) before and after CRH test helps to localize the place of ACTH secretion.

Computed tomography and MRI are used in the examination of the adrenal glands for the ectopic source of ACTH [19, 20].

The treatment of CS is based mainly on surgical methods. In case of adenoma of the pituitary, the method of treatment is transsphenoidal resection. Recently Gamma Knife, as well as linear accelerator-assisted radiosurgery state other therapeutic options. Surgical resection is also a recommended method of treatment in case of tumors producing ACTH ectopically or adrenal tumors, regardless if they are benign or aggressive [23].

Drug therapy is used in case of inoperative tumors or in the period preceding surgical treatment. Mitotan treatment is recommended in case of adrenal carcinoma. In other conditions aminoglutethimide, metyrapone, trilostane and ketoconazole are useful.

Such treatment usually ameliorates metabolic disorders caused by GC excess but in some patients does not normalize glucose intolerance or DM [11, 23].

Primary Hyperaldosteronism

Primary hyperaldosteronism (PH) is the most common endocrine cause of hypertension . Taking into account the criteria for aldosterone secretion, it is estimated that the prevalence of patients suffering from hypertension with this disease is between 4.6 and 12% [24, 25]. It is even greater in the 20% of cases of hypertension that is unresponsive to treatment [24, 25]. The majority of cases of PH are due to aldosterone-secreting adenoma or result from bilateral adrenal hyperplasia [26]. Adenoma is most common in middle-aged women (in men - three times less), while bilateral adrenal hyperplasia occurs with similar frequency in both sexes [24]. Rarely hyperaldosteronism may be a consequence of: unilateral adrenal hyperplasia, cancer of the adrenal cortex, genetic predisposition (i.e., Family hyperaldosteronism type I-III), or is caused by ectopic aldosterone secretion by the kidneys or ovarian cancer [25]. 11-beta hydroxylase deficiency and 17-alpha hydroxylase deficiency are very rare cause of PH.

Among the clinical symptoms of the disease, revealed most commonly between the 3 and 6 decade of life, are clinical symptoms due to hypertension and hypokalemia [26]. Approximately 15-25% of patients with PH develop disorders of carbohydrate metabolism, in the form of glucose intolerance or diabetes. These are caused by both, impaired insulin secretion and increased insulin resistance [27]. The observed disorder, although in most cases is a consequence of potassium deficiency, may be, to some extent also directly caused by the overproduction of mineralocorticoid [28, 29]. Compared to healthy subjects, patients with PH have a higher value of HOMA-IR index and lower QUICKI, as well as a greater increase in plasma insulin levels after a high glucose load, which proves the existence of insulin resistance [30]. Prevalence of metabolic syndrome is also greater in patients with PH (41.1%) than in idiopathic hypertension (29.6%) and also higher among them are fasting plasma glucose levels [30]. The aforementioned variations of hyperaldosteronism supports the demonstration of correlation between HOMA-IR and aldosterone concentrations in patients with hypertension [28, 29]. The diagnosis of PH requires: screening and demonstrating: increased levels of aldosterone, reduced plasma renin activity and an increased ratio of aldosterone / plasma renin activity. The next step is a confirmation of the diagnosis by the statement "rigidity" of aldosterone (in inhibition assay 0.9% sodium chloride solution with fludrocortisone test or a test with captopril) and plasma renin activity (similar values in supine and standing). It is also necessary to determine the disease requiring further treatment [24, 25]. Management of choice for patients with adenoma producing aldosterone is laparoscopic adrenalectomy. While in case of bilateral adrenal hyperplasia, recommended is conservative treatment, comprising administering aldosterone antagonists: spironolactone, in recent years eplerenone [26]. In very rare cases, hyperaldosteronism sensitive to glucocorticoids are treated with glucocorticoids (especially dexamethasone) [25]. Surgical treatment increases insulin sensitivity, whereas drug therapy stimulates secretory pancreatic β-cells [22, 24]. This effect is revealed in the first six months after the procedure or initiation of treatment [30].

Pheochromocytoma

Pheochromocytoma (PCC) is a rare tumor secreting excessive amounts of epinephrine. It is usually situated in the adrenal medulla. Only 10 - 20% of PCC arise from sympathetic tissue localized outside of the adrenals.

Almost 30-40% of PCC consist of familial cases caused by germ line mutations in one of the genes encoding succinate dehydrogenase subunits (SDH). The pattern of inheritance is autosomal dominant [31, 32].

PCC can occur commonly in MEN IIa and IIb, neurofibromatosis type I, von Hippel-Lindau and hereditary PCC/PGL syndromes. Recently the mutation in a gene predisposing to pituitary adenomas (MEN1) and PCC was described [33].

The main symptom of PCC is hypertension. Paroxysmal unprovoked hypertension with pallor, sweating and tachycardia strongly suggest the diagnosis of PCC.

Diabetes is present in one in three patients with pheochromocytoma. The diagnosis of DM in children with hypertension and normal body weight, should direct the diagnostic process to pheochromocytoma [34].

Epinephrine also disturbs glucose homeostasis by stimulation of alfa-2-adrenergic receptors which cause a decrease in insulin secretion and glucose tissue uptake, while increasing gluconeogenesis and glycogenolysis. Epinephrine additionally augments insulin resistance. Glucose intolerance and DM type 2 may be the first presentation of pheochromocytoma. In the research of Pogorzelski, DM was recorded in 31.3% of patients with PC and arterial hypertension occurred in 79.1% of them. After surgical removal of the tumor, 90% of type 2 DM and 70% of arterial hypertension subsided [35]. An improvement in glucose tolerance, and DM withdrawal, are also reported by other authors [36, 37]. Moreover, alpha–blockers administered to patients preparing for surgery, ameliorate glucose intolerance [38]. In the postoperative period patients are at risk of hypoglycemia. The recovery of insulin release leads to transient hyperinsulinemia [39].

The biochemical evaluation in PCC consists of estimation of plasma free metanephrines and/or fractionated urine metanephrines and normetanephrines. A sensitivity of these methods approaches 100% for diagnosis of a catecholamine-secreting tumor [31, 39, 40]. It is recommended to consider probability of false-positive or false-negative results, in particularly possibility of pseudopheochromocytoma, a state much more frequent than PCC [39]. Visualizing techniques include CT and MRI. CT is advised as the method of choice in imaging of the adrenal glands, whereas MRI is a better tool in searching for metastases or extra adrenal tumor. In cases of tumors difficult to localize, (123)I-metaiodobenzylguanidine scintigraphy is useful [41].

Genetic testing for SDHB mutations should be offered to patients with metastatic disease.

The method of choice in management of PCC is laparoscopic resection of the adrenal gland. Partial adrenalectomy is advised to patients with bilateral disease in hereditary PCC. Preparation for surgery should include a high-sodium diet and fluid intake to prevent postoperative hypotension. Each patient even with normal blood pressure demands treatment with an a-adrenoreceptor blocker to prevent hypertensive crisis. The most widely used blocker is phenoxybenzamine at a dose of 10-20 mg twice daily starting 10-14 days prior to surgery [31, 32].

OVARIES: PCOS

Polycystic ovary syndrome (PCOS) is a very common complex disorder. The prevalence of PCOS ranges from 6 to 20%, depending on the criteria used. The Rotterdam European Society for Human Reproduction and Embryology/American Society for Reproductive Medicine consensus statement defined PCOS as two out of three of the following: oligo-ovulation and/or anovulation, clinical and/or biochemical signs of hyperandrogenism, or polycystic ovaries, and exclusion of other etiologies [42-47].

Diseases leading to androgen excess, such as congenital adrenal hyperplasia, Cushing syndrome, virilising tumors, or disturbing functional integrity of the hypothalamo-pituitary-ovarian axis such as hyperprolactinemia, thyroid dysfunction, severe insulin resistance should be excluded.

More recently NIH Experts Panel determined four different phenotypes of PCOS 1) hyperandrogenism (clinical or biochemical) and CA (H-CA); 2) hyperandrogenism and polycystic ovaries on ultrasound (PCOm) but with ovulatory cycles (H-PCOm); 3) CA and polycystic ovaries without hyperandrogenism (CA-PCOm); and, 4) hyperandrogenism, CA and polycystic ovaries (H-CA-PCOm) [45].

Etiology

PCOS is a heterogeneous condition; genetic and environmental factors may be responsible for its development. Carbohydrate metabolism and adrenal secretion during fetal development influences PCOS appearance later in life [48-51]. There is a higher frequency of PCOS in a group of girls born small for their gestational age (SGA), than in the girls with normal birth weight, supports the importance of intrauterine conditions [52]. Besides SGA, patients with precocious pubarche, a state resembling early androgen hypersecretion, are also prone to PCOS development, and insulin resistance and visceral adiposity are observed very early in them [53, 54]. The highest incidence rate of PCOS was found in girls born SGA who develop premature adrenarche [52].

The investigation of the genetic background of PCOS did not confirm single gene mutation in the pathogenesis, and the disease seems to be polygenic or oligogenic in origin [50]. The role of the excessive androgen secretion during fetal development is confirmed and connected with early puberty, particularly early pubic hair appearance, and the clinical features of PCOS later in life [55, 56]. The activity of hypothalamo-pituitary-ovarian axis develops in the second trimester of pregnancy and hyperandrogenemia may disturb it's programming. As a result, luteinizing hormone (LH) secretion is in turn is connected with hyperinsulinemia and insulin resistant state [50]. In such a milieu, ovarian function and ovulation are disturbed [50, 52, 56]. Franks et al. reported on higher concentration of anti-Müllerian hormone (AMH) in daughters of women with PCOS in infancy, early childhood and prepuberty [50]. Moreover Maliqueo et al. confirmed lower 3b-hydroxysteroid dehydrogenase 1 and aromatase activities in the placentas of women with PCOS would augment fetal androgens exposure [57].

Diagnosis

Estimation of serum testosterone concentration is essential in the diagnostic procedure of PCOS. There is no notable fluctuation of testosterone during the menstrual cycle. The testosterone concentration in PCOS can be elevated or normal.

In case of normal testosterone concentrations, the free androgen index (FAI) should be calculated. The FAI is the ratio between total testosterone and sex hormone-binding globulin (SHBG). It is regarded as the most sensitive marker of hyperandrogenemia in PCOS [47].

Low SHBG frequently observed in PCOS, is a marker of insulin resistance and androgen excess. Women with PCOS and low SHBG are prone to metabolic syndrome and gestational diabetes [46, 47]. Recently discovered polymorphism in the gene encoding SHBG may be associated with reduced SHBG levels in PCOS, regardless of the effects of insulin resistance and obesity [58, 59].

The other methods include: gas chromatography with mass spectrometry (GC–MS) or liquid chromatography with tandem mass spectrometry (LC–MS/MS); both are very sensitive and specific but expensive and time-consuming. Fanelli at al. used LC–MS/MS method in a group of late adolescent, and young women, and observed higher T and androstendione levels in anovulatory females compared to ovulatory females [60].

In recent consensus, estimation of gonadotropins and calculation of LH/ FSH ratio is not recommended in the diagnosis of PCOS. There is no recommendation for measurement of anti-Müllerian hormone (AMH) as well. Despite it, AMH is an important marker of the number of small antral follicles. The concentration of AMH reflects ovarian reserve. Higher concentration of AMH in PCOS is connected with a greater number of follicles in the ovaries, correlating with the antral follicle count and degree of menstrual disturbances [61, 62]. Moreover the concentration of AMH correlates with the severity of both hyperandrogenism and oligoanovulation in PCOS [63].

It is also postulated that vitamin D deficiency may play a role in pathogenesis of PCOS, intensifying metabolic disturbances. Low concentration of vitamin D was found in PCOS more frequently in women with obesity, insulin resistance and glucose intolerant states [64].

Lower concentrations of vitamin D are observed in PCOS, in connection with obesity and metabolic alterations, and may affect fertility [65-68]. An interrelation between vitamin D and AMH has also been confirmed [67, 68].

Ultrasound diagnostic criteria of PCOS changed within the last decade as the result of marked improvements in technical ability of newer ultrasound scanners. It is suggested that raising the diagnostic threshold from 19 to 26 or more follicles per ovary respectively instead of 12 or more follicles measuring 2–9 mm in diameter [69-71].

In adolescent girls when the transvaginal route is not feasible, the number of follicles per ovary is difficult to count. In these circumstances, ovarian volume of at least 10 ml can be used as ultrasound criteria or the ratio of ovarian stroma to total ovary, with a cutoff value of 0.32 [72].

Metabolic Disorders

PCOS is commonly connected with insulin resistance, which occurs in 50% to 80% of women. Post-binding defect in signal transduction is responsible for the resistance. This defect results from impaired receptor kinase activity. Additionally there is also concomitant beta cell dysfunction. It is responsible for impaired insulin secretion in response to glucose.

Obesity, particularly the accumulation of fatty tissue in the abdominal/visceral region is frequently observed in PCOS and worsens the metabolic and reproductive features of PCOS [73]. It may be favored by androgen excess during each step of individual development and is connected with insulin resistance. Obese women with PCOS frequently develop type 2 diabetes mellitus (T2D), the highest risk is for those with BMI over 30 [74]. Moreover, the adipocyte function is changed in PCOS [75]. They secret lower amounts of high molecular weight (HMW) adiponectin, which is an insulin-sensitizing agent. Its activity is mediated by activation of AMP kinase [76]. In patients with PCOS HMW-adiponectin concentration is low, and dependent on the concentration of androgens and their influence on adipocyte function [77].

The adipose tissue of PCOS women differs substantially from the tissue of the others according to morphology and function [78]. Adipocytes in PCOS are hypertrophic. Impaired activity of the sympathetic system in their abdominal fat was confirmed, and reduced vascularization [79]. Hypoxia of the adipose tissue leads to the development of inflammatory reactions with an increased secretion of cytokines, chemokines, adipokines (free fatty acid (FFA), leptin, resistin and visfatin) and a decreased secretion of adiponectin [79]. In such conditions, insulin sensitivity is reduced and T2D subsequently develops [80]. The role of androgens in adipocyte hypertrophy stimulation is confirmed. They facilitate maturation of adipocytes and play a role in activation of enzymes involved in lipid and carbohydrate metabolism, and increase lipolysis [81, 82].

Recently, it was observed that the high gene expression of CD11c (ITGAX), along with tumor necrosis factor alpha (TNFa), in subcutaneous adipose tissue were significantly higher in PCOS women. TNFa has proinflammatory activity and may contribute to the pathogenesis of insulin resistance in PCOS women [83].

As the result of the above mentioned mechanisms 70% of PCOS women not only obese commonly suffer from dyslipidemia with high triglycerides and low high density lipoprotein cholesterol levels. This atherogenic pattern of lipids is observed in other states of insulin resistance.

Although PCOS is strongly connected with the risk of cardiovascular disease (CVD), mainly in adulthood, cardio-metabolic risk factors, insulin resistance, and endothelial dysfunction accompany the disease from the very beginning. Especially hyperandrogenemia negatively influence CVD [84].

Insulin resistance in the endothelial cells of arteries leads to reduced synthesis and release of nitric oxide (NO). NO released from endothelial cells is quickly inactivated. On the other hand, increased synthesis of vasoconstricting factors leads to increased vascular stiffness and impaired vasodilatory action of insulin [85]. Insulin is a potent growth factor and exerts a hypertrophic effect on the vascular endothelium and the vascular smooth muscle cells. Insulin resistance stimulates endothelin-1 (ET1) synthesis, harmful for endothelial function [86]. Moreover, in PCOS, increased secretion of free radicals and decreased total antioxidant concentration in serum, even in non-obese women, was described [87].

Oxidative stress plays a role as a link among systemic inflammation and dysfunction of endothelial cells and abnormal thecal cell action [88]. Children born from mothers with PCOS, display increased oxidative stress markers comparable to offspring from gestational diabetes [89].

The other factors which play a role in glyco-oxidative stress, are advanced glycosylation end products (AGEs). AGEs are the products of glycation or glycooxidation of proteins and lipids and play a role as proinflammatory mediators. In women with PCOS, the concentration of AGEs is elevated [90].

Christakou et al. established a close and positive correlation between serum AGEs and serum ET1 concentration, and postulated its role in endothelial dysfunction in women with PCOS [91]. Thus, increased serum AGEs in young PCOS women may play a role as a predictive marker of CV morbidity.

The positive correlation of serum AGEs with serum testosterone levels in PCOS points towards a linkage of metabolic and reproductive status [90]. The treatment efforts in PCOS depend on the patient's main problems. Modification of diet and lifestyle is recommended to each woman independent of age and weight.

It is of great importance to reduce intake of carbohydrates with a high glycemic index. Moreover, overweight and obese women should be advised to maintain a low calorie diet to lose weight. Reduction of BMI improves metabolic and hormonal profile and helps in normalization of menstrual cycle regularity. An increase in physical activity is helpful in reducing insulin resistance.

As young girls are not attempting to be pregnant and their complaints focus on the symptoms related to hyperandrogenemia, such as: acne, oily skin and hirsutism, therapy should be tailored to their needs. Oral contraceptive pills (OCP), anti-androgens and insulin sensitizers are the most frequently used.

The most widely used insulin sensitizer drug is metformin. The benefits of metformin treatment are connected with ameliorating cardiometabolic disorders, and the reproductive abnormalities in PCOS.

It improves insulin sensitivity in the liver and peripheral tissue, and ovarian steroidogenesis diminishes the concentration of inflammatory and atherogenic molecules [92]. It is suggested that AGE dietary interventions or inhibitors might represent an emerging therapeutic approach with significant applications in the metabolic and reproductive consequences of PCOS [92, 93].

PITUITARY GLAND

Growth Hormone

Growth hormone (GH) (also called somatotropin), is an anabolic hormone that is synthesized and secreted by the somatotroph cells of the anterior lobe of the pituitary gland. It is the most abundant hormone in the pituitary accounting for 25% of the gland's hormones. It is a member of the GH gene family, together with prolactin (PRL) and human placental lactogen (hPL). Binding of GH to the extracellular domain of GH receptor is the first step in the signal pathway, accounting for various biological effects of somatotropin [94].

Several factors including hormones, neurotransmitters and metabolic modulators are known to regulate GH in humans, including three major hypothalamic factors: growth hormone releasing hormone (GHRH), somatostatin (SRIF) and ghrelin [95]. The others include GH secretagogues (GHSs), thyroid hormones, glucocorticoids, and sex steroids [96].

Hypoglycemia induces GH release, probably due to somatostatin withdrawal. On the contrary, the inhibitory effect of glucose on GH release may be due to a discharge of hypothalamic somatostatin [97]. Moreover, GH secretion is controlled in the negative feedback manner by GH itself and IGF-1. The latter inhibits somatotropin release by influencing GHRH and SRIF production in the hypothalamus, and inhibition of GHRH action in the pituitary gland [98].

GH action is mediated by GH-dependent factors, among them insulin-like growth factor-1 (IGF-1) is the most potent, having both mitogenic and metabolic functions. IGF-1, along with IGF-2, belongs to a family of insulin-like growth factors (IGFs) that share close structural homology to the precursor form of insulin (pro-insulin). It acts locally in a paracrine/autocrine manner and at a distance in a hormone-like fashion. GH stimulates the synthesis and secretion of IGF-1 in the liver (in the presence of sufficient nutrient intake and elevated portal insulin levels) and growth plate. Though GH is the most potent regulator of IGF-I expression, nutrition, liver function, and sex hormones also contribute to its expression [99, 100]. IGF-1 is critical for promoting the protein anabolic effects of GH. Apart from its effects on growth and development, IGF-1 also has insulin-like effects on metabolism. However, there is growing evidence that not all GH actions need to be run through IGF-1 [101].

GH is involved in multiple and complex actions in human body. First of all, GH plays a central role in the modulation of a child's growth until the completion of puberty, even during fetal development (however it may contribute only to approximately 20% of fetal size) [102]. Somatotropin, together with IGF-1, are the main stimulators of longitudinal bone growth, as well as the acquisition of bone mass during the prepubertal period and further bone homeostasis maintenance. Moreover, it controls body composition and body metabolism throughout life. Although it is generally regarded as an anabolic factor, it has both anabolic and catabolic actions on different tissues.

A metabolic role of GH arises from three hypotheses established in the 1960s:

- The "thrifty genotype" hypothesis by Neel. In accordance with his hypothesis, evolution favored survival of individuals that were genetically capable to store a surplus of calories as fat [103].
- The "glucose fatty acid cycle" by Randle, according to which, supply of glucose to the liver promotes glucose oxidation and glucose and lipid storage, inhibiting fatty acid oxidation at the same time, while providing free fatty acids promotes fatty acid oxidation and storage, inhibiting glucose oxidation and promoting glucose storage, in case of incomplete glycogen reserves, eventually leading to insulin resistance [104].
- The "feast and famine cycle" by Rabinowitz and Zierler, according to which insulin is the major anabolic hormone storing all fuels during conditions of energy surplus and GH is the major anabolic hormone during starvation and stress, sparing glucose and protein at the expense of lipids, thereby allowing conservation of protein stores [105]. There is growing evidence that starvation and physical stress promotes GH secretion, whereas in profusion of glucose and lipid, GH release is inhibited in humans [106, 107]. The secretion of GH is maximal at puberty, which is accompanied by very high circulating IGF-1 levels. That would prove the suggestion that the activated GH/IGF-1 axis at puberty may be causally linked to the concomitant increase in insulin resistance [108].

The dominant metabolic effect of GH is stimulation of lipolysis in the adipose tissue, resulting in an increased flux of free fatty acids into the circulation, and lipid oxidation. The increments of circulating levels of FFA and ketone bodies following GH pulse is observed, indicating lipolysis and ketogenesis stimulation. Consequently, increase of FFA cause insulin resistance. Lipolysis is induced in the adipose tissue by increasing hormone-sensitive lipase (HSL) activity and promoting the differentiation of preadipocytes into adipocytes [109]. GH exerts a lipolytic effect predominantly in the visceral adipose tissue [110]. Moreover, it suppresses insulin action in the fat tissue by inhibiting glucose uptake by down-regulating glucose-transporter 1 and 4 (GLUT-1, GLUT-4) in the adipose tissue or up-regulating the activity cell's kinase pathways (phosphatidylinositol-triphosphate kinase – PI3K, and mitogen activated protein kinase – MAPK) [111].

In the muscle and liver, GH increases triglyceride (TG) uptake, by enhancing lipoprotein lipase (LPL) expression. LPL mediates the breakdown of circulating TG to FFA and their subsequent uptake into cells. The TG storage in the liver is induced by several mechanisms including suppressing lipolysis, lipid oxidation, and stimulating lipogenesis. Moreover, GH antagonizes insulin signaling in the liver, therefore stimulating glycogenolysis and mediates an increase in hepatic glucose production (HGP), primarily suppressed by insulin [112].

In muscles GH enhances LPL-mediated TG uptake, similarly to the liver. Moreover it also stimulates lipid oxidation there, as well as amino acid uptake and nitrogen retention. GH changes body composition corresponding to that of physical exercise [113, 114].

The effects of GH on carbohydrate metabolism are more complicated and presumably they are mediated indirectly, counteracting the effects of insulin on glucose. In the presence of unchanged endogenous glucose production and plasma glucose concentrations, GH decreases glucose oxidation and muscle glucose uptake, and promotes non-oxidative glucose utilization at the same time [115]. It inhibits insulin-mediated suppression of hepatic gluconeogenesis. The stimulatory effect of GH on HGP may be a result of its antagonism of insulin action and leads to insulin resistance. The other mechanism that directly induces insulin resistance is the activation of intracellular lipolysis and lipid oxidation. Moreover, somatotropin decreases insulin receptor numbers and binding affinity. Last but not least, GH controls the IGF-1 production, which has insulin-like effects on metabolism, and enhances insulin sensitivity [98].

In patients with acromegaly and GH deficiency, the metabolic effects of GH lead to distinct clinical features as presented below.

Acromegaly is a rare chronic endocrine disease caused almost invariably by a GH secreting pituitary adenoma (more than 95% of cases), albeit rarely (<2% of cases) may be due to hypothalamic tumor secreting GHRH or ectopic GHRH secretion. The prevalence varies between 40 and 130 cases per million adults and an annual incidence of 3-4 per million, with an equal number of men and women affected. However, several recent studies suggest it is underestimated. The average age at diagnosis is 40 years, although the disease may develop insidiously over many years, resulting in delayed diagnosis. Epidemiological studies demonstrated that the diagnosis is generally preceded by approximately 4–10 years of unrecognized active disease. The clinical features of acromegaly suit to the somatic and metabolic effects of prolonged exposure to excess GH and/or to local effects of an expanding pituitary tumor [116]. The most characteristic clinical features of acromegaly are coarse facial features and large hands and feet resulting from soft tissue swelling and enlargement of bones.

Furthermore, deep nasolabial furrows, prominent supraorbital ridges, prognathism and macroglossia, as well as enlargement of lips and nose, swelling of nasopharyngeal tissue, may cause sleep apnea. Up to 70% of patients develop arthropathy as a result of excess local production of IGF-1 and its action in the cartilages. High incidence of cardiovascular complications are also a problem in acromegaly, with hypertension in 30-60% of patients, as well as cardiomyopathy and cardiac hypertrophy as an effect of chronic GH excess in the heart [94].

As GH influences glucose and lipid metabolism, metabolic complications in patients with acromegaly are observed. The most frequent of these is altered glucose metabolism. Impaired glucose tolerance or diabetes will occur in up to 55% of patients with acromegaly, and the prevalence of frank diabetes in those patients varies from 19% to 56%. Acromegaly is considered more prevalent in patients with type 2 diabetes mellitus (T2DM) than in the normal population [117-119]. Kreze et al. suggested that the development of glucose intolerance appeared to be associated with a family history of diabetes and with the concomitant presence of arterial hypertension [120]. No correlation between GH levels and occurrence of diabetes was found [121].

The risk of metabolic complications increases with the level of GH, duration of the disease and age. GH excess in acromegaly induces insulin resistance as a consequence of impaired insulin ability to suppress gluconeogenesis, decreased peripheral glucose utilization and reduced insulin receptor number and binding affinity. Unfortunately increased IGF-1 levels are unable to counteract the insulin-resistance status determined by GH excess. Associations between raised GH and IGF-1 levels and especially metabolic complications of acromegaly were identified [122]. The other study indicated that IGF-1 but not GH correlates more closely than GH with insulin resistance and are associated with glucose intolerance in acromegaly [123].

The confirmation of diagnosis of acromegaly relies on the demonstration of excessive GH secretion. Baseline assessment includes fasting or random measurement. Random measurements of GH and IGF-1 are not useful, except when the result is particularly low (GH less than 0.4 μg/L, IGF-1 normal for age and sex), which confidently excludes the diagnosis. If either of these levels is not achieved, an oral glucose tolerance test according to WHO with glucose and GH measurements every 30 minutes for two hours, should be performed. If GH level exceeds 1μg/L during this test, it suggests acromegaly, however the diagnosis should be considered together with IGF-1 level and overall clinical presentation and confirmed with radiological investigations [124].

The main aim of the treatment is disease control and remission, that may be defined as reversing the symptoms and signs, control of co-morbidities and improving the long-term survival. This entails the use of trans-sphenoidal surgery (TSS), radiotherapy and/or medications. Current options include: dopamine agonists (e.g., bromocriptine, cabergoline), somatostatin analogues (e.g., Sandostatin LAR by Novartis and Lanreotide by Ipsen) as well as GH receptor agonists (pegvisomant). Insulin resistance, impaired glucose tolerance and diabetes mellitus secondary to acromegaly generally improve following treatment of the underlying disease [125, 126].

Growth hormone deficiency (GHD) results from various causes, both hereditary and acquired. Furthermore, it may be isolated or combined with deficiencies of other pituitary hormones (panhypopituitarism). A number of genetic mutations (e.g., in Prop-1 gene or Pit-1 gene) cause panhypopituitarism. Moreover, isolated GHD may be a consequence of inactivating mutations on GHRH receptor (GHRH-R) gene or GH (GH-1) gene, manifested by pituitary hypoplasia or directly affecting GH expression [127]. Also mutations in genes encoding transcription factors (OTX2, HESX1, SOX2, SOX3, PROP1 and POU1F1) cause isolated GHD. Mutations in known genes are present in up to 11% of GHD patients [128].

A number of conditions that affect the hypothalamo-pituitary axis may lead to acquired GHD. They include pituitary tumors, septo-optic dysplasia, infection, vascular causes, infiltrative diseases, or damage secondary to trauma, surgery, or irradiation and chemotherapy. In rare causes growth hormone deficiency is a result of autoimmune mechanisms as part of the spectrum of endocrine deficits in autoimmune hypophysitis [129].

In children GHD presents with growth retardation that begins shortly after birth: low growth velocity for age results in short stature. Adult onset GH deficiency may not be recognized or considered clinically important because of the lack of subjective manifestations. However, it usually presents as part of a combined pituitary hormone deficiency (panhypopituitarism). Adult GHD should be suspected if history of GHD in childhood is positive, however in approximately 20% of patients with childhood GHD a normal GH peak is observed [130].

The symptoms of GHD include perturbations in body composition (reduced skeletal muscles, reduced lean body mass and increase of fat mass), unfavorable changes in bone metabolism, as well as dyslipidemia and an increased cardiovascular risk, reduced left ventricular mass, impaired cardiac systolic function, and impaired response to peak exercise [131-133]. An increased risk of cardiovascular disease has also been reported in young overweight and obese patients [134-135]. Furthermore, presumably IGF-1 signaling regulates contractility, metabolism, hypertrophy, apoptosis, autophagy, stem cell regeneration and senescence in the heart to maintain cardiac homeostasis. IGF-1 deficiency could additionally influence the cardiovascular maintenance in GHD patients, and is associated with premature atherosclerosis, higher cardiovascular morbidity, and premature cardiovascular death [136].

Moreover, GHD usually is accompanied by an increase in visceral fat accumulation that escalates insulin resistance. GH therapy in GHD patients results in reduction of adipose tissue mass and improvement of cardiovascular profile [137, 138]. Despite lower GH secretion patterns in obese patients without GHD, the effectiveness of GH replacement therapy to reduce adipose tissue is still controversial [134, 139].

In growth hormone deficiency, hyperglycemia occurs more often than in controls, while glucose-intolerant GHD patients have decreased basal insulin secretion and hyperproinsulinemia. It is suggested that IGF-1 is essential for the development of pancreatic β cells [140, 141]. Insulin resistance in GHD patients usually is due to visceral obesity and also contributes to hyperglycemia. Growth hormone treatment increases fasting plasma insulin, which may cause transient worsening of insulin resistance. In pediatric patients, in contrast to adult ones, hypoglycemia, especially overnight, may occur in GHD. This is probably due to low glycogen concentrations in the liver and hormonal changes during puberty. When growth hormone replacement is induced, blood glucose levels return to normal [142]. The diagnosis is confirmed with low GH levels in response to stimulation tests (clonidine, glucagon, insulin, arginine and GHRH) [143, 144]. The measurement of IGF-1 as well as of IGFBP-3 levels may not distinguish between normal and GHD subjects [145].

GH deficiency treatment consists of recombinant human growth hormone, administrated subcutaneously once daily. The therapy is highly effective in accelerating growth velocity in children, together with normalization of body composition and metabolic status (lipids levels, glycemia) both in children and adults [133, 146].

Prolactin

Prolactin (PRL) is a pituitary hormone essential for regulation of mammary gland growth, lactogenesis, and galactopoiesis. However, as PRL receptors are expressed in different tissues and cells e.g., lymphoid cells, endometrium, prostate, and adipocytes, it may be involved in different physiological functions in the human body. Therefore, it is not only important for the initiation and maintenance of lactation, but also fertility, growth and development, osmoregulation, immune regulation, brain function, behavior, and metabolism [147, 148]. It was shown that in vitro PRL activates an inflammatory response with infiltration of lymphocytes, neutrophils and macrophages and mononuclear cell adhesion to the endothelial wall [149]. Moreover, numerous prolactin receptors (PRLr) were found in macrophages of atherosclerotic plaques, with numbers correlating positively with inflammation intensity. Therefore, a hypothesis was established that hyperprolactinemia implicated metabolic and inflammatory changes which account for an accelerated atherosclerotic process. The results of the following studies are controversial. Reuwer AQ et al. and Medic-Stojanoska M et al. evaluated atherothrombotic parameters in prolactinoma patients and concluded that those subjects displayed several pro-atherogenic disturbances that may be reversible after treatment [150, 151]. On the contrary, other studies did not implicate any association between PRL concentrations and risk of type 2 diabetes, as well as its relationship with obesity, insulin resistance, glucose and lipids levels [152, 153].

Thyroid Gland Disorders

Thyroid disorders and diabetes have been shown to coexist often, mutually influencing each other. Glucose homeostasis is influenced by thyroid hormones (TH), even in the euthyroid state. As a key metabolic regulator, thyroid hormones influence diverse metabolic pathways taking part not only in glucose metabolism, but also in lipogenesis, lipolysis, and thus - regulation of body weight. On the other hand, diabetes may change the thyroid function test [154].

Thyroid disorders are common in the general population, with a higher prevalence among patients with diabetes. Moreover, an increasing frequency is observed with advancing age, females compared to males and in obese patients compared to normal weight persons, and non-diabetics. The prevalence is variable among the different populations.

Whickam study, conducted in the late 1970s, showed that thyroid disorders affected 6.6% of the adult population of northern England [155]. Whereas, in the NHANES III survey, 4.6% of studied population were found to have hypothyroidism and 1.3% of subjects suffered from hyperthyroidism [156]. The meta-analysis of nine studies on the prevalence and incidence of thyroid dysfunction in Europe, published between 1975 and 2012, done by Garmendia Madariaga et al., revealed that the mean frequency of thyroid dysfunction was 3.82% (3.77%-3.86%): 3.05% (3.01%-3.09%) and 0.75% (0.73%-0.77%) for hypothyroidism and hyperthyroidism, respectively [157]. Thyroid dysfunction seems to be more frequent in diabetics, especially patients with type 1 diabetes (DM1). The prevalence of autoimmune thyroiditis in this group of patients is reported to be as high as 19.6% with female predominance (62%, $p < 0.0001$) [158].

This observation was confirmed in DCCT/EDIC participant analysis by Buschur et al., who reported higher incidence of hypothyroidism in women than in men (24.6% vs. 11.3%, $p<0.0001$) [159]. In our diabetes center, 17% of patients with DM1 were found to have elevated titers of at least one thyroid antibody [160].

Hyperthyroidism results from excess synthesis and secretion of TH by the thyroid gland, which leads to a hyper metabolic condition. The most common forms of hyperthyroidism include: diffuse toxic goiter (Graves' disease), toxic multinodular goiter (Plummer disease), and toxic adenoma. Hyperthyroidism, characterized by elevated serum thyroid hormone levels, is associated with accelerated metabolism. The early symptoms of hyperthyroidism are often subtle with changes in behavior, irritability, emotional lability, fatigue, nervousness, palpitations, tremor, insomnia, excessive perspiration, increase in appetite accompanied by no weight gain or even weight loss, and diarrhea. A deterioration in concentration may lead to decline in school performance. Other classic signs and symptoms of thyrotoxicosis include the following: heat intolerance, warm, moist, smooth skin, tachycardia or atrial arrhythmia and systolic hypertension, as well as precordial thrill, and an ejection murmur due to functional insufficiency of the mitral valve, dyspnea, muscle weakness, and menstrual disturbances (oligomenorrhea). Ophthalmopathy (e.g., periorbital edema, diplopia, or proptosis), however rare in children, suggests Graves' disease. In the developmental age, acceleration in height velocity, with advanced bone age may be observed, corresponding with a longer duration of hyperthyroidism [161-163].

As a result of excess thyroid hormones, an increase in gluconeogenesis and a reduction in insulin levels, due to accelerated insulin degradation and impaired glucose-stimulated insulin secretion, is observed. Impaired glucose tolerance (IGT), impaired fasting glucose tolerance (IFG) and diabetes may occur, with the IGT prevalence in up to 40% of patients with Graves' disease.

Diagnosis of hyperthyroidism is confirmed by thyroid hormone measurements that reveal suppressed thyroid-stimulating hormone (thyreotropin, TSH) levels and elevated free triiodothyronine (fT3) and thyroxine (fT4) levels. Patients with milder thyrotoxicosis may have elevation of fT3 levels only. Subclinical hyperthyroidism features decreased TSH and normal fT3 and fT4 levels [164].

Hypothyroidism is a common endocrine disorder resulting from deficiency of thyroid hormone. Worldwide, iodine deficiency remains the foremost cause of hypothyroidism, however in areas of adequate iodine intake, autoimmune thyroid disease (Hashimoto disease) is the most common cause. Hypothyroidism, characterized by low serum thyroid hormone levels, is associated with a reduction in metabolism. It commonly manifests as a slowing in physical and mental activity but may be asymptomatic. Symptoms and signs are often subtle and neither sensitive nor specific, they include: fatigue, loss of energy, lethargy, weight gain (with decreased appetite), cold intolerance, dry and pallor skin, coarse, brittle, straw-like hair and hair loss, sleepiness, emotional lability, depression, difficulty to concentrate, constipation, menstrual disturbances (amenorrhea and menorrhagia) and impaired fertility. Moreover, physical signs include goiter, decreased blood pressure and bradycardia, edema of lower extremities and myxedema and hyporeflexia [165].

The diagnosis is based on measuring the level of TSH, that is above the reference range. Free thyroxin or the free thyroxin index (FTI), which serves as a surrogate of the free hormone level, is decreased. Routine measurement of fT3 is not recommended. Elevated TSH (up to 10.0 mIU/L) with normal free fT4 level is considered mild or subclinical hypothyroidism [164].

Moreover, some other abnormalities may be present in hypothyroidism, including: anemia, hyponatremia, dyslipidemia, as well as hypertransaminasemia and elevated creatinine and creatinine kinase [165, 166]. Also, there is an expanding body of literature showing that low thyroid function (even in the euthyroid state) predisposes to higher glucose, insulin, HOMA-IR, and cholesterol levels, and therefore contributes to the pathogenesis of atherosclerotic cardiovascular disease [167, 168].

The thyroid hormones, triiodothyronine and thyroxin, act on almost every cell and organ in the body. Although T4 is the major thyroid secretion product, T3 is responsible for most of the biological effects, including the negative-feedback system for regulation of their secretion by thyrotropin-releasing hormone (TRH) in the hypothalamus and TSH in the pituitary gland.

Significant metabolic changes depend on thyroid status in humans. However, one needs to remember, that many of the actions of thyroid hormone as a metabolic regulator involve modulation of other metabolic signaling pathways. Thyroid receptor-mediated effects on gene transcription and translation are key in the regulation of glucose metabolism. Thyroid hormone acts predominantly through its nuclear receptors, thyroid hormone receptor (TR) α and β, which isoforms are expressed in the developmental- and tissue-dependent manner. Thyroid hormone nuclear receptors (TRs) mediate the biological activities of T3 via transcriptional regulation. Liganded TR stimulates genes positively regulated by T3, while unliganded TR binds to a thyroid hormone response unit (THRU) to repress those genes. The repressive actions of unliganded TR are important in metabolic regulation, especially in antagonizing the action of other nuclear receptors [169]. So-called non-genomic actions of thyroid hormone are not initiated by liganding of the hormone to intranuclear TR but begin at the plasma membrane or in cytoplasm. Those actions result not only in DNA-dependent effects, such as angiogenesis or cell proliferation, but also modulation of the activity of the Na/K-ATPase in the plasma membrane. Through Na/K-ATPase TH and TR regulates mitochondrial energetics and basal metabolic rate, including thermogenesis in brown adipose tissue [170].

Thyroid hormones have well-documented influence on glucose homeostasis. They affect glucose metabolism via several mechanisms, including action in adipose tissue, skeletal muscles, and pancreas. However, the liver seems to be the major target of thyroid hormones. TH regulate several genes involved in gluconeogenesis, glycogen metabolism, and insulin signaling. Hyperthyroidism may promote hyperglycemia. First of all, TH increase the concentration of plasma membrane GLUT2, which is the main glucose transporter in liver, which contributes to the increased hepatic glucose efflux and distorted glucose metabolism [171]. Additionally, T3 contributes to an increase in glucose-6-phosphatase (G6Pase) mRNA expression, that hydrolyzes glucose-6-phosphate and completes the final step in gluconeogenesis and glycogenolysis, thus glycemia homeostasis. Moreover, TH decrease the expression of Akt2 (proteinkinase B), an essential factor in the insulin signaling pathway, that also promotes glycogenesis in the liver. The decrease in Akt2 activity has been shown to decrease glycogen synthesis. This explains the antagonistic insulin effect of thyroid hormones in the liver. Moreover, an influence of T3 in the glycogenolytic and gluconeogenic effects of adrenaline and glucagon, by induction of β2-adrenergic receptor and repression of adenylate cyclase were also reported. Phosphoenolpyruvate carboxykinase (PEPCK), the enzyme catalyzing the rate-controlling step of gluconeogenesis, is also positively regulated by TH [172].

TH stimulates both lipogenesis and lipolysis, although when TH levels are elevated, lipolysis is also increased as T3 together with PPAR α (peroxisome proliferator-activated receptor α) modulate the transcription genes of enzymes involved in fatty acids oxidation: carnitine palmitoyltransferase Iα (CPT-Iα) and acyl-CoA oxidase (ACO) [173]. The net effect in hyperthyroidism is fat loss. In consequence, higher levels of free fatty acids (FFA) stimulate gluconeogenesis in the liver even more. This process is co-modulated by catecholamines that also enhance lipolysis, especially in the excess of thyroid hormones [169].

Furthermore, higher levels of TH result in augmented glycolysis and lactate formation relative to glucose oxidation in skeletal muscles, as well as non-oxidative glucose disposal. Production of lactate provides substrates for gluconeogenesis (increased Cori cycle activity) in the liver and muscles. As glucose utilization is higher in hyperthyroidism, this mechanism, together with a decrease in glycogen synthesis and an increase in glycogenolysis, keeps plasma glucose normal or slightly increased to avoid hypoglycemia [175].

Another mechanism explaining the relationship between hyperthyroidism and hyperglycemia is insulin resistance with elevated fasting and/or postprandial insulin and proinsulin levels. Moreover, during hyperthyroidism, the half-life of insulin is reduced, due to an increased rate of degradation. Hyperinsulinemia and insulin resistance in hyperthyroidism is considered to be compensatory for the increased insulin clearance [175]. Liganded TRs act to impair glucose-stimulated insulin release directly at the level of the islet, despite increased islet glucose utilization and oxidation. This effect is rescued by liganded PPARα.

Hyperthyroidism and high-fat feeding result in significant impairment of islet function. Also, enhanced FFA production in the liver increases insulin resistance [175, 176].

The intracellular action of TH is regulated by the amount of local T3 available for receptor binding. That depends on iodothyronine deiodinases activity: activating enzymes are D1 (expressed mostly in the liver, kidney, and thyroid gland) and D2 (functioning in brain, pituitary, thyroid gland, and brown adipose tissue); as well as D3 (in the skin, vascular tissue, and placenta) that is an inactivating enzyme. D2 is the primary enzyme responsible for the serum T3 level and rapid increase in intracellular T3 in specific tissues. Inactivating polymorphisms in the D2 gene have been associated with insulin resistance, type 2 diabetes, and obesity [177].

An impaired glucose tolerance in thyrotoxic subjects is a result of increased glucose absorption through the gastrointestinal tract and post-absorptive hyperglycemia, as well [178].

The increase in GH, glucagon, and catecholamine levels, associated with hyperthyroidism further contributes to the impaired glucose tolerance [154].

Also, hypothyroidism affects glucose homeostasis via several mechanisms. This disorder is associated with hypometabolism characterized by reduced resting energy expenditure. First of all, reduced hepatic gluconeogenesis and reduced lipolysis are observed. That may account for the decreased insulin requirement in hypothyroid diabetic patients and present as a recurrent hypoglycemia [179, 180]. Furthermore, hypothyroidism is associated with insulin resistance. It is due to T3-dependent reduction of acetyl-CoA carboxylase 1 (ACC1) gene expression and activity, accounting to reduced malonyl-CoA and increased expression of CPT-I and therefore, reduced glucose-induced insulin secretion [173, 174]. In hyperthyroidism however, there are some actions of TH that increase insulin sensitivity, e.g., reduction of body fat and rise of mitochondrial oxidative metabolism in skeletal muscle; two features whose defective action results in insulin resistance in type 2 diabetes and hypothyroidism [174].

Recent studies revealed that lower thyroid function, not only subclinical and overt hypothyroidism, but also low-normal thyroid function, may be implicated in the pathogenesis of cardiovascular disease and development of metabolic syndrome [167, 181-183].

Several studies have suggested that thyroid hormone and thyroid hormone receptors may play a supporting role in pancreatic and islet cell development, maturation, and function. Pancreatic islets contain TRα and TRβ which are important for normal islet proliferation and development (3). T3 acts by stimulating the islet transcription factor MAFA, and is required for the transition of islets to glucose-responsive insulin-secreting cells [184]. T3 treatment prevents streptozocin-induced diabetes and maintains proper structure, size, and consistency of islets via nongenomic activation of the AKT signaling pathway [185]. As mentioned before, hyperthyroidism and high-fat feeding accounts for significant impairment of pancreatic islet function.

We conclude that screening for thyroid disease among patients with diabetes mellitus should be routinely performed, considering the prevalence of new cases diagnosed, and the possible aggravation of classical risk factors such as hypertension and dyslipidemia, arising from undiagnosed thyroid dysfunction.

CONCLUSION

The secondary occurrence of glucose metabolism disorders in various hormonal diseases is a recurrent observation. The increased cardiovascular morbidity and mortality associated may be a consequence of increased insulin resistance that normally accompanies hormone excess. Adequate diagnosis and treatment for glucose imbalance is mandatory in these patients.

REFERENCES

[1] Batista D. L., Oldfield E. H., Keil M. F., Stratakis C. A. Postoperative testing to predict recurrent Cushing disease in children. *J. Clin. Endocrinol. Metab.*, 2009; 94: 2757–2765.

[2] Shah N. S., George J., Acharya S. V., Lila A. R., Sarathi V., Bandgar T. R. et al. Cushing disease in children and adolescents: twenty years' experience in a tertiary care center in India. *Endocr. Pract.*, 2011; 17(3): 369-376.

[3] Giordano C., Guarnotta V., Pivonello R., Amato M. C., Simeoli C., Ciresi A. et al. Is diabetes in Cushing's syndrome only a consequence of hypercortisolism? *Eur. J. Endocrinol.*, 2013; 170: 311-319.

[4] Karageorgiadis A. S., Papadakis G. Z., Biro J., Keil M. F., Lyssikatos C., Quezado M. M. et al. Ectopic adrenocorticotropic hormone and corticotropin-releasing hormone co-secreting tumors in children and adolescents causing Cushing syndrome: a diagnostic dilemma and how to solve it. *J. Clin. Endocrinol. Metab.*, 2015; 100: 141-148.

[5] Lacroix A. ACTH independent macronodular adrenal hyperplasia. *Best Pract. Res. Clin. Endocrinol. Metab.*, 2009; 23: 245–259.

[6] Dekkers O. M., Horváth-Puhó E., Jørgensen J. O., Cannegieter S. C., Ehrenstein V., Vandenbroucke J. P. et al. Multisystem morbidity and mortality in Cushing's syndrome: a cohort study. *J. Clin. Endocrinol. Metab.*, 2013; 98(6): 2277-2284.

[7] Pivonello R., De Leo M., Vitale P., Cozzolino A., Simeoli C., De Martino M. C. et al. Pathophysiology of diabetes mellitus in Cushing's syndrome. *Neuroendocrinology,* 2010; 92 Suppl. 1: 77-81.

[8] Krarup T., Krarup T., Hagen C. Do patients with type 2 diabetes mellitus have an increased prevalence of Cushing's syndrome? *Diabetes Metab. Res. Rev.,* 2012; 28(3): 219-227.

[9] Mazziotti G., Gazzaruso C., Giustina A. Diabetes in Cushing syndrome: basic and clinical aspects. *Trends Endocrinol. Metab.*, 2011; 22(12): 499-506.

[10] Hansen K. B., Vilsbøll T., Bagger J. I., Holst J. J., Knop F. K. Reduced glucose tolerance and insulin resistance induced by steroid treatment, relative physical inactivity, and high-calorie diet impairs the incretin effect in healthy subjects. *J. Clin. Endocrinol. Metab.*, 2010; 95(7): 3309-3317.

[11] Resmini E., Minuto F., Colao A., Ferone D. Secondary diabetes associated with principal endocrinopathies: the impact of new treatment modalities. *Acta Diabetol.*, 2009; 46(2): 85-95.

[12] Witek P., Zieliński G., Szamotulska K., Witek J., Zgliczyński W. Complications of Cushing's disease - prospective evaluation and clinical characteristics. Do they affect the efficacy of surgical treatment? *Endokrynol. Pol.*, 2012; 63(4): 277-285.

[13] van Raalte D. H., Ouwens D. M., Diamant M. Novel insights into glucocorticoid-mediated diabetogenic effects: towards expansion of therapeutic options? *Eur. J. Clin. Invest.*, 2009; 39(2): 81-93.

[14] Valassi E., Biller B. M., Klibanski A., Misra M. Adipokines and cardiovascular risk in Cushing's syndrome. *Neuroendocrinology,* 2012; 95(3): 187-206.

[15] Martin S. S., Qasim A., Reilly M. P. Leptin resistance: a possible interface of inflammation and metabolism in obesity-related cardiovascular disease. *J. Am. Coll. Cardiol.,* 2008; 52(15): 1201-10.

[16] Trementino L., Appolloni G., Concettoni C., Cardinaletti M., Boscaro M., Arnaldi G. Association of glucocorticoid receptor polymorphism A3669G with decreased risk of developing diabetes in patients with Cushing's syndrome. *Eur. J. Endocrinol.*, 2012; 166(1): 35-42.

[17] Manenschijn L., van den Akker E. L., Lamberts S. W., van Rossum E. F. Clinical features associated with glucocorticoid receptor polymorphisms. An overview. *Ann. N Y Acad. Sci.,* 2009; 1179: 179-198.

[18] van Rossum E. F., Koper J. W., Huizenga N. A., Uitterlinden A. G., Janssen J. A., Brinkmann A. O. et al. A polymorphism in the glucocorticoid receptor gene, which decreases sensitivity to glucocorticoids in vivo, is associated with low insulin and cholesterol levels. *Diabetes,* 2002; 51(10): 3128-3134.

[19] Bertagna X., Guignat L., Groussin L., Bertherat J. Cushing's disease. *Best Pract. Res. Clin. Endocrinol. Metab.*, 2009; 23: 607-623.

[20] Pivonello R., De Martino M. C., De Leo M., Lombardi G., Colao A.: Cushing's syndrome. *Endocrinol. Metab. Clin. North Am.*, 2008; 37: 135-149.

[21] Newell-Price J., Bertagna X., Grossman A. B., Nieman L. K.: Cushing's syndrome. *Lancet,* 2006; 367, 1605-1617.

[22] Nieman L. K., Biller B. M., Findling J. W., Newell-Price J., Savage M. O., Stewart P. M. et al. The diagnosis of Cushing's syndrome: An Endocrine Society Clinical Practice Guideline. *J. Clin. Endocrinol. Metab.,* 2008; 93: 1526–40.

[23] Munir A., Newell-Price J. Management of diabetes mellitus in Cushing's syndrome. *Neuroendocrinology*, 2010; 92 Suppl. 1:82-85.

[24] Luther J. M. Effects of aldosterone on insulin sensitivity and secretion. *Steroids,* 2014; 91: 54-60.

[25] Young W. F. Primary aldosteronism: renaissance of a syndrome. *Clin. Endocrinol.,* 2007; 66: 607-618.

[26] Mulatero P., Monticone S., Bertello C., Tizzani D., Iannaccone A., Crudo V. et al. Evaluation of primary aldosteronism. *Curr. Opin. Endocrinol. Diabetes Obes.*, 2010; 17: 188-193.

[27] Fallo F., Sonino N. Metabolic syndrome and primary aldosteronism: time for reappraisal? *J. Hum. Hypertens.*, 2010; 24: 623-624.

[28] Sechi L. A., Colussi G., Catena C. Metabolic dysfunction in primary aldosteronism. *Hypertension,* 2009; 53: e37-e42.

[29] Corry D. B., Tuck M. L. The effect of aldosterone on glucose metabolism. *Curr. Hypertens. Rep.*, 2003; 5: 106-109.

[30] Quinkler M., Born-Frontsberg E., Fourkiotis V. G. Comorbidities in primary aldosteronism. *Horm. Metab. Res.*, 2010; 42: 429-434.

[31] Waguespack S. G., Rich T., Grubbs E., Ying A. K., Perrier N. D., Ayala-Ramirez M. et al. A current review of the etiology, diagnosis, and treatment of pediatric pheochromocytoma and paraganglioma. *J. Clin. Endocrinol. Metab.,* 2010; 95: 2023–2037.

[32] Choat H., Derrevere K., Knight L., Brown W., Mack E. H. SDHB-Associated paraganglioma in a pediatric patient and literature review on hereditary pheochromocytoma-paraganglioma syndromes. *Case Rep. Endocrinol.*, 2014; 2014: 502734. doi: 10.1155/2014/502734.

[33] Dénes J., Swords F., Rattenberry E., Stals K., Owens M., Cranston T. et al. Heterogeneous genetic background of the association of pheochromocytoma/ paraganglioma and pituitary adenoma: results from a large patient cohort. *J. Clin. Endocrinol. Metab.*, 2015; 100: E531-541.

[34] La Batide–Alanore A., Chatellier G., Plouin P. F. Diabetes as a marker of pheochromocytoma in hypertensive patients. *J. Hypertens.,* 2003; 21: 1703–1707.

[35] Pogorzelski R., Toutounchi S., Krajewska E., Fiszer P., Łykowski M., Zapała Ł. et al. The effect of surgical treatment of phaeochromocytoma on concomitant arterial hypertension and diabetes mellitus in a single-centre retrospective study. *Cent. European J. Urol.,* 2014; 67: 361-365.

[36] Denisewicz K. Glucose tolerance disorders in pheochromocytoma patients. *Przeg Kardiodiab.,* [Cardio–Diabetological Review] 2009; 4: 87–89.

[37] Kazic M. R., Zivaljevic V. R., Milan Z. B., Paunovic I. R. Perioperative risk factors, morbidity, and outcome of 145 patients during pheochromocytoma resection. *Acta Chir. Belg.*, 2011; 111: 223–227.

[38] Douma S., Petidis K., Kartali N., Mahera K., Sabanis C., Zamboulis C. Pheochromocytoma presenting as diabetic ketoacidosis. *J. Diabetes Complications,* 2008; 22: 295–296.

[39] Lenders J. W. M., Duh Q.-Y., Eisenhofer G. et al. Pheochromocytoma and paraganglioma: an endocrine society clinical practice guideline. *J. Clin. Endocrinol. Metab.*, 2014; 99: 1915–1942.

[40] Eisenhofer G., Lenders J. W. M., Timmers H. et al. Measurements of plasma methoxytyramine, normetanephrine, and metanephrine as discriminators of different hereditary forms of pheochromocytoma. *Clinical Chemistry*, 2011; 57: 411–420.

[41] Szalat A., Fraenkel M., Doviner V., Salmon A., Gross D. J. Malignant pheochromocytoma: predictive factors of malignancy and clinical course in 16 patients at a single tertiary medical center. *Encocrine*, 2011; 39:160–166.

[42] The Rotterdam ESHRE/ASRM-Sponsored PCOS Consensus Workshop Group. Revised 2003 consensus on diagnostic criteria and long-term health risks related to polycystic ovary syndrome (PCOS). *Hum. Reprod.,* 2004; 19: 41–47.

[43] Conway G., Dewailly D., Diamanti-Kandarakis E., Escobar-Morreale H. F., Franks S., Gambineri A. et al.; ESE PCOS Special Interest Group. The polycystic ovary syndrome: a position statement from the European Society of Endocrinology. *Eur. J. Endocrinol.*, 2014; 171: 1-29.

[44] Fauser B. C., Tarlatzis B. C., Rebar R. W., Legro R. S., Balen A. H., Lobo R. et al. Consensus on women's health aspects of polycystic ovary syndrome (PCOS): the Amsterdam ESHRE/ASRM-Sponsored 3rd PCOS Consensus Workshop Group. *Fertil Steril.,* 2012; 97: 28–38.

[45] Ehrmann D. Final Report National Institute of Health. Evidence-based Methodology Workshop on Polycystic Ovary Syndrome December 3–5, 2012 Executive summary at http://prevention.nih.gov/workshops/2012/ pcos/resourc-es.aspx

[46] Legro R. S., Arslanian S. A., Ehrmann D. A., Hoeger K. M., Murad M. H., Pasquali R. et al. Diagnosis and treatment of polycystic ovary syndrome: an Endocrine Society clinical practice guideline. *J. Clin. Endocrinol. Metab.*, 2013; 98: 4565–4592.

[47] Dunaif A., Fauser B. C. Renaming PCOS – a two state solution. *J. Clin. Endocrinol. Metab.*, 2013; 98: 4325–4328.

[48] Dumesic D. A., Richards J. S. Ontogeny of the ovary in polycystic ovary syndrome. *Fertil Steril.*, 2013; 100: 23–38.

[49] Dumesic D. A., Abbott D. H., Padmanabhan V. Polycystic ovary syndrome and its developmental origins. *Rev. Endocrine and Metabolic Disorders,* 2007; 8: 127–141.

[50] Franks S., Berga S. L. Does PCOS have developmental origins? *Fertil Steril.,* 2012; 97: 2-6.

[51] Welt C. K., Carmina E. Lifecycle of polycystic ovary syndrome (PCOS): from in utero to menopause. *J. Clin. Endocrinol. Metab.*, 2013; 98: 4629–4638.

[52] de Zegher F., Lopez-Bermejo A., Ibáñez L. Adipose tissue expandability and the early origins of PCOS. *Trends Endocrinol. Metabol.*, 2009; 20: 418–423.

[53] Ibáñez L., de Zegher F., Potau N. Premature pubarche, ovarian hyperandrogenism, hyperinsulinism and the polycystic ovary syndrome: from a complex constellation to a simple sequence of prenatal onset. *J. Endocrinol. Invest.*, 1998; 21: 558–566.

[54] Meas T., Chevenne D., Thibaud E., Léger J., Cabrol S., Czernichow P. et al. Endocrine consequences of premature pubarche in post-pubertal Caucasian girls. *Clin. Endocrinol.*, 2002; 57: 101–106.

[55] Ezeh U., Yildiz B. O., Azziz R. Referral bias in defining the phenotype and prevalence of obesity in polycystic ovary syndrome. *J. Clin. Endocrinol. Metab.*, 2013; 98: E1088–E1096.

[56] Abbott D. H., Barnett D. K., Bruns C. M., Dumesic D. A. Androgen excess fetal programming of female reproduction: a developmental aetiology for polycystic ovary syndrome? *Hum. Reprod. Update*, 2005; 11: 357–374.

[57] Maliqueo M., Lara H. E., Sanchez F., Echiburu B., Crisosto N., Sir-Petermann T. Placental steroidogenesis in pregnant women with polycystic ovary syndrome. *Eur. J. Obstet., Gynecol. Reprod. Biol.,* 2013; 166: 151–155.

[58] Hacihanefioglu B., Aybey B., Hakan Ozon Y., Berkil H., Karsidag K. Association of anthropometric, androgenic and insulin-related features with polymorphisms in exon 8 of SHBG gene in women with polycystic ovary syndrome. *Gynecol. Endocrinol.*, 2013; 29: 361–364.

[59] Wickham E. P., Ewens K. G., Legro R. S., Dunaif A., Nestler J. E., Strauss J. F. Polymorphisms in the SHBG gene influence serum SHBG levels in women with polycystic ovary syndrome. *J. Clin. Endocrinol. Metab.,* 2011; 96: E719–E727.

[60] Fanelli F., Gambineri A., Belluomo I., Repaci A., Di Lallo V. D., Di Dalmazi G. et al. Androgen profiling by liquid chromatography–tandem mass spectrometry (LC–MS/MS) in healthy normal-weight ovulatory and anovulatory late adolescent and young women. *J. Clin. Endocrinol. Metab.,* 2013; 98: 3058–3067.

[61] Homburg R., Ray A., Bhide P., Gudi A., Shah A., Timms P. et al. The relationship of serum anti-Mullerian hormone with polycystic ovarian morphology and polycystic ovary syndrome: a prospective cohort study. *Hum. Reprod.*, 2013; 28: 1077–1083.

[62] Pawelczak M., Kenigsberg L., Milla S., Liu Y. H., Shah B. Elevated serum anti-Mullerian hormone in adolescents with polycystic ovary syndrome: relationship to ultrasound features. *J. Pediatr. Endocrinol. Metab.*, 2012; 25: 983–989.

[63] Dewailly D., Lujan M. E., Carmina E., Cedars M. I., Laven J., Norman R. J. et al. Definition and significance of polycystic ovarian morphology: a task force from the Androgen Excess and Polycystic Ovary Syndrome Society. *Hum. Reprod. Update*, 2014; 20: 334–352.

[64] Thomson R. L., Spedding S., Brinkworth G. D., Noakes M., Buckley J. D. Seasonal effects on vitamin D status influence outcomes of lifestyle intervention in overweight and obese women with polycystic ovary syndrome. *Fertil Steril.*, 2013; 99: 1779–1785.

[65] Li H. W., Brereton R. E., Anderson R. A., Wallace A. M., Ho C. K. Vitamin D deficiency is common and associated with metabolic risk factors in patients with polycystic ovary syndrome. *Metabolism,* 2011; 60: 1475–1481.

[66] Krul-Poel Y. H., Snackey C., Louwers Y., Lips P., Lambalk C. B., Laven J. S. et al. The role of vitamin D in metabolic disturbances in polycystic ovary syndrome: a systematic review. *Eur. J. Endocrinol.,* 2013; 169: 853–865.

[67] Wojtusik J., Johnson P. A. Vitamin D regulates anti-Mullerian hormone expression in granulosa cells of the hen. *Biol. Reprod.*, 2012; 86: 91. doi: 10.1095/biolreprod.111.094110.

[68] Dicken C. L., Israel D. D., Davis J. B., Sun Y., Shu J., Hardin J. et al. Peripubertal vitamin D(3) deficiency delays puberty and disrupts the estrous cycle in adult female mice. *Biol. Reprod.*, 2012; 87: 51. doi: 10.1095/biolreprod.111.096511.

[69] Dewailly D., Gronier H., Poncelet E., Robin G., Leroy M., Pigny P. et al. Diagnosis of polycystic ovary syndrome (PCOS): revisiting the threshold values of follicle count on ultrasound and of the serum AMH level for the definition of polycystic ovaries. *Hum. Reprod.,* 2011; 26: 3123–3129.

[70] Lujan M. E., Jarrett B. Y., Brooks E. D., Reines J. K., Peppin A. K., Muhn N. et al. Updated ultrasound criteria for polycystic ovary syndrome: reliable thresholds for elevated follicle population and ovarian volume. *Hum. Reprod.*, 2013; 28: 1361–1368.

[71] Deb S., Campbell B. K., Clewes J. S., Raine-Fenning N. J. Quantitative analysis of antral follicle number and size: a comparison of two-dimensional and automated three-dimensional ultrasound techniques. *Ultrasound Obstet. Gynecol.*, 2010; 35: 354–360.

[72] Fulghesu A. M., Angioni S., Frau E., Belosi C., Apa R., Mioni R. et al. Ultrasound in polycystic ovary syndrome – the measuring of ovarian stroma and relationship with circulating androgens: results of a multicentric study. *Human Reprod.*, 2007; 22: 2501–2508.

[73] Lim S. S., Norman R. J., Davies M. J., Moran L. J. The effect of obesity on polycystic ovary syndrome: a systematic review and meta-analysis. *Obesity Rev.*, 2013; 14: 95–109.

[74] Morgan C. L., Jenkins-Jones S., Currie C. J., Rees D. A. Evaluation of adverse outcome in young women with polycystic ovary syndrome versus matched, reference controls: a retrospective, observational study. *J. Clin. Endocrinol. Metab.*, 2012; 97: 3251–3260.

[75] Mannerås-Holm L., Leonhardt H., Kullberg J., Jennische E., Odén A., Holm G. et al. Adipose tissue has aberrant morphology and function in PCOS: enlarged adipocytes and low serum adiponectin, but not circulating sex steroids, are strongly associated with insulin resistance. *J. Clin. Endocrinol. Metab.*, 2011; 96: E304–E311.

[76] Yamauchi T., Kamon J., Minokoshi Y., Ito Y., Waki H., Uchida S. et al. Adiponectin stimulates glucose utilization and fatty-acid oxidation by activating AMP-activated protein kinase. *Nat. Med.*, 2002; 8: 1288–1295.

[77] O'Connor A., Phelan N., Tun T. K., Boran G., Gibney J., Roche H. M. High-molecular-weight adiponectin is selectively reduced in women with polycystic ovary syndrome independent of body mass index and severity of insulin resistance. *J. Clin. Endocrinol. Metab.*, 2010; 95: 1378–1385.

[78] Yildiz B. O. Polycystic ovary syndrome: is obesity a symptom? *Women's Health*, 2013; 9: 505–507.

[79] Villa J., Pratley R. Adipose tissue dysfunction in polycystic ovary syndrome. *Curr. Diab. Reports*, 2011; 11: 179–184.

[80] Repaci A., Gambineri A., Pasquali R. The role of low-grade inflammation in the polycystic ovary syndrome. *Mol. Cell. Endocrinol.,* 2011; 15: 30–41.

[81] Cortón M., Botella-Carretero J. I., Benguría A., Villuendas G., Zaballos A., San Millán J. L. et al. Differential gene expression profile in omental adipose tissue in women with polycystic ovary syndrome. *J. Clin. Endocrinol. Metab.,* 2007; 92: 328–337.

[82] Xu X. F., De Pergola G., Bjorntorp P. Testosterone increases lipolysis and the number of b-adrenoceptors in male rat adipocytes. *Endocrinology,* 1991; 128: 379–382.

[83] Tao T., Li S., Zhao A., Zhang Y., Liu W. Expression of the CD11c gene in subcutaneous adipose tissue is associated with cytokine level and insulin resistance in women with polycystic ovary syndrome. *Eur. J. Endocrinol.*, 2012; 167: 705–713.

[84] Carmina E., Campagna A. M., Lobo R. A. Emergence of ovulatory cycles with aging in women with polycystic ovary syndrome (PCOS) alters the trajectory of cardiovascular and metabolic risk factors. *Hum. Reprod.,* 2013; 28: 2245–2252.

[85] Kelly C. J., Speirs A., Gould G. W., Petrie J. R., Lyall H., Connell J. M. Altered vascular function in young women with polycystic ovary syndrome. *J. Clin. Endocrinol. Metab*., 2002; 87: 742–746.

[86] Diamanti-Kandarakis E., Alexandraki K., Piperi C., Protogerou A., Katsikis I., Paterkis T. et al. Inflammatory and endothelial markers in women with polycystic ovary syndrome. *Eur. J. Clin. Invest*., 2006; 36: 691–697.

[87] Macut D., Simic T., Lissounov A., Pljesa-Ercegovac M., Bozic I., Djukic T. et al. Insulin resistance in non-obese women with polycystic ovary syndrome: relation to byproducts of oxidative stress. *Exp. Clin.Endocrinol. Diabetes,* 2011; 119: 451–455.

[88] Kodaman P. H., Duleba A. J. Statins: do they have potential in the treatment of polycystic ovary syndrome? *Semin. Reprod. Med*., 2008; 26: 127–138.

[89] Boutzios G., Livadas S., Piperi C., Vitoratos N., Adamopoulos C., Hassiakos D. et al. Polycystic ovary syndrome offspring display increased oxidative stress markers comparable to gestational diabetes offspring. *Fertil Steril*., 2013; 99: 943–950.

[90] Diamanti-Kandarakis E., Katsikis I., Piperi C., Kandaraki E., Piouka A., Papavassiliou A. G. et al. Increased serum advanced glycation end-products is a distinct finding in lean women with polycystic ovary syndrome. *Clin. Endocrinol*., 2008; 69: 634–641.

[91] Christakou C., Economou F., Livadas S., Piperi C., Adamopoulos C., Marinakis E. et al. Strong and positive association of endothelin-1 with AGEs in PCOS: a causal relationship or a bystander? *Hormones,* 2011; 10: 292–297.

[92] Diamanti-Kandarakis E., Christakou C. D., Kandaraki E., Economou F. N. Metformin: an old medication of new fashion: evolving new molecular mechanisms and clinical implications in polycystic ovary syndrome. *Eur. J. Endocrinol*., 2010; 162: 193–212.

[93] Tantalaki E., Piperi C., Livadas S., Kollias A., Adamopoulos C., Koulouri A. et al. Impact of dietary modification of advanced glycation end products (AGEs) on the hormonal and metabolic profile of women with polycystic ovary syndrome (PCOS). *Hormones,* 2014; 13: 65–73.

[94] Ayuk J., Sheppard M. C. Growth hormone and its disorders. *Postgrad. Med. J.,* 2006; 82(963): 24-30.

[95] Benso A., Gramaglia E., Olivetti I., Tomelini M., Gigliardi V. R., Frara S. et al. The GH-releasing effect of acylated ghrelin in normal subjects is refractory to GH acute auto-feedback but is inhibited after short-term GH administration inducing IGF1 increase. *Eur. J. Endocrinol*., 2013; 168(4): 509-514.

[96] Tannenbaum G. S., Bowers C. Y. Interactions of growth hormone secretagogues and growth hormone-releasing hormone/somatostatin. *Endocrine,* 2001; 14(1): 21-27.

[97] Shibasaki T., Hotta M., Masuda A., Imaki T., Obara N., Demura H. et al. Plasma GH responses to GHRH and insulin-induced hypoglycemia in man. *J. Clin. Endocrinol. Metab*., 1985; 60(6): 1265-1267.

[98] Kato Y., Murakami Y., Sohmiya M., Nishiki M. Regulation of human growth hormone secretion and its disorders. *Intern. Med*., 2002; 41(1): 7-13.

[99] Ohlsson C., Mohan S., Sjögren K., Tivesten A., Isgaard J., Isaksson O. et al. The role of liver-derived insulin-like growth factor-I. *Endocr. Rev.,* 2009; 30(5): 494-535.

[100] Møller N., Jørgensen J. O. Effects of growth hormone on glucose, lipid, and protein metabolism in human subjects. *Endocr. Rev*., 2009; 30(2): 152-177.

[101] Mauras N., Haymond M. W. Are the metabolic effects of GH and IGF-I separable? *Growth Horm. IGF Res*., 2005; 15(1): 19-27.

[102] Waters M. J., Kaye P. L. The role of growth hormone in fetal development. *Growth Horm. IGF Res.*, 2002; 12(3): 137-46.

[103] Neel J. V., Weder A. B., Julius S. Type II diabetes, essential hypertension, and obesity as "syndromes of impaired genetic homeostasis": the "thrifty genotype" hypothesis enters the 21st century. *Perspect. Biol. Med.,* 1998; 42(1): 44-74.

[104] Hue L., Taegtmeyer H. The Randle cycle revisited: a new head for an old hat. *Am. J. Physiol. Endocrinol. Metab.*, 2009; 297(3): E578-E591.

[105] Sakharova A. A., Horowitz J. F., Surya S., Goldenberg N., Harber M. P., Symons K. et al. Role of growth hormone in regulating lipolysis, proteolysis, and hepatic glucose production during fasting. *J. Clin. Endocrinol. Metab.*, 2008; 93(7): 2755-2759.

[106] Pena-Bello L., Pertega-Diaz S., Outeiriño-Blanco E., Garcia-Buela J., Tovar S., Sangiao-Alvarellos S. et al. Effect of oral glucose administration on rebound growth hormone release in normal and obese women: the role of adiposity, insulin sensitivity and ghrelin. *PLoS One,* 2015; 10(3): e0121087.

[107] Cornford A. S., Barkan A. L., Hinko A., Horowitz J. F. Suppression in growth hormone during overeating ameliorates the increase in insulin resistance and cardiovascular disease risk. *Am. J. Physiol. Endocrinol. Metab.,* 2012; 303(10): E1264-E1272.

[108] Moran A., Jacobs D. R. Jr., Steinberger J., Cohen P., Hong C. P., Prineas R. et al. Association between the insulin resistance of puberty and the insulin-like growth factor-I/growth hormone axis. *J. Clin. Endocrinol. Metab.,* 2002; 87(10): 4817-4820.

[109] Yang H. L., Sun C., Sun C., Qi R. L. Effect of suppressor of cytokine signaling 2 (SOCS2) on fat metabolism induced by growth hormone (GH) in porcine primary adipocyte. *Mol. Biol. Rep.*, 2012; 39(9): 9113-9122.

[110] Kim K. R., Nam S. Y., Song Y. D., Lim S. K., Lee H. C., Huh K. B. Low-dose growth hormone treatment with diet restriction accelerates body fat loss, exerts anabolic effect and improves growth hormone secretory dysfunction in obese adults. *Horm. Res.,* 1999; 51(2): 78-84.

[111] Vijayakumar A., Novosyadlyy R., Wu Y., Yakar S., LeRoith D. Biological effects of growth hormone on carbohydrate and lipid metabolism. *Growth Horm. IGF Res.*, 2010; 20(1): 1-7.

[112] Sos B. C., Harris C., Nordstrom S. M., Tran J. L., Balázs M., Caplazi P. et al. Abrogation of growth hormone secretion rescues fatty liver in mice with hepatocyte-specific deletion of JAK2. *J. Clin. Invest.,* 2011; 121(4): 1412-1423.

[113] Vijayakumar A., Wu Y., Buffin N. J., Li X., Sun H., Gordon R. E. et al. Skeletal muscle growth hormone receptor signaling regulates basal, but not fasting-induced, lipid oxidation. *PLoS One,* 2012; 7(9): e44777.

[114] Chikani V., Ho K. K. Action of GH on skeletal muscle function: molecular and metabolic mechanisms. *J. Mol. Endocrinol.*, 2013; 52(1): R107-R123.

[115] Mauras N., Martinez V., Rini A., Guevara-Aguirre J. Recombinant human insulin-like growth factor I has significant anabolic effects in adults with growth hormone receptor deficiency: studies on protein, glucose, and lipid metabolism. *J. Clin. Endocrinol. Metab.,* 2000; 85(9): 3036-3042.

[116] Adelman D. T., Liebert K. J., Nachtigall L. B., Lamerson M., Bakker B. Acromegaly: the disease, its impact on patients, and managing the burden of long-term treatment. *Int. J. Gen. Med.*, 2013; 6: 31-38.

[117] Suda K., Fukuoka H., Iguchi G., Hirota Y., Nishizawa H., Bando H. et al. The prevalence of acromegaly in hospitalized patients with type 2 diabetes. *Endocr. J.*, 2015; 62(1): 53-59.

[118] Colao A., Baldelli R., Marzullo P., Ferretti E., Ferone D., Gargiulo P. et al. Systemic hypertension and impaired glucose tolerance are independently correlated to the severity of the acromegalic cardiomyopathy. *J. Clin. Endocrinol. Metab.,* 2000; 85(1): 193-199.

[119] Resmini E., Minuto F., Colao A., Ferone D. Secondary diabetes associated with principal endocrinopathies: the impact of new treatment modalities. *Acta Diabetol.*, 2009; 46(2): 85-95.

[120] Kreze A., Kreze-Spirova E., Mikulecky M. Risk factors for glucose intolerance in active acromegaly. *Braz. J. Med. Biol. Res.*, 2001; 34(11): 1429-1433.

[121] Biering H., Knappe G., Gerl H., Lochs H. Prevalence of diabetes in acromegaly and Cushing syndrome. *Acta Med. Austriaca*, 2000; 27(1): 27-31.

[122] Jayasena C. N., Comninos A. N., Clarke H., Donaldson M., Meeran K., Dhillo W. S. The effects of long-term growth hormone and insulin-like growth factor-1 exposure on the development of cardiovascular, cerebrovascular and metabolic co-morbidities in treated patients with acromegaly. *Clin. Endocrinol.*, (Oxf). 2011; 75(2): 220-225.

[123] Niculescu D., Purice M., Coculescu M. Insulin-like growth factor-I correlates more closely than growth hormone with insulin resistance and glucose intolerance in patients with acromegaly. *Pituitary,* 2013; 16(2): 168-74.

[124] Katznelson L., Atkinson J. L., Cook D. M., Ezzat S. Z., Hamrahian A. H., Miller K. K.; AACE Acromegaly Task Force. American Association of Clinical Endocrinologists Medical Guidelines for Clinical Practice for the Diagnosis and Treatment of Acromegaly - 2011 update: executive summary. *Endocr. Pract.*, 2011; 17(4): 636-646.

[125] Vilar L., Valenzuela A., Ribeiro-Oliveira A. Jr., Gómez Giraldo C. M., Pantoja D., Bronstein M. D. Multiple facets in the control of acromegaly. *Pituitary,* 2014; 17 Suppl. 1: S11-S17.

[126] Andersen M. Management of endocrine disease: GH excess: diagnosis and medical therapy. *Eur. J. Endocrinol.*, 2013; 170(1): R31-R41.

[127] Corazzini V., Salvatori R. Molecular and clinical aspects of GHRH receptor mutations. *Endocr. Dev.,* 2013; 24: 106-17.

[128] Alatzoglou K. S., Webb E. A., Le Tissier P., Dattani M. T. Isolated growth hormone deficiency (GHD) in childhood and adolescence: recent advances. *Endocr. Rev.,* 2014; 35(3): 376-432.

[129] De Bellis A., Sinisi A. A., Pane E., Dello Iacovo A., Bellastella G., Di Scala G. et al.; Italian Autoimmune Hypophysitis Network Group. Involvement of hypothalamus autoimmunity in patients with autoimmune hypopituitarism: role of antibodies to hypothalamic cells. *J. Clin. Endocrinol. Metab.*, 2012; 97(10): 3684-3690.

[130] Gelwane G., Garel C., Chevenne D., Armoogum P., Simon D., Czernichow P. et al. Subnormal serum insulin-like growth factor-I levels in young adults with childhood-onset nonacquired growth hormone (GH) deficiency who recover normal GH secretion may indicate less severe but persistent pituitary failure. *J. Clin. Endocrinol. Metab.*, 2007; 92(10): 3788-3795.

[131] Molitch M. E., Clemmons D. R., Malozowski S., Merriam G. R., Vance M. L.; Endocrine Society. Evaluation and treatment of adult growth hormone deficiency: an

Endocrine Society clinical practice guideline. *J. Clin. Endocrinol. Metab.*, 2011; 96(6): 1587-1609.

[132] Reed M. L., Merriam G. R., Kargi A. Y. Adult growth hormone deficiency - benefits, side effects, and risks of growth hormone replacement. *Front Endocrinol.*, (Lausanne). 2013; 4: 64.

[133] Höybye C., Christiansen J. S. Growth hormone replacement in adults - current standards and new perspectives. *Best Pract. Res. Clin. Endocrinol. Metab.*, 2015; 29(1): 115-123.

[134] Slattery M., Bredella M. A., Stanley T., Torriani M., Misra M. Effects of recombinant human growth hormone (rhGH) administration on body composition and cardiovascular risk factors in obese adolescent girls. *Int. J. Pediatr. Endocrinol.*, 2014; 2014(1): 22.

[135] Bredella M. A., Gerweck A. V., Lin E., Landa M. G., Torriani M., Schoenfeld D. A. et al. Effects of GH on body composition and cardiovascular risk markers in young men with abdominal obesity. *J. Clin. Endocrinol. Metab.*, 2013; 98(9): 3864-3872.

[136] Ren J., Anversa P. The insulin-like growth factor I system: physiological and pathophysiological implication in cardiovascular diseases associated with metabolic syndrome. *Biochem. Pharmacol.*, 2015; 93(4): 409-417.

[137] Elbornsson M., Götherström G., Bosæus I., Bengtsson B. Å., Johannsson G., Svensson J. Fifteen years of GH replacement improves body composition and cardiovascular risk factors. *Eur. J. Endocrinol.*, 2013; 168(5): 745-753.

[138] Beauregard C., Utz A. L., Schaub A. E., Nachtigall L., Biller B. M., Miller K. K. et al. Growth hormone decreases visceral fat and improves cardiovascular risk markers in women with hypopituitarism: a randomized, placebo-controlled study. *J. Clin. Endocrinol. Metab.*, 2008; 93(6): 2063-2071.

[139] Bredella M. A., Gerweck A. V., Lin E., Landa M. G., Torriani M., Schoenfeld D. A. et al. Effects of GH on body composition and cardiovascular risk markers in young men with abdominal obesity. *J. Clin. Endocrinol. Metab.*, 2013; 98(9): 3864-3872.

[140] Oliveira C. R., Salvatori R., Barreto-Filho J. A., Rocha I. E., Mari A., Pereira R. M. et al. Insulin sensitivity and β-cell function in adults with lifetime, untreated isolated growth hormone deficiency. *J. Clin. Endocrinol. Metab.*, 2012; 97(3): 1013-1019.

[141] Vicente T. A., Rocha I. E., Salvatori R., Oliveira C. R., Pereira R. M., Souza A. H. et al. Lifetime congenital isolated GH deficiency does not protect from the development of diabetes. *Endocr. Connect.*, 2013; 2(2): 112-117.

[142] Johnstone H. C., McNally R. J., Cheetham T. D. The impact of fasting and treatment omission on susceptibility to hypoglycemia in children and adolescents with GH and cortisol insufficiency. *Clin. Endocrinol.*, (Oxf). 2008; 69(3): 436-442.

[143] Binder G., Huller E., Blumenstock G., Schweizer R. Auxology-based cut-off values for biochemical testing of GH secretion in childhood. *Growth Horm. IGF Res.*, 2011; 21(4): 212-218.

[144] Ghigo E., Aimaretti G., Corneli G. Diagnosis of adult GH deficiency. *Growth Horm. IGF Res.*, 2008; 18(1): 1-16.

[145] Pawlikowska-Haddal A., Cohen P., Cook D. M. How useful are serum IGF-I measurements for managing GH replacement therapy in adults and children? *Pituitary,* 2012; 15(2): 126-134.

[146] Ross J. L., Lee P. A., Gut R., Germak J. Increased height standard deviation scores in response to growth hormone therapy to near-adult height in older children with delayed

skeletal maturation: results from the ANSWER Program. *Int. J. Pediatr. Endocrinol.*, 2015; 2015(1): 1.

[147] Trott J. F., Schennink A., Petrie W. K., Manjarin R., VanKlompenberg M. K., Hovey R. C. Triennial Lactation Symposium: Prolactin: The multifaceted potentiator of mammary growth and function. *J. Anim. Sci.,* 2012; 90(5): 1674-1686.

[148] Cejkova P., Fojtikova M., Cerna M. Immunomodulatory role of prolactin in diabetes development. *Autoimmun. Rev.*, 2009; 9(1): 23–27.

[149] Suarez A. L., López-Rincón G., Martínez Neri P. A., Estrada-Chávez C. Prolactin in inflammatory response. *Adv. Exp. Med. Biol.*, 2015; 846: 243-264.

[150] Reuwer A. Q., Sondermeijer B. M., Battjes S., van Zijderveld R., Stuijver D. J., Bisschop P. H. et al. Microcirculation and atherothrombotic parameters in prolactinoma patients: a pilot study. *Pituitary,* 2012; 15(4): 472-481.

[151] Medic-Stojanoska M., Icin T., Pletikosic I., Bajkin I., Novakovic-Paro J., Stokic E. et al. Risk factors for accelerated atherosclerosis in young women with hyperprolactinemia. *Med. Hypotheses*, 2015; 84(4): 321-326.

[152] Ernst B., Thurnheer M., Schultes B. Basal serum prolactin levels in obesity - unrelated to parameters of the metabolic syndrome and unchanged after massive weight loss. *Obes. Surg*., 2009; 19(8): 1159-1162.

[153] Balbach L., Wallaschofski H., Völzke H., Nauck M., Dörr M., Haring R. Serum prolactin concentrations as risk factor of metabolic syndrome or type 2 diabetes? *BMC Endocr. Disord*., 2013; 13: 12.

[154] Hage M., Zantout M. S., Azar S. T. Thyroid disorders and diabetes mellitus. *J. Thyroid Res.,* 2011; 2011: 439463.

[155] Tunbridge W. M., Evered D. C., Hall R., Appleton D., Brewis M., Clark F. et al. The spectrum of thyroid disease in a community: the Whickham survey. *Clin. Endocrinol.*, (Oxf). 1977; 7(6): 481-493.

[156] Hollowell J. G., Staehling N. W., Flanders W. D., Hannon W. H., Gunter E. W., Spencer C. A. et al. Serum TSH, T(4), and thyroid antibodies in the United States population (1988 to 1994): National Health and Nutrition Examination Survey (NHANES III). *J. Clin. Endocrinol. Metab.,* 2002; 87(2): 489-499.

[157] Garmendia Madariaga A., Santos Palacios S., Guillén-Grima F., Galofré J. C. The incidence and prevalence of thyroid dysfunction in Europe: a meta-analysis. *J. Clin. Endocrinol. Metab*., 2014; 99(3): 923-931.

[158] Warncke K., Fröhlich-Reiterer E. E., Thon A., Hofer S. E., Wiemann D., Holl R. W., DPV Initiative of the German Working Group for Pediatric Diabetology; German BMBF Competence Network for Diabetes Mellitus. Polyendocrinopathy in children, adolescents, and young adults with type 1 diabetes: a multicenter analysis of 28,671 patients from the German/Austrian DPV-Wiss database. *Diabetes Care*, 2010; 33: 2010–2012.

[159] Buschur E., Sarma A. V., Pietropaolo M., Dunn R. L., Braffett B. H., Cleary P. A. et al.; DCCT/EDIC Research Group. Self-reported autoimmune disease by sex in the diabetes control and complications trial/epidemiology of diabetes interventions and complications (DCCT/EDIC) study. *Diabetes Care*, 2014; 37(2): e28-e29.

[160] Zubkiewicz-Kucharska A., Seifert M., Chrzanowska J., Noczyńska A. Analysis of the results of genetic testing for celiac disease in children and adolescents with diabetes type 1. *Clinical Diabetology*, in press.

[161] Vargas-Uricoechea H., Sierra-Torres C. H. Thyroid hormones and the heart. *Horm. Mol. Biol. Clin. Investig.,* 2014; 18(1): 15-26.

[162] Bossowski A. T., Reddy V., Perry L. A., Johnston L. B., Banerjee K., Blair J. C. et al. Clinical and endocrine features and long-term outcome of Graves' disease in early childhood. *J. Endocrinol. Invest.*, 2007; 30(5): 388-392.

[163] Kaguelidou F., Carel J. C., Léger J. Graves' disease in childhood: advances in management with antithyroid drug therapy. *Horm. Res.,* 2009; 71(6): 310-317.

[164] Boucai L., Hollowell J. G., Surks M. I. An approach for development of age-, gender-, and ethnicity-specific thyrotropin reference limits. *Thyroid,* 2011; 21(1): 5-11.

[165] Devdhar M., Ousman Y. H., Burman K. D. Hypothyroidism. *Endocrinol. Metab. Clin. North Am.,* 2007; b36(3): 595-615.

[166] Garduño-Garcia J., Alvirde-Garcia U., López-Carrasco G., Padilla Mendoza M. E., Mehta R., Arellano-Campos O. et al. TSH and free thyroxine concentrations are associated with differing metabolic markers in euthyroid subjects. *Eur. J. Endocrinol.*, 2010; 163(2): 273-278.

[167] van Tienhoven-Wind L. J., Dullaart R. P. Low-normal thyroid function and the pathogenesis of common cardio-metabolic disorders. *Eur. J. Clin. Invest.,* 2015; doi: 10.1111/eci.12423.

[168] Triolo M., Kwakernaak A. J., Perton F. G., de Vries R., Dallinga-Thie G. M., Dullaart R. P. Low normal thyroid function enhances plasma cholesteryl ester transfer in Type 2 diabetes mellitus. *Atherosclerosis,* 2013; 228(2): 466-471.

[169] Liu Y. Y., Brent G. A. Thyroid hormone crosstalk with nuclear receptor signaling in metabolic regulation. *Trends Endocrinol. Metab.*, 2010; 21(3): 166-173.

[170] Cheng S. Y., Leonard J. L., Davis P. J. Molecular aspects of thyroid hormone actions. *Endocr. Rev.*, 2010; 31(2): 139-170.

[171] Mokuno T., Uchimura K., Hayashi R., Hayakawa N., Makino M., Nagata M., Kakizawa H., Sawai Y., Kotake M., Oda N., Nakai A., Nagasaka A., Itoh M. Glucose transporter 2 concentrations in hyper- and hypothyroid rat livers. *J. Endocrinol.*, 1999; 160(2): 285-289.

[172] Feng X., Jiang Y., Meltzer P., Yen P. M. Thyroid hormone regulation of hepatic genes in vivo detected by complementary DNA microarray. *Mol. Endocrinol.,* 2000; 14(7): 947-955.

[173] Oppenheimer J. H., Schwartz H. L., Lane J. T., Thompson M. P. Functional relationship of thyroid hormone-induced lipogenesis, lipolysis, and thermogenesis in the rat. *J. Clin. Invest.,* 1991; 87: 125–132.

[174] Mullur R., Liu Y. Y., Brent G. A. Thyroid hormone regulation of metabolism. *Physiol. Rev.*, 2014; 94(2): 355-382.

[175] Dimitriadis G. D., Raptis S. A. Thyroid hormone excess and glucose intolerance. *Exp. Clin. Endocrinol. Diabetes*, 2001;109 Suppl. 2: S225-S239.

[176] Daniele G., Eldor R., Merovci A., Clarke G. D., Xiong J., Tripathy D., Taranova A., Abdul-Ghani M., DeFronzo R. A. Chronic reduction of plasma free fatty acid improves mitochondrial function and whole-body insulin sensitivity in obese and type 2 diabetic individuals. *Diabetes,* 2014; 63(8): 2812-2820.

[177] Estivalet A. A., Leiria L. B., Dora J. M., Rheinheimer J., Bouças A. P., Maia A. L., Crispim D. D2 Thr92Ala and PPARγ2 Pro12Ala polymorphisms interact in the

modulation of insulin resistance in type 2 diabetic patients. *Obesity,* (Silver Spring). 2011; 19(4): 825-832.

[178] Foss M. C., Paccola G. M., Saad M. J., Pimenta W. P., Piccinato C. E., Iazigi N. Peripheral glucose metabolism in human hyperthyroidism. *J. Clin. Endocrinol. Metab.*, 1990; 70 (4): 1167-1172.

[179] Kumar A. Recurrent hypoglycemia in type 2 diabetic patient due to hypothyroidism. *J. Diabetes Metab. Disord.*, 2015; 14: 13.

[180] Mohn A., Di Michele S., Di Luzio R., Tumini S., Chiarelli F. The effect of subclinical hypothyroidism on metabolic control in children and adolescents with type 1 diabetes mellitus. *Diabet Med.,* 2002; 19(1): 70-73.

[181] Nada A. M. Effect of treatment of overt hypothyroidism on insulin resistance. *World J. Diabetes*, 2013; 4(4): 157-161.

[182] Gyawali P., Takanche J. S., Shrestha R. K., Bhattarai P., Khanal K., Risal P., Koju R. Pattern of thyroid dysfunction in patients with metabolic syndrome and its relationship with components of metabolic syndrome. *Diabetes Metab. J.,* 2015; 39(1): 66-73.

[183] Garduño-Garcia Jde J., Camarillo Romero E., Loe Ochoa A., Romero-Figueroa S., Huitron Bravo G., Torres García R., Montenegro-Morales P., Mendieta-Zerón H. Thyroid function is associated with insulin resistance markers in healthy adolescents with risk factors to develop diabetes. *Diabetol. Metab. Syndr.*, 2015; 7: 16.

[184] Aguayo-Mazzucato C., Zavacki A. M., Marinelarena A., Hollister-Lock J., El Khattabi I., Marsili A., Weir G. C., Sharma A., Larsen P. R., Bonner-Weir S. Thyroid hormone promotes postnatal rat pancreatic β-cell development and glucose-responsive insulin secretion through MAFA. *Diabetes,* 2013; 62(5): 1569-1580.

[185] Verga Falzacappa C., Mangialardo C., Madaro L., Ranieri D., Lupoi L., Stigliano A., Torrisi M. R., Bouchè M., Toscano V., Misiti S. Thyroid hormone T3 counteracts STZ induced diabetes in mouse. *PLoS One,* 2011;6(5):e19839.

In: Alterations in Glucose Homeostasis in Children
Editor: E. Otto-Buczkowska
ISBN: 978-1-63483-861-0

Chapter 11

POLYGLANDULAR AUTOIMMUNE SYNDROMES IN CHILDREN WITH TYPE 1 DIABETES MELLITUS

Iwona Beń-Skowronek*
Department of Pediatric Endocrinology and Diabetology,
Medical University of Lublin, Lublin, Poland

ABSTRACT

Polyglandular autoimmune syndromes (PAS) are conditions characterized by the association of two or more organ-specific disorders. Diagnosis should be considered in every patient in case of coexistence of two or more autoimmune endocrinopathies. On the basis of the clinical picture, they are divided into three different types according to Neufeld and Blizzard's classification.

Type I PAS (APECED – autoimmune polyendocrinopathy – candidiasis – ectodermal dystrophy) is a monogenic autoimmune syndrome, which is caused by defects in the AIRE gene located on chromosome 21. Its major components include candidiasis of the skin and mucous membranes, hypoparathyroidism, and Addison's disease. PAS I is extremely rarely connected with diabetes type 1.

Type II PAS is defined as a combination of autoimmune adrenal insufficiency with autoimmune thyroid disease and/or type 1 diabetes mellitus. It is characterized by obligatory occurrence of autoimmune Addison's disease in combination with thyroid autoimmune diseases and/or type 1 diabetes mellitus (also known as insulin-dependent diabetes mellitus, or T1D). The prevalence of PAS II increases gradually in the first decade of life and reaches the highest values between 25-40 years of age. Type III PAS is the most common syndrome in children with diabetes mellitus. PAS III is composed of autoimmune thyroid diseases associated with endocrinopathy other than adrenal insufficiency (e.g., premature gonad failure, type 1 diabetes mellitus). Thyroiditis usually occurs first. PAS type III can be further classified into the following 3 subcategories:

A. Autoimmune thyroiditis with immune-mediated diabetes mellitus (T1D) – also known as polyglandular autoimmune syndrome type 3 variant
B. Autoimmune thyroiditis with pernicious anemia

* Corresponding author: Iwona Beń-Skowronek. E-mail: skowroneki@interia.pl.

C. Autoimmune thyroiditis with vitiligo and/or alopecia and/or other organ-specific autoimmune disease [1]

This syndrome is associated with organ-specific autoimmune diseases (coeliac disease, hypogonadism and myasthenia gravis), organ-non-specific or systemic autoimmune diseases (sarcoidosis, Sjögren syndrome, rheumatoid arthritis), and other diseases (gastric carcinoid tumor, malabsorption due to exocrine pancreatic deficiency). PAS II and III syndromes exhibit polygenic inheritance and are connected with the HLA system. Frequent coexistence of diabetes mellitus and thyroid diseases requires monitoring and estimation of thyroid hormones and levels of anti-thyroid antibodies in patients with type 1 diabetes mellitus.

Prevalence of type 1 diabetes mellitus is 4-18% in PAS I and 60% PAS II. Clinical manifestations of T1DM in PAS I appear in patients about 20 years of age. It is estimated that about one-fourth of the patients with a single organ-specific autoimmune disease may have, or develop, other autoimmune diseases during their life. Typical PAS occurs usually around the third and fourth decades of life but first manifestations of autoimmune endocrinopathies can be present in childhood.

Keywords: PAS, autoimmune thyroid disease, Hashimoto's thyroiditis, Addison's disease, coeliac disease

CLASSIFICATION

Polyglandular autoimmune syndromes (PAS) are conditions characterized by the association of two or more organ-specific disorders. Diagnosis should be considered in every patient in the case of coexistence of two or more autoimmune endocrinopathies. On the basis of the clinical picture, they are divided into three different types according to Neufeld and Blizzard's classification [1].

PATHOGENESIS

Cell-mediated immune processes play a pivotal role in the immunopathogenesis of PAS. Naive CD4+ T cells, after contacting with antigens presented on professional antigen-presenting cells, differentiate into effector Th1, Th2, Th17, regulatory or Treg cells. Th1 cells produce interferon gamma (IFN-γ) and tumor necrosis factor alpha (TNF-α) and regulate antigen presentation and immunity against intracellular pathogens. Th2 cells, which produce interleukin IL-4, IL-5, and IL-13, mediate humoral responses and are important mediators of allergic diseases. Th17 cells express IL-17, IL-17F, IL-21, and IL-22 (and IL-26 in humans) and participate in inflammation and autoimmunity processes.

T regulatory cells (Tregs) express the Forkhead box P3 (Foxp3) transcription factor and mediate immune suppression by secretion of the transforming growth factor-β (TGF-β) and IL-10 [2, 3]. Differentiation of Th cells is controlled by the innate immune system, which provides the T cell receptor (TCR) and co-stimulatory signals as well as an appropriate cytokine microenvironment that ultimately leads to the preferential induction of one specific cell lineage over the other.

	Components	Genetics
PAS I	APECED – autoimmune polyendocrinopathy – candidiasis – ectodermal dystrophy	Defects in the AIRE gene located on chromosome 21
PAS II	Combination of autoimmune adrenal insufficiency with autoimmune thyroid disease and/or type 1 diabetes mellitus	Polygenic inheritance connected with chromosome 6
PAS III	Combination of autoimmune thyroid diseases associated with endocrinopathy other than adrenal insufficiency (e.g., premature gonad failure, type 1 diabetes mellitus)	Polygenic inheritance connected with chromosome 6
PAS III A	Autoimmune thyroiditis with immune-mediated diabetes mellitus	
PAS III B	Autoimmune thyroiditis with pernicious anemia	
PAS III C	Autoimmune thyroiditis with vitiligo and/or alopecia and/or other organ-specific autoimmune disease	

Lymphocyte infiltrations of various glands are usually associated with functional loss of epithelial cells. The cellular defect in PAS may be associated with abnormal balances in cytokine production by T cells. A polarized Th2 response is associated with Graves' disease (GD) and Th1 with type 1 diabetes mellitus.

A hypothesis for the pathogenesis of PAS as "two hits" has been presented. Accordingly, a genetically predisposed person might develop an autoimmune process after initiation by an infectious agent. Animal models of the immunopathogenesis of PAS are consistent with a viral infection theory as well as a suppressor effect theory. The molecular mimicry is characterized by an immune response to an environmental agent that cross-reacts with a host antigen, resulting in disease. In an animal model, mice infected with reovirus type 1 developed autoimmune polyendocrinopathy [4]. In PAS I, Th2 immune responses to self-antigens and defective protective Th1 responses against invasion of Candida albicans yeast were observed [5, 6]. PAS I is caused by a loss of central tolerance [6].

The precise etiology of PAS II and III is unknown. A further hypothesis refers to Tregs, which are implicated in self-tolerance and autoimmunity. As previously stated, the adaptive immune system not only consists of immune-stimulatory CD4+ helper or effector T cells and cytotoxic CD8+ T cells but is also regulated by immunosuppressive T cells. Three classes of immunosuppressive CD4+ T cells are known to date: induced Th3, Tr1 cells, and the naturally present CD4+CD25+FoxP3+ Tregs [7, 8].

Tregs suppress the activation of CD4+ and CD8+ cells in vitro and in vivo and in this way prevent autoimmune disease, although the exact mechanism of suppression is unknown, and cell contact is crucial, at least in vitro. Depletion or dysfunction of Tregs leads to autoimmunity in mice [7].

In a murine model, depletion of thymically derived CD4+CD25+ Tregs resulted in a syndrome that is similar to human PAS with multiple endocrinopathies [8]. Defective Treg function in patients with PAS has important implications for autoimmunity in general. The results indicate that once the molecular mechanisms of suppressor function are better delineated, manipulations of human Tregs may allow improvement of immunomodulatory strategies in diseases with impaired suppressor function.

Another category of immunosuppressive cells are the Tregs Tr1/Th3, and knowledge of the development and functions of such immunoregulatory cells may elucidate the etiology for developing autoimmunity [7, 8]. Recent studies on animal models of diabetes and human Tregs have shown that TNF-α impairs the ability of these cells to prevent the disease. Increased susceptibility of Tregs to the modulatory effects of TNF-α involves signaling through TNFR2 that is expressed on the surface of these cells. It seems that TNF-α neutralization may rescue regulatory T cells and restore their function in several autoimmune and inflammatory diseases. This review describes recent data concerning regulatory T cells in the context of inflammation that is present during diabetes type 1. It describes how TNF contributes to the pathogenesis of type 1 diabetes, what the impact of this cytokine on regulatory T cell population is, and therapeutic effects that result from neutralization thereof in several inflammatory and autoimmune diseases [9].

The pathological processes lead to the preclinical phase of PAS, with production of organ-specific antibodies and progressive immune-mediated destruction of endocrine tissues. In the clinical phase, major organ destruction occurs due to the autoimmune activity that is primarily characterized by chronic inflammatory infiltration of lymphocytes . Destruction of endocrine glands causes their secretory insufficiency [9].

The role of apoptosis in immunodestruction has been associated with dysregulation of apoptotic signaling pathways. Dysfunction of the Fas apoptotic pathway or production of soluble factors including soluble Fas and soluble Fas ligand may be involved in the pathogenesis of endocrine diseases. The activity of deoxyribonuclease 1 (DNase 1) was lowered in patients with PAS compared to healthy subjects [9]. It catalyzes DNA hydrolysis by cleaving double-stranded DNA. The deficiency in DNase 1 may result in reduced or delayed removal of DNA from nuclear antigens and, thereby, may promote disease susceptibility to autoimmune disorders. Antigen complexes containing DNA can initiate the manifestation of endocrine autoimmunity by stimulation of an innate immune signaling pathway via toll-like receptors (TLRs). The activation of TLR9 promotes the production of IL-12 and a shift towards a Th1 immune response, activities that are accompanied by the breakdown of tolerance. Antigen complexes can initiate the production of autoantibodies by their ability to activate B-cells without a second signal from T-helper cells.

This "single handed" B-cell activation is mediated by antigen recognition with cell-surface antigen receptors and the additional co-stimulatory activation of TLR9 by antigen-associated self DNA [6-9].

GENETICS

Type I PAS (APECED – autoimmune polyendocrinopathy – candidiasis – ectodermal dystrophy) is a monogenic autoimmune syndrome, which is caused by defects in the AIRE gene located on chromosome 21 [7, 10].

The inheritance of PAS II and PAS III is complex, with genes on chromosome 6 playing a predominant role. In humans, this chromosome contains the major histocompatibility loci. Within some families, susceptibility to autoimmune endocrine disease appears to be inherited as an autosomal dominant form associated with a specific HLA haplotype.

Nevertheless, family members may manifest different diseases, though the more common the disease in the general population, the higher its prevalence in affected families.

Environmental factors may have an important influence on the development of autoimmune diseases, but the exposure to environmental pathogens does not always lead to disease. With respect to genetics, PAS II is supposed to be a polygenic disease with autosomal dominant inheritance and incomplete penetrance. Also, familial clustering provided evidence for a genetic predisposition.

Many of the PAS II component diseases are associated with HLA alleles B8 and DR3. The DR antigen has two glycoprotein chains coded for by the DR loci of chromosome 6. The alleles HLA-B8 and DR3, which are coded for by separate genes, are found together on the same chromosome more often than one would predict from their frequency in the general population [7-9].

The MHC class I related gene A (MICA), located on chromosome 6 within the HLA region, is an additional locus associated with PAS. The MICA exon 5 microsatellite is of special interest, because this trinucleotide (GCT) repeat lies in the transmembrane region encoding alanine residues. It is characterized by five common alleles (MIC-A4, -A5, -A5.1, A6 and A9); allele A5.1 shows an additional 1-bp insertion (GGCT). The MICA molecule is a membrane-bound glycoprotein. It binds to NKG2D, which is an activating receptor of natural killer (NK) cells. Combination of MIC-A5 and HLA-DR3-DQ2 and/or HLA-DR4-DQ8 represents a strong genetic marker for T1D with PAS [7, 9].

Further susceptible genes encompass the cytotoxic T lymphocyte-associated antigen 4 (CTLA-4) gene, a strong inhibitor of T cells (40). CTLA-4 is a T lymphocyte surface antigen that interacts with B7-1 (CD80) and B7-2 (CD86) of antigen presenting cells. It inhibits T cell activation by reducing IL-2 production and IL-2 receptor expression and by arresting T cells at the G1 phase of the cell cycle [7, 9].

Of further interest is the recent finding that the protein tyrosine phosphatase non-receptor 22 (PTPN22) gene might contribute to susceptibility to PAS II. This gene encodes an intracellular phosphatase with negative regulatory effects on T cell activation. A functional single nucleotide polymorphism (C1858T) in the PTPN22 gene is associated with several autoimmune diseases [7, 9]. Due to the higher prevalence, previous studies investigated associations of PTPN22 C1858T with AITD and T1D as single diseases. With respect to T1D and AITD as separate entities, for each a less than twofold increased risk (OR) for PTPN22 T allele carriers was reported [10]. LYP encoded by PTPN22 is expressed in lymphocytes where it binds to the protein kinase, a suppressor of the lymphocyte-specific protein tyrosine kinase and the protein tyrosine kinase p59fyn, which are associated with the TCR signaling pathway [7-9].

CLINICAL SIGNS AND SYMPTOMS

Type I PAS (APECED – autoimmune polyendocrinopathy – candidiasis – ectodermal dystrophy) – its major components include candidiasis of the skin and mucous membranes, hypoparathyroidism, and Addison`s disease. PAS I is extremely rarely connected with diabetes type 1 in children [9, 10].

Type II PAS is defined as a combination of autoimmune adrenal insufficiency with autoimmune thyroid disease and/or type 1 diabetes mellitus. It is characterized by obligatory occurrence of autoimmune Addison's disease in combination with thyroid autoimmune diseases and/or type 1 diabetes mellitus. PAS II prevalence increases gradually in the first decade of life and reaches the highest values between 25-40 years of age. The coexistence of adrenal failure with either autoimmune thyroid disease (AITD) and/or type 1 diabetes (T1D) is defined as Schmidt's or Carpenter's syndrome [9, 11]. PAS II presenting in childhood is extremely rare; a case with hypothyroidism, followed by diabetic ketoacidosis, and adrenal insufficiency has been reported.

Type III PAS is the most common syndrome in children with diabetes mellitus. PAS III is composed of autoimmune thyroid diseases associated with endocrinopathy other than adrenal insufficiency (e.g., premature gonad failure, type 1 diabetes mellitus). In some patients, the presence of one autoimmune endocrine disease is associated with an increased risk of developing autoimmunity to other tissues detected by increased autoantibodies level. Type 1 diabetes mellitus is a very frequent component disorder of PAS II and is often its first symptom. According to Dittmar and Kahaly, 21% patients with T1D showed multiple glandular involvement, 4% had vitiligo and 2% had immune gastritis [9]. The screening of 15000 consecutive subjects with endocrine disorders revealed a PAS prevalence of 1%.

There is often a long time-interval between the manifestation of the first and second component of PAS II disease: frequent disease combinations are T1D/AITD (41%), T1D/ vitiligo (9.9%), T1D/AITD/pernicious anemia (5.3%), hypogonadism/alopecia (5.3%), and T1D/Addison's disease (3.3%) [9].

Thyroid Disorders and Type 1 Diabetes Mellitus

The simultaneous occurrence of immune hypothyroidism and type 1 diabetes mellitus is often accompanied by hypoglycemia due to increased insulin sensitivity. Hypothyroid children may develop growth disorders caused by chronic hypoglycemia and due to hypothyroidism itself. Proper treatment with L-thyroxine and insulin guarantees normal growth and metabolic status.

Substitution therapy with levothyroxine leads to increased insulin demand [9, 11-18, 20-29]. Thyroid autoantibodies without clinical disease manifestation are noted in nearly 50% of the PAS patients, and the time interval between the detection of the autoantibodies and presentation of the disease is longer for AITD than for other immunopathies. Thus, high titers of autoantibodies against TPO and Tg do not automatically precede the development of clinical AITD [11-29].

Frequent coexistence of diabetes mellitus and thyroid diseases requires monitoring and estimation of thyroid hormones and levels of anti-thyroid antibodies in patients with type 1 diabetes mellitus. Prevalence of type 1 diabetes mellitus is 4-18% in PAS I and 60% in PAS II. Clinical manifestations of type 1 diabetes mellitus in PAS I appear in patients about 20 years of age. It is estimated that about one-fourth of patients with a single organ-specific autoimmune disease may have or develop other autoimmune diseases during their life. Typical PAS occurs usually around the third and fourth decades of life but first manifestations of autoimmune endocrinopathies can be present in childhood [9].

The exact international prevalence of PAS III is unknown. Polish observation indicated that in a cohort of 461 patients with type 1 diabetes mellitus autoimmune PAS III A was diagnosed in 11,1% (8,2% girls and 2,6% boys) [19].

An extensive retrospective study carried out by Kordonouri et al. revealed that children with a positive antibody titer were older, developed the disease at an older age, and their disease lasted longer [20]. At the time of the diabetes diagnosis, the proportion of patients with a positive anti-thyroid antibody titer ranged from 2 to over 30% [20-26]. Hypothyroidism was diagnosed in 10% of children with a positive anti-thyroid antibody titer, whereas ultrasound changes, i.e., thyroid enlargement, disturbances in the echostructure, and/or nodular lesions, were observed in 50% [20-26]. Clinical hypothyroidism may affect 1-5% of children with type 1 diabetes mellitus (International Society for Pediatric and Adolescent Diabetes – ISPAD Consensus), and subclinical hypothyroidism 5-10% (15,8% of patients with the positive anti-thyroid antibody titer – Kordonuri), 5-10% (Glastras), 7,3% (Mantovani), or 15% (Szalecki) [9, 23, 27-29]. PAS III A was often diagnosed in the group of the youngest children under 5 years of age. The ISPAD Consensus 2009 recommends screening according to the following scheme: in the case of autoimmune thyroid diseases, assays of the TSH concentration and antibodies should be performed at the diagnosis of diabetes and followed by biennial tests [28]. In screening examinations, the ISPAD recommends assessment of thyroid function in the case of presence of goiter, a decreased growth rate, thyroid disease symptoms, and elevated antibody titers [28]. In turn, the Australasian Pediatric Endocrine Group suggests measurement of anti-thyroid antibody titers (anti-Tg and anti-TPO) every 2-3 years, and assessment of the TSH concentration every year, rather than biennially [26]. The American Diabetes Association (ADA) acknowledges that presence of anti-thyroid antibodies raises the risk of AITD – however, screening is not recommended [26, 27]. In turn, it is recommended that anti-thyroid peroxidase antibodies and TSH and fT4 levels should be assayed in children at diagnosis and repeated every 1-2 years. In patients with a positive antibody titer, assessment of thyroid function should take place every 6-12 months [26]. According to Kordonuri and Mantovani, antibodies (or rather anti-TPO) should be assayed annually [20, 22, 23].

In the case of a positive antibody titer, the concentrations of TSH and fT4 should be determined and thyroid ultrasound performed. The risk group includes girls, patients affected by diabetes type 1 for over 9 years, patients over 12 years of age, and patients with another autoimmune disease. Prophylactic treatment with thyroxine in euthyroid patients with a positive antibody reduces the thyroid volume, prevents development of goiter and hypothyroidism by decreasing the TSH concentration and thyroid antigen expression, and leads to a decline in the antibody concentration and lower levels of lymphocytic infiltration in the thyroid [29, 31-35]. Hyperthyroidism can affect 0,5% of children with diabetes type 1 (in the US: 0,2-0,4%) [26]. Clinical or subclinical hyperthyroidism is accompanied by deterioration of metabolic control in diabetes – its frequent first manifestation is hyperglycemia, an effect of increased gluconeogenesis and glycogenolysis and increased glucose autoantibody sorption.

Additionally, reduced insulin sensitivity can be found, i.e., higher receptor affinity accompanied by abnormal post-receptor signal transmission [26, 27, 30]. Treatment of hyperthyroidism with thyreostatic drugs or radioiodine leads to sufficient diabetes control.

In contrast, hyperthyroidism is accompanied in 50% of the cases by glucose intolerance and in 3% of the cases by overt diabetes.

Impaired glucose tolerance is caused by decreased insulin sensitivity and hepatic storage of glycogen, whereas both secretion of glucagon and intestinal glucose sorption of autoantibodies are enhanced. In diabetic children with concomitant Graves' disease and the risk of fluctuating thyroid functional status, treatment of type 1 diabetes mellitus is much more difficult, which might call for a decision of definitive treatment, especially when remission is not obtained after an anti-thyroid drug course of reasonable duration [9, 36].

ADDISON'S DISEASE AND TYPE 1 DIABETES MELLITUS

Concomitant presence of Addison's disease and type 1 diabetes mellitus leads to hypoglycemia due to decreased gluconeogenesis and increased insulin sensitivity [9, 11]. In adrenal autoimmunity, autoantibodies are directed against antigens identified as P450 cytochromes, i.e., enzymes involved in the synthesis of steroid hormones.

These are 21-hydroxylase (OH), 17-alpha-OH, and the side-chain cleavage enzyme. The 21-OH has been identified as a major target antigen in Addison's disease and has been detected in 85% of patients with PAS and adrenal failure. Autoantibodies against steroidal enzymes (21-OH) are of high prognostic value and will help identify patients at risk for developing Addison's disease. This might prevent delayed diagnosis of adrenal failure [9, 11].

Subjects with pathological ACTH test results and increased levels of basal plasma ACTH require close clinical follow-up with repetition of the test every six months.

A replacement therapy with hydrocortisone or cortisone acetate should be considered in the case of undercurrent stressful events. Hypoglycemic episodes and a decreasing insulin requirement in a type 1 diabetic can be one of the earliest signs of the development of adrenal failure. Replacement of levothyroxin without previous or simultaneous adrenal steroid replacement in a hypothyroid patient with Addison's disease can precipitate an adrenal crisis. Replacement of levothyroxin increases the cortisol turnover rate in the liver, and this may tax a failing adrenal gland [5, 6, 9].

IMMUNE GASTRITIS AND PERNICIOUS ANEMIA AND TYPE 1 DIABETES MELLITUS

Other disorders like immune gastritis and pernicious anemia may occur in PAS. Immune gastritis, eventually leading to pernicious anemia, is an organ-specific autoimmune disease characterized by pathological lesions, affecting the fundus and body of stomach, which are typified by gastric mucosal atrophy, selective loss of parietal cells from the gastric mucosa, submucosal lymphocyte infiltration, as well as circulating gastric parietal cell autoantibodies. Subsequently, pernicious anemia, which is considered to be the most common cause of vitamin B12 deficiency, may develop. In this case, autoantibodies to the intrinsic factor, itself a secretory product of gastric chief cells, are found in the circulation and in gastric secretions [9].

COELIAC DISEASE AND TYPE 1 DIABETES MELLITUS

Coeliac disease is one of the most common genetic disorders. Its incidence in the European and American population ranges from 0.03 to 0.04% and exhibits variation [37-48]. The highest incidence rates are found in north European countries: Sweden – 2.4/1000 births, Great Britain – 1.49/1000, and Finland – 0,86/1000 [37-39]. Promotion of breast-feeding as well as delayed introduction of gluten-containing products into the diet has changed the clinical manifestation of the disease. Classic forms of coeliac disease have given way to its atypical forms. The frequency of coexistence of both disorders increases with patients' age and duration of diabetes [9]. In their investigations conducted on similar populations in terms of numbers, Crone and collaborators have reported 5% and Sanchez-Albisua 3.2% of patients with diagnoses coeliac disease among diabetes type 1 patients [44, 45]. In Poland, Myśliwiec reported coeliac disease in 5.7% of children with newly diagnosed diabetes type 1 [46]. In patients affected by diabetes type 1, coeliac disease is usually oligosymptomatic or asymptomatic – therefore, it is difficult to identify the risk group requiring regular screening examination among all patients [39-41]. A reverse phenomenon is characteristic as well, i.e., higher prevalence of diabetes type 1 in coeliac disease patients and greater frequency of diabetes-specific antibodies in this group of patients [39-41]. The findings about the gluten-free diet used for children with diabetes type 1 that clearly reduces the risk of autoimmunity are similar [37-41]. As in other autoimmune diseases, a specific HLA system located on the short arm of chromosome 6 in the 6p23 region was observed. The strongest association with coeliac disease is exhibited by centromere-associated HLA class II genes (DQ2 and DR3), which are present in 80-90% of patients. Telomere-associated HLA class I (B8 and A1) genes are responsible for development of coeliac disease to a lesser extent. It has been demonstrated that genes DQA1*0501 and DQB1*0201 are characteristic for coeliac disease [50]. Positive anti-transglutaminase antibodies have been detected in as many as 1/3 of diabetic patients with HLA DQ2, compared with <2% diabetic patients without HLA DQ2 or DQ8 [49]. Sumnik et al. showed that HLA DQ2 was present in 80% of patients affected by diabetes type 1 and coeliac disease and in 49% of patients affected by diabetes type 1 without coeliac disease [51]. Coeliac disease patients who do not express HLA DQ2 usually have genotype DQ8. Therefore, patients with diabetes type 1 with HLA DQ2 or DQ8 are susceptible to coeliac disease [43, 44, 49, 50]. In the case of coeliac disease, the ISPAD Consensus 2009 recommends assessment of the antibody titer at diagnosis of diabetes, followed by annual assessment for five years, and next – every two years [28].

Strict use of the gluten-free diet in patients with diabetes type 1 is still a controversial issue. In patients exhibiting even mild clinical symptoms of coeliac disease, the gluten-free diet is indispensable in order to prevent worsening of the symptoms [51, 52, 55]. When untreated, coeliac disease may lead to changes in insulin requirement and recurrent hypoglycemia, weight and/or growth disorders, maturation disorders, alimentary anemia, osteopenia, osteoporosis, and neurological disorders. The most severe complications of coeliac disease comprise gastrointestinal cancer – lymphomas and tumors. Therefore, despite being inconvenient and restrictive, the gluten-free diet should be introduced in all patients affected by diabetes type 1 and coeliac disease and, given the protective effect of the gluten-free diet on pancreatic β cells and improved insulin secretion, in patients with newly diagnosed diabetes [53, 58].

OTHER AUTOIMMUNE DISEASES ANDTYPE 1 DIABETES MELLITUS

PAS III C is sometimes associated with other autoimmune diseases. In half of the cases, autoimmune arthritis preceded the development of T1D. Only a few papers have described coexistence of JIA and T1D in children [56-60]. Due to use of steroid therapy, there was always difficulty in diabetes control and treatment. Currently, biological therapy in patients with JIA is disputable, since there are reports on development of diabetes type 1 after administration of etanercept [59, 60].

SCREENING

Approximately one in seven first-degree relatives of patients with PAS have an unrecognized endocrine disorder, usually the relatively common immune thyroiditis. According to the literature, genetic screening is useful for PAS I and less for type II, because one would likely identify the alleles associated with the disease that have already been identified, e.g., DR3 and DR4, in patients with type 1 diabetes. Below we propose the screening for frequent autoimmune diseases in children with type 1 diabetes mellitus.

Many of the endocrine disorders of PAS are adequately treated with hormonal replacement therapy if the disease is recognized early.

	Immune screening	Frequency	If positive immune screening – hormonal investigations	Frequency
Autoimmune thyroid disorders	TPO Ab TG Ab Optionally TSI Ab	Every year	TSH fT4, fT3 thyroid ultrasound	Every year
Adrenal disorders	Antibodies against 21-hydroxylase	Every 2-3 years	basal and stimulated cortisol basal ACTH	Every year
Coeliac disease	Antibodies against transglutaminase	Every year	Gastroscopy with histopathological investigations of the intestinal mucosa	–

REFERENCES

[1] Neufeld, M., Blizzard, R. M. Polyglandular autoimmune disease. In: Pinchera, A., Doniach, D., Fenzi, D. F., Baschieri, L., eds. *Autoimmune aspects of endocrine disorders*. London, UK: Academic Press; 1980: 357-65.

[2] Zhu, J., Paul, W. E. CD4 T cells: fates, functions, and faults. *Blood*. 2008; 112: 1557-1569.

[3] Martinez, G. J., Nurieva, R. I., Yang, X. O., Dong, C. Regulation and function of proinflammatory TH17 cells. *Annals of the New York Academy of Sciences*. 2008; 1143: 188-211.
[4] Onodera, T., Toniolo, A., Ray, U. R., Jenson, A. B., Knazek, R. A., Notkins, A. L. Virus-induced diabetes mellitus. Polyendocrinopathy and autoimmunity. *J. Exp. Med.* 1981; 153: 1457-1473.
[5] Anderson, M. S. Autoimmune endocrine diseases. *Current Opinion in Immunology* 2002; 14: 760-764.
[6] Devendra, D., Eisenbarth, G. S. Immunologic endocrine disorders. *Journal of Allergy and Clinical Immunology* 2003; 111: S624-S636.
[7] Eisenbarth, G. S., Gottlieb, P. A. Autoimmune polyendocrine syndromes. *N. Engl. J. Med.* May 13 2004; 350(20): 2068-79.
[8] Shevach, E. M. Regulatory T cells in autoimmmunity. *Annual Review of Immunology* 2000; 18: 423-449.
[9] Dittmar, M., Kahaly, G. J. Polyglandular Autoimmune Syndromes: Immunogenetics and Long-Term Follow-Up. *The Journal of Clinical Endocrinology and Metabolism* 2013; 88(7): 2983-2992.
[10] Heino, M., Peterson, P., Kudoh, J., Shimizu, N., Antonarakis, S. E., Scott, H. S. et al. APECED mutations in the autoimmune regulator (AIRE) gene. *Hum. Mutat.* 2001; 18: 205-211.
[11] Betterle, C., Lazzarotto, F., Presotto, F. Autoimmune polyglandular syndrome Type 2: the tip of an iceberg? *Clin. Exp. Immunol.* Aug. 2004; 137(2): 225-33.
[12] Bahceci, M., Tuzcu, A., Pasa, S., Ayyildiz, O., Tuzcu, S. Polyglandular autoimmune syndrome type III accompanied by common variable immunodeficiency. *Gynecol. Endocrinol.* Jul. 2004; 19(1): 47-50.
[13] Turkoglu, Z., Kavala, M., Kolcak, O., Zindanci, I., Can, B. Autoimmune polyglandular syndrome-3C in a child. *Dermatol. Online J.* 2010; 16(3): 8.
[14] Kahaly, G. J. Polyglandular autoimmune syndromes. *Eur. J. Endocrinol.* Jul. 2009; 161 (1): 11-20.
[15] Lorini, R., Gastaldi, R., Traggiai, C., Perucchin, P. P. Hashimoto`s thyroiditis. *Pediatric Endocrinol. Rev.* 2003; Suppl. 2: 205-211.
[16] Karagüzel, G., Simşek, S., Değer, O., Okten, A. Screening of diabetes, thyroid, and coeliac diseases-related autoantibodies in a sample of Turkish children with type 1 diabetes and their siblings. *Diabetes Res. Clin. Pract.* 2008 May; 80(2): 238-43.
[17] Kakleas, K., Paschali, E., Kefalas, N. et al. Factors for thyroid autoimmunity in children and adolescents with type 1 diabetes mellitus. *Ups. J. Med. Sci.* 2009; 114: 214-220.
[18] Karavanaki, K., Kakleas, K., Paschali, E., Kefalas, N., Konstantopoulos, I., Petrou, V. et al. Screening for Associated Autoimmunity in Children and Adolescents with Type 1 Diabetes Mellitus (T1DM). *Horm. Res.* 2009; 71: 201-206.
[19] Ben-Skowronek, I., Michalczyk, A., Piekarski, R., Wysocka, B., Banecka, B. Polyglandular autoimmune syndromes type III in children with type 1 diabetes mellitus. *Ann. Agric. Environ. Med.* 2013; 20(1): 140-146.
[20] Kordonouri, O., Klinghammer, A., Lang, E. B., Grüters-Kieslich, A., Grabert, M., Holl, R. W. Thyroid autoimmunity in children and adolescents with type 1 diabetes: a multicenter survey. *Diabetes Care*. 2002; 25: 1346-1350.

[21] Piątkowska, E., Szalecki, M. Autoimmune thyroiditis in children and adolescents with type 1 diabetes. *Ped. End. Diab. and Metab.* 2011; 17(4): 173-177.

[22] Kordonouri, O., Hartmann, R., Deiss, D., Wilms, M., Gruters-Kieslich, A. Natural course of autoimmune thyroiditis in type 1 diabetes: association with gender, age, diabetes duration, and puberty. *Arch. Dis. Child.* 2005; 90: 411-414.

[23] Mantovani, R. M., Mantovani, L. M., Dias, V. M. Thyroid autoimmunity in children and adolescents with type 1 diabetes mellitus prevalence and risk factors. *J. Pediatr. Endocrinol. Metab.* 2007; 20: 669-675.

[24] Kondrashova, A., Viskari, H., Haapala, A. M., Seiskari, T., Kulmala, P., Ilonen, J. et al. Serological evidence of thyroid autoimmunity among schoolchildren in two different socioeconomic environments. *J. Clin. Endocrinol. Metab.* 2008; 93: 729-734.

[25] Brix, T. H., Knudsen, G. P., Kristiansen, M., Kyvik, K. O., Orstavik, K. H., Hegedüs, L. High frequency of skewed X -chromosome inactivation in females with autoimmune thyroid disease: a possible explanation for the female predisposition to thyroid autoimmunity. *J. Clin. Endocrinol. Metab.* 2005; 90: 5949-5953.

[26] Barker, J. M. Clinical review: Type 1 diabetes-associated autoimmunity: natural history, genetic associations, and screening. *J. Clin. Endocrinol. Metab.* 2006; 91: 1210-1217.

[27] Glastras, S. J., Craig, M. E., Verge, C. F., Chan, A. K., Cusumano, J. M., Donaghue, K. C. The role of autoimmunity at diagnosis of type 1 diabetes in the development of thyroid and coeliac disease and microvascular complications. *Diabetes Care.* 2005; 28: 2170-2175.

[28] ISPAD Clinical Practice Consensus Guidelines Compendium 2009. *Pediatr. Diabetes.* 2009; 10 (Suppl. 12): 1-210.

[29] Szalecki, M. Screening for thyroid disease in the pediatric insulin dependent diabetic population in Kielce Province. *Horm. Res.* 1998; 50: 147.

[30] Kordonouri, O., Hartmann, R., Riebel, T., Liesenkoetter, K. P. Early treatment with L - thyroxine in children and adolescents with type 1 diabetes, positive thyroid antibodies, and thyroid gland enlargement. *Pediatr. Diabetes.* 2007; 8: 180-184.

[31] Karges, B., Muche, R., Knerr, I., Ertelt, W., Wiesel, T., Hub, R. et al. Levothyroxine in euthyroid autoimmune thyroiditis and type 1 diabetes: a randomized, controlled trial. *J. Clin. Endocrinol. Metab.* 2007; 92: 1647-1652.

[32] Padberg, S., Heller, K., Usadel, K. H., Schumm-Draeger, P. M. One-year prophylactic treatment of euthyroid Hashimoto's thyroiditis patients with levothyroxine: is there a benefit? *Thyroid.* 2001; 11: 249-255.

[33] Villano, M. J., Huber, A. K., Greenberg, D. A. et al. Autoimmune thyroiditis and diabetes: dissecting the joint genetic susceptibility in a large cohort of multiplex families. *J. Clin. Endocrinol. Metab.* 2009; 94(4): 1458-66.

[34] Huber, A., Menconi, F., Corathers, S. et al. Joint genetic susceptibility to type 1 diabetes and autoimmune thyroiditis: from epidemiology to mechanisms. *Endocr. Rev.* 2008; 29(6): 697-725.

[35] Dittmar, M., Kahaly, G. J. Genetics of the autoimmune polyglandular syndrome type 3 variant. *Thyroid.* 2010; 20(7): 737-43.

[36] Otoda, T., Sakurai, M., Ota, T., Kurita, S., Misu, H., Shimizu, A. et al. Two cases of type 1 diabetes mellitus accompanied by Graves' disease. *Je. Japan Diab. Soc.* 2008; 51 (4): 323-327.

[37] Hill, I., Fasano, A., Schwartz, R. The prevalence of coeliac disease in at-risk groups of children in the United States. *J. Pediatr.* 2000; 136: 86-90.

[38] Maki, M., Mustalahti, Kokkonen, J. Prevalence of coeliac disease among children in Finland. *N. Engl. J. Med.* 2003; 348: 2517-2524.

[39] Hansen, D., Bennedback, F. N., Hansen, L. K. High prevalence of coeliac disease in Danish children with type 1 diabetes mellitus. *Acta Peadiatr.* 2001; 90(11): 1217-1220.

[40] Ludvigsson, I. F., Ludvigson, J., Ekbom, A. Coeliac disease and risk of subsequent type 1 diabetes: a general population cohort study of children and adolescents. *Diabetes Care* 2006; 29: 2483-2488.

[41] Li Voon Chong, J. S., Leong, K. S., Wallymahmed, M. Is coeliac disease more prevalent in young adult with coexisting type 1 diabetes mellitus and autoimmune thyroid disease compared with those with type 1 diabetes mellitus alone? *Diabet. Med.* 2002; 19: 334-337.

[42] Picarelli, A., Sabbatella, L., Di Tola, M., Vetrano, S., Casale, C., Porowska, B. et al. Anti-endomysial antibody of IgG1 isotype detection strongly increases the prevalence of coeliac disease in patients affected by type 1 diabetes mellitus. *Clin. Exp. Immunol.* 2005; 142: 11-15.

[43] Arentz-Hansen, A., Fleckenstein, B., Moberg, O. HLA DQ2 and DQ8 signatures of gluten T cell epitopes in coeliac disease. *Clin. Invest.* 2006; 116: 2226-2232.

[44] Crone, J., Rami, B., Huber, W. D., Granditsch, G., Schober, E. Prevalence of coeliac disease and follow up to EMA in children with type 1 diabetes mellitus. *J. Pediatr. Gastroenterol. Nutr.* 2003; 37: 67-71.

[45] Sanchez-Albisua, I., Wolf, J., Neu, A., Geiger, H., Wäscher, I., Stern, M. Coeliac disease in children with Type 1 diabetes mellitus: the effect of the gluten-free diet. *Diabet. Med.* 2005; 22: 1079-1082.

[46] Myśliwiec, M., Balcarska, A., Stopiński, J. Prognostic factors of coeliac disease occurrence in type 1 diabetes mellitus children. *Endokr. Diabetol. Chor. Przem. Mati. Wieku Rozw.* 2006; 12: 281-285.

[47] Giletti, P. M., Giletti, H. R., Israel, D. M. High prevalence of coeliac disease in patients with type 1 diabetes detected by antibodies to endomysium and tissue transglutaminase. *Can. J. Gastroenterol.* 2003; 15: 297-301.

[48] Mathieu, S., Stassen, A., Paquot, N., Scheen, A. J. Type 1 diabetes and coeliac disease. *Rev. Med. Liege.* 2006; 61: 637-642.

[49] Galli-Tsianopoulou, A., Nousia-Arvanitakis, S., Dracoulacos, D. Autoantibodies predicting diabetes mellitus type I in coeliac disease. *Horm. Res.* 1999; 52, 119-124.

[50] Doolan, A., Donaghue, K., Fairchild, J. F. Use of HLA typing in diagnosing coeliac disease in patients with type 1 diabetes. *Diabetes Care.* 2005; 28: 806-809.

[51] Sumnik, Z., Koluskova, S., Cinek, O. HLA-DQA180201 positivity predisposes to coeliac disease in Czech diabetic children. *Acta Paediatr.* 2007; 89: 1426-1430.

[52] Calvo, G., Rica, J., Vittoria, J. C. Conserved extend haplotypes discriminate HLA-DR3 -homozygous Basque patients with type 1 diabetes mellitus and coeliac disease. *Genes Immun.* 2006; 7: 550-554.

[53] Hansed, D., Brock-Jacobsen, D., Lund, E. Clinical benefit of gluten-free diet in type 1 diabetic children with screening-detected coeliac disease: a population-based screening study with 2 years' follow-up. *Diabetes Care.* 2006; 29: 2452-2456.

[54] Acerini, C. L., Ahmed, M. I., Ross, K. M. Coeliac disease in children and adolescents with IDDM: clinical characteristic and response to gluten-free diet. *Diabet. Med.* 1998; 15: 38-44.

[55] Fuchtenbush, M., Ziegler, A. G., Hummel, M. Elimination of dietary gluten and development of type 1 diabetes high risk subjects. *Rev. Diabet. Study*. 2004; 1: 39-41.

[56] Pohjankoski, H., Kautiainen, H., Korppi, M., Savolainen, A. Simultaneous juvenile idiopathic arthritis and diabetes mellitus type 1 - a Finnish Nationwide study. *J. Rheumatol*. 2012; 39(2): 377-381.

[57] Rudolf, M. C., Genel, M., Tamborlane, W. V. Jr, Dwyer, J. M. Juvenile rheumatoid arthritis in children with diabetes mellitus. *J. Pediatr*. 1981; 99(4): 519-524.

[58] Nolfi-Donegan, D., Viswanathan, A., Wood, R. et al. Case report of a child with recently diagnosed diabetes mellitus type 1 and subsequent systemic arthritis. *The Internet Journal of Pediatrics and Neonatology* ISSN: 1528-8374.9.

[59] Tack, C. J., Kleijwegt, F. S., Van Riel, P. L., Roep, B. O. Development of type 1 diabetes in a patient treated with anti-TNF-alpha therapy for active rheumatoid arthritis. *Diabetologia*. 2009; 52: 1442-1444.

[60] Bloom, B. J. Development of diabetes mellitus during etanercept therapy in a child with systemic-onset juvenile rheumatoid arthritis. *Arthritis Rheum*. 2000; 43: 2606-2608.

In: Alterations in Glucose Homeostasis in Children
Editor: E. Otto-Buczkowska
ISBN: 978-1-63483-861-0

Chapter 12

OVARIAN FUNCTION AND PUBERTY IN GIRLS WITH TYPE 1 DIABETES MELLITUS

***E. Wender-Ożegowska*[1,*] *and K. Ożegowska*[2]**

[1]Department of Obstetrics and Women's Diseases, Faculty of Gynecology, Obstetrics and Oncological Gynecology, Poznań University of Medical Sciences, Poznań, Poland

[2]Department of Infertility and Reproductive Endocrinology, Faculty of Gynecology, Obstetrics and Oncological Gynecology, Poznań University of Medical Sciences, Poznań, Poland

ABSTRACT

Insulin plays a pivotal role in carbohydrate metabolism but also a regulatory role in the ovarian function. Ovarian granulosa cells, as well as theca lutein and stromal cells, are susceptible to both, insulin deficiency or excess. Delayed puberty (thelarche, pubarche and menarche) continues to be observed in adolescent girls with diabetes, even despite good metabolic control. Also, amenorrhea, hyperandrogenism, and symptoms consistent with polycystic ovary syndrome (PCOS) are typical manifestations of the disease. These disorders are believed to be caused by deficiency of natural insulin and, consequently, low concentrations of insulin-like growth factors (absence of puberty), but symptoms of hyperandrogenism may also be noted during intensive insulin therapy with high doses of exogenous insulin. The paper deals with puberty-related problems, menstrual disorders, and disturbances in the ovarian function in young women with type 1 diabetes.

Keywords: Type 1 diabetes, puberty, hyperandrogenism, polycystic ovary syndrome

[*] Corresponding author: E. Wender-Ożegowska. E-mail: ewaoz@post.pl.

INTRODUCTION

Diabetes influences the process of puberty in a specific way. Recent decades have witnessed the secular trend in healthy girls, manifested as earlier menarche and other elements of puberty compared to the previous centuries [1, 2] (Figure 1). The beginnings of diabetes treatment consisted in conventional insulin therapy, often with high levels of glycemia in the affected patients, elevated risk for acidosis, and fear of hypoglycemia in attempts to correct the doses of insulin regimen.

All of the above mentioned factors were the reason why patients had poor metabolic control and, in case of children and teenagers, caused serious disturbances, including a 2-year delay of the first menarche. Consequently, young girls with type 1 diabetes had a significantly delayed menarche, what was largely correlated with insufficient glycemic control [1]. These disorders were known as hypogonadism triggered by insulin deficiency.

Insulin deficiency results in the deficiency of insulin-like growth factors, indispensable stimulators in the production of the growth factor, which in turn stimulates the hypothalamic LH-peak that is vital for the onset of puberty.

Owing to advances in the management of the disease, i.e., intensive insulin therapy and personal insulin pumps, recent years have brought a notably improved metabolic control in diabetic children, what translates into an improved pubertal process, more similar to the normal healthy peers [3]. A study from Germany reported that thelarche in young girls with diabetes was delayed by approximately 6 months as compared to normal healthy controls. Regardless, the age of the onset of puberty in both groups was considered as normal [4]. These authors also reported that earlier menarche was observed in girls with higher body mass index (BMI) and lower levels of glycated hemoglobin, whereas delayed menarche correlated not only with the level of glycated hemoglobin but also the duration of diabetes [5]. Delayed menarche and slightly delayed thelarche have also been observed in young girls with diabetes [6, 7]. Importantly, delayed menarche applies to girls who were diagnosed with diabetes before the first menstruation [8, 9]. Some studies showed also a delay of between two and six months to reach the final stages of breast development and pubic hair growth among girls with type 1 diabetes [4].

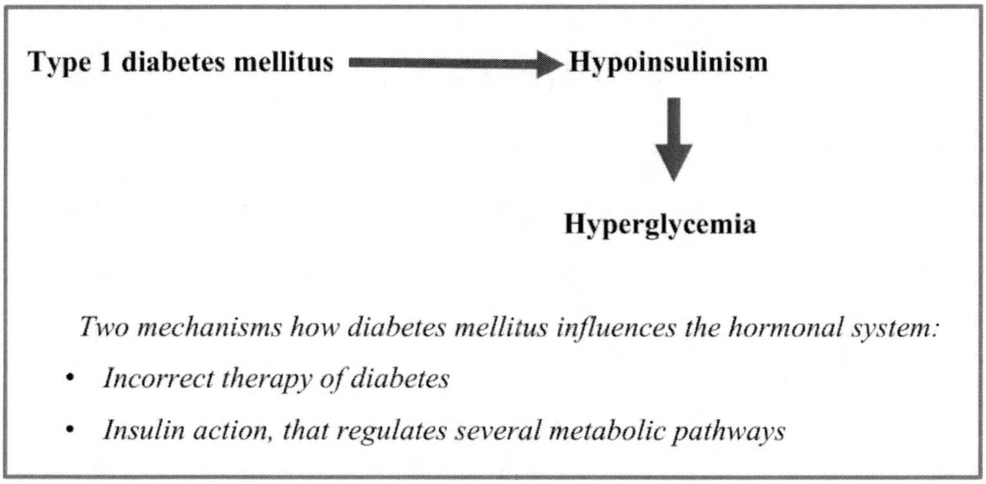

Figure 1. Influence of diabetes on the hormonal system.

Delayed menarche in diabetic girls is also associated with irregular menstrual cycles and lower mineral bone density later in life [10-12]. Schweiger et al. noticed that incident cases of hypoglycemia and acidosis do not significantly influence the time of menarche in girls with type 1 diabetes [9]. BMI and ethnic background have a significant impact on the time of first menstruation. Black girls have been observed to menstruate one year earlier than their Caucasian peers. A higher BMI among diabetic girls is associated with an earlier onset of puberty and age of menarche, similar to that observed in healthy girls [13-15] (Figure 2).

In case of teenagers with diabetes, apart from delayed menarche, longer menstrual cycles as compared to normal healthy women are typically observed. According to the recommendations of The American College of Obstetrics and Gynecology (ACOG) and The American Academy of Pediatrics (AAP), the duration of a normal menstrual cycle varies from 21 to 45 days during the pubertal period of 5 years since the menarche [16, 17].

Codner found that mean duration of the menstrual cycle in a group of adolescents with diabetes was 48.0 ± 38.9 days, as compared to normal healthy teenagers, whose cycles typically lasted 32.7 ± 8.6 days [18]. These authors also observed that diabetic girls manifest significantly greater variability of the menstrual cycles during a 6-month observation period than normal healthy adolescents (81% of the diabetic subjects had at least one disturbed menstrual cycle as compared to 59% of the healthy controls, OR = 5.9). Other authors observed a similar frequency of anovulatory cycles during the first year after menarche in T1D and healthy girls. However, during the second year, control girls experience a higher frequency of ovulatory cycles than T1D girls, suggesting that anovulatory cycles may cause menstrual abnormalities [7]. Menstruation-related disturbances, similarly to the onset of menarche, depend on the degree of metabolic control in the affected patients. Each event of a 1% increase in glycated hemoglobin has been proven to prolong the menstrual cycle by 1.5 days, what results in oligomenorrhoea in the group of diabetic women as compared to their healthy peers [19]. Longer menstrual cycles and higher variability in menstrual lengths are frequently observed in these patients. Although these symptoms become less severe after adolescence, menstrual irregularities should be included in the list of critical problems during adolescence for girls with T1D [18].

Even a couple of weeks of carbohydrate changes may cause the following complications among young type 1 diabetic women:

- Disturbances in neurosecretive hypothalamic GnRh production
- Impaired gonadal function
- Delayed menarche
- Disturbances in menstrual cycles: oligomenorrhea, polimenorrhea, amenorrhea secundaria
- Prolonged follicular phase
- Ovulation despite irregular cycles

Figure 2. Complications among type 1 diabetic women.

Other disturbances reported in girls with type 1 diabetes include hyperandrogenism, symptoms resembling PCOS, and earlier menopause in the future. Also, hypogonadism which accompanies poor glycemic control results in lower estradiol levels, what may favor the development of cardiovascular diseases in the perimenopausal period.

Notably, higher total and free testosterone levels that have been observed in T1D girls by the end of puberty may be associated with a higher risk of developing microalbuminuria or retinopathy [20, 21]. An ovarian source of this hyperandrogenism has been documented both, in adults and pubertal girls [22].

Nevertheless, the rate of ovulatory cycles is similar in teenagers with type 1 diabetes and poor metabolic control and healthy adolescents, even despite longer and irregular metabolic cycles in the former group [18]. The proportion of girls with anovulatory cycles was similar in both groups (19.4 and 19.2% in the T1D and control groups, respectively), although 35% of the girls with T1D had at least one cycle of oligomenorrhoea. Regression analyses showed that the presence of T1D did not significantly affect the ovulatory function. These data are consistent with studies evaluating the etiology of infertility, which have shown that T1D is a rare cause for consultation in infertility clinics [18].

In summary, frequent hormonal changes in adolescents with type 1 diabetes include higher serum levels of testosterone and androstenedione, with normal concentrations of SHBG, gonadotropin, estradiol and DHEA-S.

Normal SHBG levels result from the fact that insulin administered via the subcutaneous route is not subject to first-pass liver metabolism. Thus, free androgen binding is analogous to that in the general population, resulting in slightly intensified hirsutism, explained in part by the protection conferred by the normal sex-hormone-binding globulin (SHBG) levels observed in T1D women [22].

According to the literature, hyperandrogenism correlated with PCOS on ultrasound is the most common PCOS phenotype among women with type 1 diabetes, but the number of follicles and ovarian volume is intermediate between healthy controls and non-diabetic PCOS women [23]. T1D onset before the first menstruation has been observed to predispose to PCOS symptoms, as well as higher BMI, increased visceral fat mass, insulin resistance, and disturbances between GH and IGF [8, 24].

The Pathogenesis of Menstrual Disturbances

Apart from its main effectors, insulin receptors are also located in the hypothalamus, pituitary gland, ovaries, and the uterus. Thus, all disturbances in insulin concentrations implicate abnormal function of these organs.

Also, in young patients with insulin deficiency sudden weight loss before the onset of the disease, metabolic disturbances or acidosis may manifest as hypogonadism-like symptoms due to the fact that insulin deficiency impacts the decreased synthesis of the growth hormone as well as gonadotropin-releasing hormone (GnRH), which is responsible for, among others, the onset of puberty [25].

Decreased expression of kisspeptin is one of the factors which most probably influence the abnormal secretion of GnRH in diabetes during the pubertal period as it participates in the induction of GnRH synthesis, what may result in disturbed LH pulsatility [26, 27] (Figure 3).

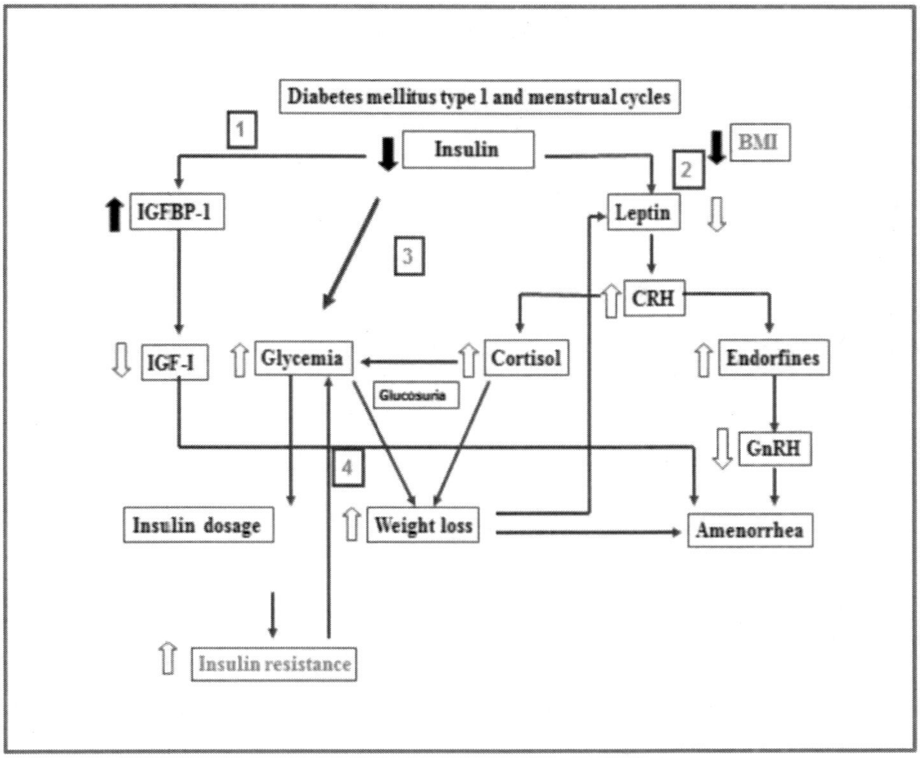

Figure 3. Mechanism of menstrual disturbances in type 1 diabetic adolescents.

On the other hand, owing to the availability of insulin pumps and intensive insulin therapy, metabolic control has greatly improved, what results in less frequent incidence of hypogonadism among adolescents with diabetes, but may induce hyprandrogenism and symptoms mirroring PCOS. In normal conditions, insulin produced by the pancreas enters the portal venous system and its metabolism is regulated by the liver. Depending on insulin concentration, the liver regulates carbohydrate metabolism, eliminating even up to 70% of the insulin produced by the pancreas [28].

In case of subcutaneous administration, the route through the portal venous system is omitted, resulting in high insulin levels in the peripheral circulation and excessive effect on a number of receptors, including in the ovaries. Insulin, affecting the insulin receptors located in thecal, granulosa, and stromal cells, acts like co-gonadotropin, both for Lutropin (LH) and Follicle-stimulating hormone (FSH). In the thecal cells, insulin contributes to synthesis activation of a number of enzymes responsible for steroidogenesis (CYP17, cytochrome p450, 3β steroid dehydrogenase), inducing hyperandrogenism which is a common occurrence in young diabetic girls [23, 29, 31]. Codner demonstrated higher androstenedione and dehydroepiandrosterone sulfate (DHEA-S) concentrations in a group of diabetic adolescents as compared to controls [31].

Folliculotropic activity of insulin is manifested as increased recruitment and growth of pre-ovulatory follicles, lowered apoptosis and atresia, resulting in PCOS-like ovaries [28, 29, 32, 33]. Insulin resistance, a symptom typical for type 2 diabetes and diagnosed more frequently in girls than in boys, is yet another factor that might disturb the menstrual cycle in adolescents with diabetes [34].

Insulin resistance may intensify symptoms of hyperandrogenism, leading to disturbed menstrual cycles, similarly to hyperglycemia, by increasing glucotoxicity and insulin resistance [34]. Hyperinsulinemia leads to higher sensitivity of the insulin and IGF-1 binding receptors in the ovary, increasing steroid synthesis and intensifying PCOS symptoms.

The levels of the anti-Müllerian hormone in young diabetic girls and women are not elevated, which is a common occurrence in PCOS. That fact might support the thesis that insulin acts in a way which is similar to FSH, and is responsible for follicle maturation and, consequently, ovulation. Numerous authors emphasize that ovulation occurs in adolescents with type 1 diabetes, despite PCOS-like ovaries [9].

The pathogenesis of disturbed menstruation in young girls with type 1 diabetes ought to include eating disorders, elements of depression, omission of insulin injection, and the consequent considerable weight fluctuations, what results in impaired metabolic control [35].

Conclusion

Adolescence and puberty is a critical period and many of the problems faced by female patients with diabetes and are of a greater magnitude than in males. The explanation for this difference is very complex and seems to be related to psychological factors and to interactions between estrogen levels, lack of insulin, insulin therapy, and the growth hormone and IGF axis. It may be the reason why girls with type 1 diabetes show abnormalities in their pubertal development and gonadal function during the second decade of life.

During the last few years, the reproductive manifestations observed in girls with T1D have evolved from signs caused by insulin deficiency and hypogonadism to those caused by insulin excess and insulin resistance. Current insulin treatment modalities seem to be associated with pubertal delay and menstrual disturbances. In addition, some type 1 diabetic girls may develop hyperandrogenism, which may evolve towards PCOS in adulthood. It could be speculated that insulin delivering in a more physiologic way, and the consequent better balance between insulin deficiency and insulin excess, in type 1 diabetic adolescents will reduce the current problems observed in pubertal T1D girls.

Also, due to the fact that ovulation is present in girls with T1D during adolescence, regardless of their metabolic control and the presence of menstrual irregularities, the importance of pregnancy prevention in all adolescents with T1D ought to be emphasized.

References

[1] Parent, A. S., Teilmann, G., Juul, A., Skakkebaek, N. E., Toppari, J., Bourguignon, J. P. The timing of normal puberty and the age limits of sexual precocity: variations around the world, secular trends, and changes after migration. *Endocr. Rev*. 2003; 24: 668–693.

[2] Wyshak, G., Frisch, R. E. Evidence for a secular trend in age of menarche. *N. Engl. J. Med*. 1982; 306: 1033–1035.

[3] Cameron, F. J., Wherrett, D. K. Care of diabetes in children and adolescents: controversies, changes, and consensus. *Lancet*. 2015; 385: 2096–106.

[4] Rohrer, T., Stierkorb, E., Heger, S., Karges, B., Raile, K., Schwab, K. O. et al. Diabetes-Patienten-Verlaufsdaten (DPV) Initiative. Delayed pubertal onset and development in German children and adolescents with type 1 diabetes: cross-sectional analysis of recent data from the DPV diabetes documentation and quality management system. *J. Endocrinol.* 2007; 157: 647–653.

[5] Strotmeyer, E. S., Steenkiste, A. R., Foley, T. P. Jr, Berga, S. L., Dorman, J. S. Menstrual cycle differences between women with type 1 diabetes and women without diabetes. *Diabetes Care*. 2003; 26: 1016–1021.

[6] Codner, E., Barrera, A., Mook-Kanamori, D., Bazaes, R. A., Unanue, N., Gaete, X. et al. Ponderal gain, waist-to-hip ratio, and pubertal development in girls with type-1 diabetes mellitus. *Pediatr. Diabetes*. 2004; 5: 182-9.

[7] Codner, E., Cassorla, F. Puberty and ovarian function in girls with type 1 diabetes mellitus. *Horm. Res*. 2009; 71: 12–21.

[8] Strotmeyer, E. S., Steenkiste, A. R., Foley, T. P. Jr, Berga, S. L., Dorman, J. S. Menstrual cycle differences between women with type 1 diabetes and women without diabetes. *Diabetes Care*. 2003; 26: 1016–1021.

[9] Schweiger, B., Klingensmith, G. J., Snell-Bergeon, J. K. Menarchal timing in type 1 diabetes through the last 4 decades. *Diabetes Care*. 2010; 33: 2521–2523.

[10] Kjaer, K., Hagen, C., Sando, S. H., Eshoj, O. Epidemiology of menarche and menstrual disturbances in an unselected group of women with insulin-dependent diabetes mellitus compared to controls. *J. Clin. Endocrinol. Metab*. 1992; 75: 524–529.

[11] Anai, T., Miyazaki, F., Tomoyasu, Matsuo, T. Risk of irregular menstrual cycles and low peak bone mass during early adulthood associated with age at menarche. *Pediatr. Int*. 2001; 43: 483–488.

[12] Ulrich, U., Pfeifer, T., Buck, G., Lauritzen, C. Osteopenia in primary and secondary amenorrhea. *Horm. Metab. Res*. 1995; 27: 432–435.

[13] Schweiger, B. M., Snell-Bergeon, J. K., Roman, R., McFann, K., Klingensmith, G. J. Menarche delay and menstrual irregularities persist in adolescents with type 1 diabetes. *Reproductive Biology and Endocrinology*. 2011; 9: 61.

[14] Adcock, C. J., Perry, L. A., Lindsell, D. R., Taylor, A. M., Holly, J. M., Jones, J. et al. Menstrual irregularities are more common in adolescents with type 1 diabetes: association with poor glycaemic control and weight gain. *Diabet. Med*. 1994; 11: 465–470.

[15] Danielson, K. K., Palta, M., Allen, C., D'Alessio, D. J. The association of increased total glycosylated hemoglobin levels with delayed age at menarche in young women with type 1 diabetes. *J. Clin. Endocrinol. Metab*. 2005; 90: 6466–6471.

[16] ACOG Committee Opinion No. 349. Menstruation in girls and adolescents: using the menstrual cycle as a vital sign. *Obstet. Gynecol*. 2006; 108: 1323–1328.

[17] American Academy of Pediatrics, Committee on Adolescence, American College of Obstetricians and Gynecologists, Committee on Adolescent Health Care. Menstruation in girls and adolescents: using the menstrual cycle as a vital sign. *Pediatrics*. 2006; 118: 2245–2250.

[18] Codner, E., Soto, N., Merino, P. M. Contraception and pregnancy in adolescents with type 1 diabetes: a review. *Pediatric Diabetes*. 2012; 13: 108–123.

[19] Gaete, X., Vivanco, M., Eyzaguirre, F. C., López, P., Rhumie, H. K., Unanue, N. et al. Menstrual cycle irregularities and their relationship with HbA1c and insulin dose in adolescents with type 1 diabetes mellitus. *Fertil. Steril*. 2010; 94: 1822–1826.

[20] Amin, R., Schultz, C., Ong, K., Frystyk, J., Dalton, R. N., Perry, L. et al. Low IGF-I and elevated testosterone during puberty in subjects with type 1 diabetes developing microalbuminuria in comparison to normoalbuminuric control subjects: the Oxford Regional Prospective Study. *Diabetes Care*. 2003; 26: 1456–1461.

[21] Haffner, S. M., Klein, R., Dunn, J. F., Moss, S. E., Klein, B. E. Increased testosterone in type I diabetic subjects with severe retinopathy. *Ophthalmology*. 1990; 97: 1270–1274.

[22] Roldan, B., Escobar-Morreale, H. F., Barrio, R., de la Calle, H., Alonso, M., Garcia-Robles, R. et al. Identification of the source of androgen excess in hyperandrogenic type 1 diabetic patients. *Diabetes Care*. 2001; 24: 1297–1299.

[23] Codner, E., Iniguez, G., Villarroel, C., Lopez, P., Soto, N., Sir-Petermann, T. et al. Hormonal profile in women with polycystic ovarian syndrome with or without type 1 diabetes mellitus. *J. Clin. Endocrinol. Metab*. 2007; 92: 4742–4746.

[24] Adcock, C. J., Perry, L. A., Lindsell, D. R., Taylor, A. M., Holly, J. M., Jones, J. et al. Menstrual irregularities are more common in adolescents with type 1 diabetes: association with poor glycaemic control and weight gain. *Diabet. Med*. 1994; 11: 465-470.

[25] Bruning, J. C., Gautam, D., Burks, D. J., Gillette, J., Schubert, M., Orban, P. C. et al. Role of brain insulin receptor in control of body weight andreproduction. *Science*. 2000; 289: 2122–2125.

[26] Castellano, J. M., Navarro, V. M., Roa, J., Pineda, R., Sánchez-Garrido, M. A., García-Galiano, D. et al. Alterations in Hypothalamic KiSS-1 System in Experimental Diabetes: Early Changes and Functional Consequences. *Endocrinology*. 2009; 150: 784–794.

[27] Sherman, L. S., Rogers, D. G., Gabbay, K. H. Pulsatility of luteinizing hormone during puberty is dependent on recent glycemic control. *Adolesc. Paediatr. Gynecol*. 1991; 4: 87–92.

[28] Polonsky, K. S., Given, B. D., Hirsch, L., Shapiro, E. T., Tillil, H., Beebe, C. et al. Quantitative study of insulin secretion and clearance in normal and obese subjects. *J. Clin. Invest*. 1988; 81: 435–441.

[29] Poretsky, L., Kalin, M. F. The gonadotropic function of insulin. *Endocr. Rev*. 1987; 8: 132–141.

[30] Poretsky, L., Cataldo, N. A., Rosenwaks, Z., Giudice, L. C. The insulin-related ovarian regulatory system in health and disease. *Endocr. Rev*. 1999; 20: 535–582.

[31] Baillargeon, J. P., Nestler, J. E. Commentary: polycystic ovary syndrome: a syndrome of ovarian hypersensitivity to insulin? *J. Clin. Endocrinol. Metab*. 2006; 91: 22–24.

[32] Codner, E., Iniguez, G., Villarroel, C., Lopez, P., Soto, N., Sir-Petermann, T. et al. Hormonal profile in women with polycystic ovarian syndrome with or without type 1 diabetes mellitus. *J. Clin. Endocrinol. Metab*. 2007; 92: 4742–4746.

[33] Hsueh, A. J., Billig, H., Tsafriri, A. Ovarian follicle atresia: a hormonally controlled apoptotic process. *Endocr. Rev*. 1994; 15: 707–724.

[34] Kezele, P. R., Nilsson, E. E., Skinner, M. K. Insulin but not insulin-like growth factor-1 promotes the primordial to primary follicle transition. *Mol. Cell. Endocrinol.* 2002; 192: 37–43.

[35] Szadkowska, A., Pietrzak, I., Mianowska, B., Bodalska-Lipinska, J., Keenan, H. A. et al. Insulin sensitivity in type 1 diabetic children and adolescents. *Diabet. Med.* 2008; 25: 282–288.

In: Alterations in Glucose Homeostasis in Children
Editor: E. Otto-Buczkowska
ISBN: 978-1-63483-861-0

Chapter 13

SECONDARY DIABETES

E. Otto-Buczkowska*[*] *and A. Chwalba
Medical Specialist Centre in Gliwice,
Gliwice, Poland

ABSTRACT

The pancreas is a dual organ with exocrine and endocrine functions. The interrelationship of the endocrine-exocrine parts of the pancreas is a complex one. Secondary diabetes may occur in association with endocrine or non-endocrine disorders.

The most common endocrine diseases associated with glucose intolerance are those involving overproduction of counterregulatory hormones. Non-endocrine conditions associated with glucose intolerance or diabetes fall into three general categories: pancreatic diseases, drug-induced diabetes and genetic syndromes.

Cystic fibrosis (CF) is the most common recessive autosomal disease. Cystic fibrosis-related diabetes (CFRD) is considered the most common complication associated with CF. It can appear as early as infancy or adolescence. Many drugs and hormones can impair insulin secretion and action. They may not cause diabetes by themselves, but they may precipitate diabetes in individuals with insulin resistance or alterations of blood glucose homeostasis.

The therapy for secondary diabetes centers on the correction of the underlying disturbance. Patients with fasting hyperglycemia should be treated with an understanding of the pathophisiologic basis of their diabetes.

Keywords: Secondary diabetes, deficient insulin secretion, insulin resistance, endocrine disorders, overproduction of counterregulatory hormones, genetic syndromes, pancreatic diseases, drug-induced diabetes

---

[*] Corresponding author: E. Otto-Buczkowska. E-mail: em.buczkowski@pro.onet.pl.

Introduction

Secondary diabetes is a disorder of glucose tolerability that is associated with other illnesses. The question of the secondary diabetes diagnosis and treatment has been given particular prominence in relation to tougher offset criteria of carbohydrates metabolism.

There are many morbidities that may be accompanied by a disorder of glucose homeostasis with different degrees of severity, among them the following can be listed:

- endocrinopathy
- pancreatic diseases
- medication-induced diabetes
- diabetes associated with genetic syndromes [1]

Endocrinopathy

Certain types of diabetes arise during endocrinological diseases proceeding with overproduction of contrregulatory hormones, e.g., growth hormone, cortisol, glucagon, epinephrine which antagonize insulin action. There are differential mechanisms leading to glucose intolerability in disorders proceeding with hormones overproduction [2-7].

Increased concentrations of hormones may influence glucose production by stimulation of glycogenolysis and/or gluconeogenesis. They can also suppress glucose utilization by affecting insulin secretion or its effect. A number of endocrine diseases with overexpression of hormones can be associated with secondary diabetes, e.g., acromegaly, Cushing's syndrome, glucagonoma, pheochromocytoma, somatostatinoma, hyperthyroidism, primary hiperaldosteronism, hyperprolactinemia. These conditions will be discussed in chapter 10.

Diseases of the Exocrine Pancreas

Any process that diffusely injures the pancreas can cause diabetes. Among such the following can be listed: pancreatitis, pancreatectomy, trauma and pancreatic carcinoma. Mechanisms that lead to diabetes in this type of disorders are numerous [8-11]. Diabetes secondary to pancreatic diseases is commonly referred to as pancreatogenic diabetes or type 3c diabetes mellitus. In contrast to the management of type 1 or type 2 diabetes mellitus, the endocrinopathy in type 3c is very complex.

Pancreatitis

Hyperglycemia in acute pancreatitis is usually temporary but can be severe. Chronic pancreatitis can be associated with diabetes in 60% of patients. Patients with pancreatitis demonstrate variable diminution of blood insulin level. This fall of insuline level correlates with an increase of pancreatic enzymes levels in acute pancreatitis and with decrease of exocrine function in chronic pancreatitis. Carbohydrate intolerability is usually a symptom of a decrease in insulin secretion. Patients with chronic pancreatitis often demonstrate normal basic blood insuline level but reduced insulin response to arginine and glucagon stimulation.

Experimental studies suggest that this abnormal response and, in effect, deficiency of the circulating insulin, can be correlated with fibrous changes (acinar fibrosis).

Insulin shortage might not be the factor responsible for progress of secondary dabetes in pancreatitis. Insulin resistance of the liver is also influential and can be enhanced by pancreatic polypeptide deficiency.

Many patients are misdiagnosed. The course of the disease is complicated by additional comorbidities such as maldigestion and concomitant qualitative malnutrition. The failure to correctly diagnose type 3 diabetes mellitus leads to a failure to implement an appropriate medical therapy [12, 13].

Pancreatectomy

The development or worsening of diabetes after pancreatic resection is extremely common, with different types of resections conveying different risks for disease progression.

Partial or complete pancreatic resection can be dictated by a neoplastic process, polycystic pancreas disease or can be a result of treatment of recurring inflammations.

The incidence of new-onset diabetes mellitus after pancreatic resection increases as the follow-up period after surgery becomes longer and is related to the progression of underlying disease, the type of surgery, and the extent of resection. The pathophysiology of pancreatogenic diabetes is related to pancreatic hormone deficiency and the altered responses of the liver and peripheral organs to lower than normal hormone levels [14].

The reduction of exocrine and endocrine pancreatic functions occurs rarely if the organ is removed completely or in more than 75%. If the endocrine secretion of all types of cells is reduced, it causes not only the deficiency of insulin secretion but also the shortage of pancreatic glucagon, somatostatine and pancreatic peptide (PP). Because of reduced pancreatic peptide level, despite insulin shortage, the risk of katoacidosis occurrence in these patients is definitely lower.

The evident reduction of insulin secretion is a main cause of postoperative diabetes. According to studies by Burkhart et al., patients with distal pancreatectomies (DP) are at a greater risk for the development of new-onset postoperative diabetes when compared to pancreaticoduodenectomies (PD) [15].

Patients with preoperative diabetes are more likely to experience worsening of their disease after PD as compared to DP.

The evaluation of pancreatectomy influence on postoperative course of diabetes was also performed by other investigators [16].

Hemochromatosis

Iron homeostasis is closely related to many metabolic processes. Perturbations in iron homeostasis are associated with several conditions, including hyperglycemia and diabetes. Hereditary hemochromatosis, transfusion-dependent thalassemia, and excess heme iron intake can increase the risk of developing diabetes [17].

Clinically, diabetes or glucose intolerance occurs in 75-90% of patients with primary hemochromatosis and in more than 65% patients with secondary hemochromatosis. Diabetes in primary as well as in secondary hemochromatosis frequently requires insulin administration and ketosis can occur in its course.

There are three factors that contribute to abnormal glucose tolerability in hemochromatosis: cirrhosis, iron precipitations in the pancreas and potential coexistence of primary diabetes. Presence of cirrhotic changes increases the probability of abnormal glucose metabolism.

Iron content in the liver, ferritin level in plasma or the size of liver damage correlate with the disturbance of glucose homeostasis. Family history of diabetes is a very reliable prognostic factor of glucose intolerance occurrence in patients with primary hereditary diabetes. Insulin resistance of the liver plays an important role in both types of hemochromatosis.

The insufficiency of the first phase of insulin secretion is observed even without significant iron deposits in pancreatic islets. Glucose and insulin tolerance tests as well as euglycemic clamp tests suggest that in this syndrome, the deficit of insulin secretion is more important than insulin resistance. This diminution of insulin secretion may be preceded by insulin resistance and hyperinsulinemia. Tightly controlled iron metabolism is essential for insulin secretion and insulin sensitivity.

Cystic Fibrosis (CF)

Cystic fibrosis-related diabetes (CFRD) is the most common comorbidity in the CF population. The first definitive reports of glucose intolerance or diabetes in CF are from 1955. The combination of CF and related diabetes (CFRD) has a negative impact on survival.

Recognition of glycemic abnormalities through aggressive screening has led to improvements in nutritional status, pulmonary function, and mortality rates [18].

CFRD is a unique type of diabetes. These patients have clinical features of both type 1 and type 2 diabetes. Although CFRD is different from type 1 and 2 DM, the development of diabetes-induced complications is similar to other types of diabetes.

However, hyperglycemia and/or insulin deficiency in people with CFRD may exacerbate underlying CF problems, such as decreased pulmonary function and weight loss [19].

The etiology of CFRD is complex, likely consisting of a combination of insulin deficiency and insulin resistance. A genetic predisposition towards the development of diabetes is an additional risk factor in CF patients.

Hepatogenous Diabetes

It is due to the hepatocellular functional loss and insulin resistance. Increased insulin resistance is frequently associated with chronic liver disease and is a pathophysiological feature of hepatogenous diabetes.

Impaired glucose tolerance is common in cirrhosis hepatis [20]. Pancreatic congestion present in cirrhotic patients and insulin secretion were correlated and could be a key factor underlying the development of hepatogenous diabetes in cirrhotic patients [21].

Diabetes is frequently diagnosed in patients with cirrhosis and represents an important risk factor for morbidity and mortality [22]. Pharmacological therapy is limited due to hepatotoxicity and the risk of hypoglycemia [23].

Non-alcoholic fatty liver disease (NAFLD) is a chronic liver disease which increasingly affects children [24]. Type 2 diabetes mellitus (T2DM), insulin resistance (IR), obesity, metabolic syndrome and NAFLD are particularly closely related.

MEDICATION-INDUCED DIABETES

Many drugs and hormones can impair insulin secretion and action. They might not cause diabetes by themselves, but they may precipitate diabetes in individuals with insulin resistance or alterations of blood glucose homeostasis [25, 26]. These substances may unveil pre-existing concealed defects of insulin secretion or insulin resistance and lead to glucose intolerance in previously undiagnosed patients or an exacerbation of diabetes, if they are administered to patients with pre-diagnosed diabetes.

Steroids are a group of diabetogenic drugs that are often used, especially in adolescent patients [27]. Interesting observations were presented by Yucel et al., who ascertained that asthmatic children on low doses of inhaled steroids had had higher HbA1c values than healthy children [28]. Kari et al. reported steroid resistant nephrotic syndrome with almost simultaneous onset of IDDM in young children [29].

The diabetogenic effect of steroids has been known for a long time, however the hyperglycemia caused by those drugs is usually not very high and rarely leads to ketosis. There are various mechanisms of adrenal steroid influence on glucose homeostasis.

An inflow of substrates for gluconeogenesis is correlated with a catabolic effect of adrenal steroids in muscle tissue (glucogenic amino acids) and in adipose tissue (glycerol, fatty acids). During treatment, there is a decrease of peripheral glucose consumption. The influence of adrenal steroids on β cell functions must not be ignored.

It is worth to mention that glucocorticoids also work synergistically with other gluconeogenic hormones, e.g., epinephrine or glucagon.

In many situations, drugs from this group are administered ad hoc or chronically in small doses. In such situations, generally, disorders of glucose homeostasis are not severe and may self-adjust. In some cases, adrenal steroids are applied in big doses and for long periods of time. Then, serious chronic disorders of carbohydrate metabolism have to be considered. They may occur as clinically overt diabetes, which is not so unusual, or essential disorders of glucose tolerance.

As mentioned before, because of the necessity of macro- and microangiopathy preventive treatment, these states absolutely require monitoring and treatment implementation.

Infections

Certain viruses have been associated with β cell destruction. Diabetes occurs in patients with congenital rubella, although most of these patients have HLA and immune markers characteristic of type 1 diabetes. In addition, Coxsackie virus B, cytomegalovirus, adenovirus, and mumps have been implicated in inducing certain types of the disease.

Diabetes in Patients with Malignant Tumors

Disturbances of carbohydrate metabolism are fairly common is patients with malignancy [30, 31]. Lengyel et al. presented an overview of literature concerning pathophysiological and clinical connections between diabetes and malignant diseases [32].

Malignant diseases and therapies used in their treatment often have an impact on carbohydrate metabolism. Some of the medications used in therapy, such as steroids or parenteral nutrition, may raise blood glucose levels. Exposure to high doses of corticosteroids increases the likelihood of developing diabetes [33].

Complications of Secondary Diabetes

For the last 20 years, the question whether vascular complications of diabetes are directly caused by metabolism disorders present in diabetic patients or are they secondary to other genetic faults has been discussed. Patients with secondary glucose intolerance in diseases with an overproduction of hormones typically do not show vascular complications, although clinically apparent diabetes occurs in about 20% of such patients. That is because survival of a patient who is not treated for the primary disease for a period of time sufficient for progression of chronic diabetic complications is unlikely. Among patients with pancreatitis and patients after pancreatectomy, the fluorescein angiography shows retinopathy in 7-30%. Typically, it is of a concealed type (background) and the proliferative retinopathy is very rare in this group. In patients with hemochromatosis and diabetes, retinopathy occurs in 20-30%. Therapeutic decrease of iron deposits is correlated with progression in carbohydrate tolerance in more than 40% of these patients. It has not been explained yet whether iron overload has independent influence on the progress of microangiopathy.

Treatment of Secondary Diabetes

Treatment of the primary disease is essential in secondary diabetes. In cases when treatment is not or can not be performed, patients with hyperglycemia should be treated according to the patophysiological background of their diabetes.

For patients with acromegaly or Cushing syndrome, who are patients who have preserved insulin secretion, diet treatment or oral antidiabetic drug treatment can be effective. Sometimes, in this group, patients require insulin therapy to achieve normoglycemia.

Patients with pheochromocytoma or glucagonoma, who have severe disorders of insulin secretion, require insulin therapy to normalize glycemia.

Insulin replacement is indispensable for patients after pancreatectomy and patients with chronic pancreatitis. It may also be necessary for patients with acute pancreatitis.

Insulin treatment is also the medical therapy recommended for CFRD [34-36]. In recent years, research has been conducted on the use of oral drugs in the treatment of CFRD. The latest Cochrane report titled "Insulin and oral agents for managing CFRD" stated that further studies are needed to establish whether there is clear benefit from the use of hypoglycemic agents [37].

In steroid therapy, the mode of treatment depends on the intensity of disorders. In extreme cases, insulin therapy is necessary – in others diet treatment or potential oral antidiabetic drugs may be sufficient. The decision should be based on the nature of glucose tolerance disorder.

Conclusion

The objective of treatment has to be the achievement of metabolic balance. The introduction of an effective treatment must not be postponed, under any circumstances. Therapy has to be chosen individually, considering the relevance of nutritional habits and lifestyle revision.

References

[1] American Diabetes Association. Diagnosis and classification of diabetes mellitus. *Diabetes Care*. 2014; 37 Suppl. 1: S81-90.

[2] Balbach, L., Wallaschofski, H., Völzke, H., Nauck, M., Dörr, M., Haring, R. Serum prolactin concentrations as risk factor of metabolic syndrome or type 2 diabetes? *BMC Endocr. Disord*. 2013; 13: 12.

[3] Colao, A., De Block, C., Gaztambide, M. S., Kumar, S., Seufert, J., Casanueva, F. F. Managing hyperglycemia in patients with Cushing's disease treated with pasireotide: medical expert recommendations. *Pituitary*. 2014; 17: 180-186.

[4] Lekva, T., Bollerslev, J., Sahraoui, A., Scholz, H., Bøyum, H., Evang, J. A. et al. Thioredoxin interacting protein is a potential regulator of glucose and energy homeostasis in endogenous Cushing's syndrome. *PLoS One*. 2013; 8: e64247.

[5] Peppa, M., Koliaki, C., Nikolopoulos, P., Raptis, S. A. Skeletal muscle insulin resistance in endocrine disease. *J. Biomed. Biotechnol*. 2010; 2010: 527850.

[6] Resmini, E., Minuto, F., Colao, A., Ferone, D. Secondary diabetes associated with principal endocrinopathies: the impact of new treatment modalities. *Acta Diabetol*. 2009; 46: 85-95.

[7] Shimon, I. Screening for Cushing's syndrome: Is it worthwhile? *Pituitary*. 2015; 18: 201-205.

[8] Chakrabarti, S. K., Biswas, D., Chaudhury, S., Jain, R., Kataria, M., Ghosh, S. Diabetes due to recurrent pancreatitis secondary to hypercalcemia due to primary hyperparathyroidism. *Indian J. Endocrinol. Metab*. 2013; 17(Suppl. 1): S212-214.

[9] Chen, N., Unnikrishnan, I. R., Anjana, R. M., Mohan, V., Pitchumoni, C. S. The complex exocrine-endocrine relationship and secondary diabetes in exocrine pancreatic disorders. *J. Clin. Gastroenterol*. 2011; 45: 850-861.

[10] Choudhuri, G., Lakshmi, C. P., Goel, A. Pancreatic diabetes. *Trop. Gastroenterol*. 2009; 30: 71-75.

[11] Ewald, N., Bretzel, R. G. Diabetes mellitus secondary to pancreatic diseases (Type 3c) – are we neglecting an important disease? *Eur. J. Intern. Med*. 2013; 24: 203-206.

[12] Ewald, N., Hardt, P. D. Diagnosis and treatment of diabetes mellitus in chronic pancreatitis. *World J. Gastroenterol*. 2013; 19: 7276-7281.

[13] Rickel, M. R., Bellin, M., Toledo, F. G., Robertson, R. P., Andersen, D. K., Chari, S. T. et al. Detection, evaluation and treatment of diabetes mellitus in chronic pancreatitis: recommendations from PancreasFest 2012. *Pancreatology*. 2013; 13: 336-342.

[14] Maeda, H., Hanazaki, K. Pancreatogenic diabetes after pancreatic resection. *Pancreatology*. 2011; 11: 268-276.

[15] Burkhart, R. A., Gerber, S. M., Tholey, R. M., Lamb, K. M., Somasundaram, A., McIntyre, C. A. et al. Incidence and Severity of Pancreatogenic Diabetes After Pancreatic Resection. *J. Gastrointest. Surg*. 2015; 19: 217-225.

[16] Hirata, K., Nakata, B., Amano, R., Yamazoe, S., Kimura, K., Hirakawa, K. Predictive factors for change of diabetes mellitus status after pancreatectomy in preoperative diabetic and nondiabetic patients. *J. Gastrointest. Surg*. 2014; 18: 1597-603.

[17] Wang, X., Fang, X., Wang, F. Pleiotropic actions of iron balance in diabetes mellitus. *Rev. Endocr. Metab. Disord*. 2015; 16: 15-23.

[18] Nathan, B. M., Laguna, T., Moran, A. Recent trends in cystic fibrosis-related diabetes. *Curr. Opin. Endocrinol. Diabetes Obes*. 2010; 17: 335-341.

[19] Hardin, D. S. Pharmacotherapy of diabetes in cystic fibrosis patients. *Expert Opin. Pharmacother*. 2010 Apr; 11(5): 771-8.

[20] Kawaguchi, T., Taniguchi, E., Itou, M., Sakata, M., Sumie, S., Sata, M. Insulin resistance and chronic liver disease. *World J. Hepatol*. 2011; 3: 99-107.

[21] Kuroda, T., Hirooka, M., Koizumi, M., Ochi, H., Hisano, Y., Bando, K. et al. Pancreatic congestion in liver cirrhosis correlates with impaired insulin secretion. *J. Gastroenterol*. 2014 Oct. 5.

[22] Jeon, H. K., Kim, M. Y., Baik, S. K., Park, H. J., Choi, H., Park, S. Y. et al. Hepatogenous diabetes in cirrhosis is related to portal pressure and variceal hemorrhage. *Dig. Dis. Sci*. 2013; 58: 3335-3341.

[23] Gundling, F., Seidl, H., Strassen, I., Haller, B., Siegmund, T., Umgelter, A. et al. Clinical manifestations and treatment options in patients with cirrhosis and diabetes mellitus. *Digestion*. 2013; 87: 75-84.

[24] Firneisz, G. Non-alcoholic fatty liver disease and type 2 diabetes mellitus: the liver disease of our age? *World J. Gastroenterol*. 2014; 20: 9072-9089.

[25] Regelmann, M. O., Greig, F., Kerkar, N., Rapaport, R. Risk factors for medication-induced diabetes mellitus. *J. Pediatr*. 2012; 160: 355; author reply 356.

[26] Yeshayahu, Y., Koltin, D., Hamilton, J., Nathan, P. C., Urbach, S. Medication-induced diabetes during induction treatment for ALL, an early marker for future metabolic risk? *Pediatr. Diabetes*. 2015; 16: 104-108.

[27] Koltin, D., Sung, L., Naqvi, A., Urbach, S. L. Medication induced diabetes during induction in pediatric acute lymphoblastic leukemia: prevalence, risk factors and characteristics. *Support. Care Cancer*. 2012; 20: 2009-2015.

[28] Yucel, O., Eker, Y., Nuhoglu, C., Ceran, O. Hemoglobin A1c levels in children with asthma using low dose inhaled corticosteroids. *Indian Pediatr*. 2009; 46: 300-303.

[29] Kari, J. A., El-Desoky, S. M., Mokhtar, G., Jalalah, S. M. Simultaneous onset of steroid resistant nephrotic syndrome and IDDM in two young children. *BMJ Case Rep*. 2010.

[30] Bizzarri, C., Bottaro, G., Pinto, R. M., Cappa, M. Metabolic syndrome and diabetes mellitus in childhood cancer survivors. *Pediatr. Endocrinol. Rev*. 2014; 11: 365-373.

[31] Godsland, I. F. Insulin resistance and hyperinsulinaemia in the development and progression of cancer. *Clin. Sci*. 2010; 118: 315-332.

[32] Lengyel, Z., Boér, K., Halászlaki, C., Németh, Z. Diabetes in patients with malignant tumors. *Magy Onkol*. 2013; 57: 177-181.

[33] Majhail, N. S., Challa, T. R., Mulrooney, D. A., Baker, K. S., Burns, L. J. Hypertension and diabetes mellitus in adult and pediatric survivors of allogeneic hematopoietic cell transplantation. *Biol. Blood Marrow Transplant*. 2009; 15: 1100-1107.

[34] Bridges, N. Diabetes in cystic fibrosis. *Paediatr. Respir. Rev*. 2013; 14 Suppl. 1: 16-18.

[35] Hameed, S., Jaffé, A., Verge, C. F. Cystic fibrosis related diabetes (CFRD) – the end stage of progressive insulin deficiency. *Pediatr. Pulmonol*. 2011; 46: 747-760.

[36] Onady, G. M., Stolfi, A. Insulin and oral agents for managing cystic fibrosis-related diabetes. *Cochrane Database Syst. Rev*. 2013; 7: CD 004730.

[37] Ballmann, M., Hubert, D., Assael, B. M., Kronfeld, K., Honer, M., Holl, R. W.; CFRD Study group. Open randomised prospective comparative multi-centre intervention study of patients with cystic fibrosis and early diagnosed diabetes mellitus. *BMC Pediatr*. 2014; 14: 70.

In: Alterations in Glucose Homeostasis in Children
Editor: E. Otto-Buczkowska
ISBN: 978-1-63483-861-0

Chapter 14

CYSTIC FIBROSIS-RELATED DIABETES

E. Otto-Buczkowska** *and A. Chwalba
Medical Specialist Centre in Gliwice, Gliwice, Poland

ABSTRACT

Diabetes mellitus has evolved as a complication because of increased longevity of patients with cystic fibrosis (CF). CF-related diabetes (CFRD) is associated with increased morbidity and mortality – therefore, prompt diagnosis and aggressive management are important. CFRD develops insidiously.

Cystic fibrosis patients demonstrate a spectrum of pancreatic β cell abnormalities. The major alteration in insulin secretion and insulin sensitivity of CF patients arises from the slowly declining β cell function resulting in delayed and reduced responsiveness to hyperglycemia. Insulin secretion is significantly decreased in CF-related diabetes versus patients with impaired glucose tolerance (CF-IGT) and normal glucose tolerance (CF-NGT). In CF patients with normal glucose tolerance, the disruption of β cell function may be compensated by increased insulin sensitivity.

In cystic fibrosis patients, it is important to evaluate the kinetics of insulin secretion and insulin sensitivity.

CFRD shares features of both type 1 and type 2 diabetes. The primary cause of CFRD is insulin deficiency, especially first-phase insulin secretion.

Insulinopenia plays a prominent role in the pathogenesis of glucose intolerance and subsequent development of CFRD, but patients also show a variable degree of insulin resistance that may worsen in some clinical conditions – such as malnutrition, steroid therapy and/or chronic pulmonary infection. Insulin therapy is the preferred treatment.

Keywords: CF-related diabetes, CFRD, impaired glucose tolerance, insulin resistance, epidemiology of CFRD, treatment of CFRD

* Corresponding author: E. Otto-Buczkowska. E-mail: em.buczkowski@pro.onet.pl.

INTRODUCTION

The improved life expectancy of patients affected by cystic fibrosis (CF) has resulted in an increasing prevalence of CF-related diabetes (CFRD). The nature of the glucose homeostasis disorder can be various and depends on potential disturbance of insulin secretion as well as on peripheral insulin resistance [1-4].

The most common comorbidity associated with CF is cystic fibrosis-related diabetes (CFRD). It occurs in about 2% of children, 19% of adolescents and 40–50% of adults [5].

During the last few years, the average survival rate of patients with cystic fibrosis (CF) has significantly improved. Complications such as cystic fibrosis related diabetes mellitus (CFRD) develop with age, and the median age at diagnosis is 21 years. The prevalence of CFRD increases progressively [6]. Children with CF have benefited from advances in medical and nutritional treatments, and this can be observed from the improvement in the survival of these patients. Because CFRD is often asymptomatic, annual screening for CFRD in CF patients is recommended at 10 years of age and onward, using the oral glucose tolerance test. Early detection of CFRD is important since weight loss, protein catabolism, lung function decline, and mortality are increased in CFRD [7, 8].

This combination of diseases is genetically determined [9-13].

The first definitive reports of glucose intolerance or diabetes in CF are from 1955. The combination of CF and related diabetes (CFRD) has a negative impact on survival. CFRD is now the most common complication of CF.

PATHOGENESIS OF CFRD

CFRD pathogenesis has not been explained yet. In order to understand these mechanisms, experimental investigations have been conducted. Recently published studies by Uc et al. show that early pathogenesis of CFRD is determined by functional islet abnormalities, independent of structural islet loss [14].

In experimental studies, the abnormalities in islets of Langerhans function in CF were diagnosed even in the neonatal period [15].

Battezzati et al. in their studies proved that in CF patients, β cell function declines slowly but, for a certain period of time, this regression is compensated by increased insulin sensitivity which suppresses the occurrence of hyperglycemia [16].

Interesting research on the role of lymphocyte T-helper 17 (Th17) dysfunction in the pathogenesis of CFRD was recently presented [17]. The Th17 pathway is involved in CF lung inflammation and β cell destruction. Waugh et al., based on an analysis of literature, ascertained that CF-related diabetes (CFRD) is caused by the damage of the pancreas which, over time, loses its capacity to produce a sufficient amount of insulin [18].

EPIDEMIOLOGY OF CFRD

Comprehensive literature overview regarding the frequency of CFRD occurrence in children and adolescents was presented by O'Riordan et al. [19].

It was formerly claimed that diabetes in CF patients unveils rarely before the age of 10 years. However, CRFD is diagnosed in more and more younger children and even infants [20]. Casas et al. described a diagnosis of diabetes in a 2-year-old child, successfully treated with insulin [21]. Siahanidou et al. published a case of a neonate with CF who presented with hyperglycemia in the diabetic range and marked insulinopenia. Insulinopenia and impaired glucose tolerance were permanent findings at 6 and 15 months of age [22].

The nature of glucose homeostasis disorder may be various and dependent on potential disorders of insuline secretion as well as on peripheral insulin sensitivity [23-25].

Insulin deficiency is the primary defect in CFRD, but insulin resistance and impairment of the enteroinsular axis play contributory roles [4]. The results of studies on the occurrence of impaired glucose tolerance (CF-IGT) and CF-related diabetes in adolescent patients were presented by other authors [26-28].

Schmid et al., based on the analysis of oral glucose tolerance tests (OGTT) performed between 2001 and 2009 on CF patients, ascertained that the disorder of glucose tolerance, regardless of the level of intensity (IFG or IGT as well as INDET: indeterminate glucose tolerance - patients with a 1-hour glucose level >11.1 mmol/L), were predictors of future CFRD. Authors noticed that OGTT in patients with CF should include 1-hour post-challenge value [32].

Patients with CF have reduced insulin secretion during an OGTT, regardless of exocrine pancreatic status [33]. In CFRD diagnostics, continuous glucose monitoring systems (CGMS) are more and more often recommended as a useful tool for early detection of hyperglycemia in a CF patient [4, 34]. In opinion of these authors, the standard oral glucose tolerance test (OGTT) may appear not sensitive enough to detect early abnormalities in blood glucose regulation for patients with CF.

Similar phenomena were studied by other investigators who state that CGMS is a reliable and valid device for the evaluation of glucose metabolism in children with CF [35].

Based on retrospective studies, Martín-Frías et al. concluded that in order to diagnose CF related diabetes or impaired glucose tolerance, the screening for abnormal glucose tolerance (AGT) should be performed on prepubertal CF patients [36].

Waugh et al. noticed that CFRD diagnosis criteria have to be more strict in relation to the ones applied in diagnostics of other types of diabetes [37].

In opinion of other investigators, CFRD should be defined according to the level at which pulmonary harm occurs, and not by the same thresholds of PG as are used for T1D and T2D.

TREATMENT OF CFRD

It is generally accepted that early insulin therapy is the treatment by choice in CFRD. It allows to maintain correct body weight as well as to protect lung tissue [21, 38]. Van Sambeek et al. believe that while current clinical care guidelines for cystic fibrosis-related diabetes target HbA1c ≤7% to limit long-term microvascular damage, more stringent glycemic control (HbA1c ≤6.5%) may further reduce the short-term pulmonary complications [39]. According to accepted rules, CFRD patients typically require 0.5–0.8 units insulin per kg body weight per day when they are in their usual state of health. Much more may be required during stress [40].

In recent years, there has been an increasing amount of research on the use of oral drugs in the treatment of CFRD [41, 42].

REFERENCES

[1] Ode, K. L., Frohnert, B., Laguna, T., Phillips, J., Holme, B., Regelmann, W. et al. Oral glucose tolerance testing in children with cystic fibrosis. *Pediatr. Diabetes*. 2010; 11: 487-492.

[2] Rana, M, Munns, C. F., Selvadurai, H., Donaghue, K. C., Craig, M. E. Cystic fibrosis-related diabetes in children – gaps in the evidence? *Nat. Rev. Endocrinol*. 2010; 6: 371-378.

[3] Zirbes, J., Milla, C. E. Cystic fibrosis related diabetes. *Paediatr. Respir. Rev*. 2009; 10: 118-123.

[4] Brennan, A. L., Beynon, J. Clinical updates in cystic fibrosis-related diabetes. *Semin. Respir. Crit. Care Med*. 2015 Apr; 36(2): 236-50.

[5] Scheuing, N., Holl, R. W., Dockter, G., Fink, K., Junge, S., Naehrlich, L. et al. Diabetes in cystic fibrosis: multicenter screening results based on current guidelines. *PLoS One*. 2013; 8: e81545.

[6] Lek, N., Acerini, C. L. Cystic fibrosis related diabetes mellitus – diagnostic and management challenges. *Curr. Diabetes Rev*. 2010; 6: 9-16.

[7] Konrad, K., Thon, A., Fritsch, M., Fröhlich-Reiterer, E., Lilienthal, E., Wudy, S. A. et al. Comparison of cystic fibrosis-related diabetes with type 1 diabetes based on a German/Austrian Pediatric Diabetes Registry. *Diabetes Care*. 2013; 36: 879-886.

[8] Noronha, R. M., Calliari, L. E., Damaceno, N., Muramatu, L. H., Monte, O. Update on diagnosis and monitoring of cystic fibrosis-related diabetes mellitus (CFRD). *Arq. Bras. Endocrinol. Metabol*. 2011; 55: 613-621.

[9] Adler, A. I., Shine, B. S., Chamnan, P., Haworth, C. S., Bilton, D. Genetic determinants and epidemiology of cystic fibrosis-related diabetes: results from a British cohort of children and adults. *Diabetes Care*. 2008; 31: 1789-1794.

[10] Blackman, S. M., Hsu, S., Vanscoy, L. L., Collaco, J. M., Ritter, S. E., Naughton, K. et al. Genetic modifiers play a substantial role in diabetes complicating cystic fibrosis. *J. Clin. Endocrinol. Metab*. 2009; 94: 1302-1309.

[11] Hardin, D. S. A review of the management of two common clinical problems found in patients with cystic fibrosis: cystic fibrosis-related diabetes and poor growth. *Horm. Res*. 2007; 68 Suppl. 5: 113-116.

[12] O'Riordan, S. M., Robinson, P. D., Donaghue, K. C., Moran, A. ISPAD Clinical Practice Consensus. Management of cystic fibrosis-related diabetes. *Pediatr. Diabetes*. 2008; 9: 338-344.

[13] Preumont, V., Hermans, M. P., Lebecque, P., Buysschaert, M. Glucose homeostasis and genotype-phenotype interplay in cystic fibrosis patients with CFTR gene deltaF508 mutation. *Diabetes Care*. 2007; 30: 1187-1192.

[14] Uc, A., Olivier, A. K., Griffin, M. A., Meyerholz, D. K., Yao, J., Abu-El-Haija, M. et al. Glycaemic regulation and insulin secretion are abnormal in cystic fibrosis pigs despite sparing of islet cell mass. *Clin. Sci.* (Lond.). 2015; 128: 131-142.

[15] Olivier, A. K., Yi, Y., Sun, X., Sui, H., Liang, B., Hu, S. et al. Abnormal endocrine pancreas function at birth in cystic fibrosis ferrets. *J. Clin. Invest.* 2012; 122: 3755-3768.

[16] Battezzati, A., Mari, A., Zazzeron, L., Alicandro, G., Claut, L., Battezzati, P. M. et al. Identification of insulin secretory defects and insulin resistance during oral glucose tolerance test in a cohort of cystic fibrosis patients. *Eur. J. Endocrinol.* 2011; 165: 69-76.

[17] Ziai, S., Coriati, A., Gauthier, M. S., Rabasa-Lhoret, R., Richter, M. V. Could T cells be involved in lung deterioration and hyperglycemia in cystic fibrosis. *Diabetes Res. Clin. Pract.* 2014; 105: 22-29.

[18] Waugh, N., Royle, P., Craigie, I., Ho, V., Pandit, L., Ewings, P. et al. Screening for cystic fibrosis-related diabetes: a systematic review. *Health Technol. Assess.* 2012; 16: iii-iv, 1-179.

[19] O'Riordan, S. M., Dattani, M. T., Hindmarsh, P. C. Cystic fibrosis-related diabetes in childhood. *Horm. Res. Paediatr.* 2010; 73: 15-24.

[20] Lombardi, F., Raia, V., Spagnuolo, M. I., Nugnes, R., Valerio, G., Ciccarelli, G. et al. Diabetes in an infant with cystic fibrosis. *Pediatr. Diabetes.* 2004; 5: 199-201.

[21] Casas, L., Berry, D. R., Logan, K., Copeland, K. C., Royall, J. A. Cystic fibrosis related diabetes in an extremely young patient. *J. Cyst. Fibros.* 2007; 6: 247-249.

[22] Siahanidou, T., Mandyla, H., Doudounakis, S., Anagnostakis, D. Hyperglycaemia and insulinopenia in a neonate with cystic fibrosis. *Acta Paediatr.* 2005; 94: 1837-1840.

[23] Ode, K. L., Frohnert, B., Laguna, T., Phillips, J., Holme, B., Regelmann, W. et al. Oral glucose tolerance testing in children with cystic fibrosis. *Pediatr. Diabetes.* 2010; 11: 487-492.

[24] Rana, M., Munns, C. F., Selvadurai, H., Donaghue, K. C., Craig, M. E. Cystic fibrosis-related diabetes in children – gaps in the evidence? *Nat. Rev. Endocrinol.* 2010; 6: 371-378.

[25] Zirbes, J., Milla, C. E. Cystic fibrosis related diabetes. *Paediatr. Respir. Rev.* 2009; 10: 118-123.

[26] Elder, D. A., Wooldridge, J. L., Dolan, L. M., D'Alessio, D. A. Glucose tolerance, insulin secretion, and insulin sensitivity in children and adolescents with cystic fibrosis and no prior history of diabetes. *J. Pediatr.* 2007; 151: 653-658.

[27] Marshall, B. C., Butler, S. M., Stoddard, M., Moran, A. M., Liou, T. G., Morgan, W. J. Epidemiology of cystic fibrosis-related diabetes. *J. Pediatr.* 2005; 146: 681-687.

[28] Tofé, S., Moreno, J. C., Máiz, L., Alonso, M., Escobar, H., Barrio, R. Insulin-secretion abnormalities and clinical deterioration related to impaired glucose tolerance in cystic fibrosis. *Eur. J. Endocrinol.* 2005; 152: 241-247.

[29] Kelly, A., Moran, A. Update on cystic fibrosis-related diabetes. *J. Cyst. Fibros.* 2013; 12: 318-331.

[30] O'Shea, D., O'Connell, J. Cystic fibrosis related diabetes. *Curr. Diab. Rep.* 2014; 14: 511.

[31] Wickens-Mitchell, K. L., Gilchrist, F. J., McKenna, D., Raffeeq, P., Lenney, W. The screening and diagnosis of cystic fibrosis-related diabetes in the United Kingdom. *J. Cyst. Fibros.* 2014; 13: 589-592.

[32] Schmid, K., Fink, K., Holl, R. W., Hebestreit, H., Ballmann, M. Predictors for future cystic fibrosis-related diabetes by oral glucose tolerance test. *J. Cyst. Fibros*. 2014; 13: 80-85.

[33] Wooldridge, J. L., Szczesniak, R. D., Fenchel, M. C., Elder, D. A. Insulin secretion abnormalities in exocrine pancreatic sufficient cystic fibrosis patients. *J. Cyst. Fibros*. 2015 Mar. 5. pii: S1569-1993(15) 00049-1.

[34] Khammar, A., Stremler, N., Dubus, J. C., Gross, G., Sarles, J., Reynaud, R. Value of continuous glucose monitoring in screening for diabetes in cystic fibrosis. *Arch. Pediatr*. 2009; 16: 1540-1546.

[35] Schiaffini, R., Brufani, C., Russo, B., Fintini, D., Migliaccio, A., Pecorelli, L. et al. Abnormal glucose tolerance in children with cystic fibrosis: the predictive role of continuous glucose monitoring system. *Eur. J. Endocrinol*. 2010; 162: 705-710.

[36] Martín-Frías, M., Lamas Ferreiro, A., Enes Romero, P., Cano Gutiérrez, B., Barrio Castellanos, R. Abnormal glucose tolerance in prepubertal patients with cystic fibrosis. *An. Pediatr*. (Barc.). 2012; 77: 339-343.

[37] Waugh, N., Royle, P., Craigie, I., Ho, V., Pandit, L., Ewings, P. et al. Screening for cystic fibrosis-related diabetes: a systematic review. *Health Technol. Assess*. 2012; 16: iii-iv, 1-179.

[38] Bridges, N. Diabetes in cystic fibrosis. *Paediatr. Respir. Rev*. 2013; 14 Suppl. 1: 16-18.

[39] Van Sambeek, L., Cowley, E. S., Newman, D. K., Kato, R. Sputum glucose and glycemic control in cystic fibrosis-related diabetes: a cross-sectional study. *PLoS One*. 2015 Mar. 24; 10(3): e0119938.

[40] ISPAD Consensus Guidelines 2014. *Pediatric Diabetes* 2014; 15(Suppl. 20): 65-76.

[41] Ballmann, M., Hubert, D., Assael, B. M., Kronfeld, K., Honer, M., Holl, R. W. Open randomised prospective comparative multi-centre intervention study of patients with cystic fibrosis and early diagnosed diabetes mellitus. *BMC Pediatr*. 2014; 14: 70.

[42] Onady, G. M., Stolfi, A. Insulin and oral agents for managing cystic fibrosis-related diabetes. *Cochrane Database Syst. Rev*. 2013 Jul. 26; 7: CD004730.

In: Alterations in Glucose Homeostasis in Children
Editor: E. Otto-Buczkowska ISBN: 978-1-63483-861-0

Chapter 15

COELIAC DISEASE IN TYPE 1 DIABETES MELLITUS

A. Szaflarska-Popławska[*]
Department of Pediatric Endoscopy and Gastrointestinal Function Testing, Collegium Medicum in Bydgoszcz, Nicolaus Copernicus University in Toruń, Toruń, Poland

ABSTRACT

Coeliac disease (CD) occurs in patients with type 1 diabetes mellitus (T1D) with the prevalence that varies from 1.6% to 12.3% compared with 0.2-1% in the general population. More than one in twenty patients with T1D have histopathologically confirmed CD. Long-term CD seems to be an independent risk factor for microvascular complications and death in patients with T1D.

There are several HLA and non-HLA loci that overlap between T1D and CD. The risk of developing both conditions increases with the diagnosis of T1D at age <4 years. Diabetes is usually diagnosed first, CD precedes diabetes onset only in 10-25% of cases.

The majority of patients with CD and T1D are either asymptomatic or present only mild symptoms. The gastrointestinal presentation of CD occurs only in 10% of T1D cases. There is no consensus on screening patients with T1D for coeliac disease. CD diagnosis requires the small intestinal biopsy confirmation. Gluten-free diet should be proposed only in patients with mucosal atrophy. The effects of gluten-free diet on growth, nutritional status and metabolic control in CD/T1D patients are still controversial. The adherence to gluten-free diet by children with both conditions seems to be low.

Keywords: Coeliac disease, type 1 diabetes, gluten-free diet, screening recommendations

[*] Corresponding author: A. Szaflarska-Popławska. E-mail: aszaflarska@wp.pl.

Introduction

Coeliac disease (CD) is a systemic disease of immunological etiology, caused by intake of gluten and gluten-related prolamines, which occurs in patients genetically predisposed to it. It is distinguished by diverse clinical course and presence of specific antibodies, haplotype HLA DQ2 and/or DQ8 and enteropathy [1].

The prevalence of CD in Europe and United States is estimated at 0.2-1% in the overall population [2]. Coeliac disease has so far been considered a disease affecting the Caucasian race. However, it seems that it can also affect populations in China and the Indian subcontinent, which is most probably associated with changes in dietary habits (intake of wheat) and global increase in incidence rate of autoimmune diseases [3].

The clinical spectrum of CD is extremely wide, from completely asymptomatic to a complex of severe disturbances of intestinal absorption of numerous nutrients. CD is diagnosed both in patients with gastrological and non-gastrological symptoms, as well as asymptomatic patients, diagnosed due to screening tests in the overall population or high-risk groups [3]. In most cases (according to estimated data from the United States it is over 95% of all patients), CD remains undiagnosed. CD can have its onset at any age. It is currently diagnosed at the age of 45 years (on average), 10-12 years after first clinical signs (on average) [4].

Clinical Types of CD

According to the definitions from Oslo, we can currently differentiate four clinical types of CD. Classic type of CD involving intestinal malabsorption can be diagnosed if symptoms such as diarrhea, steatorrhea, body mass loss or disturbances of growing are observed.

Patients presenting with symptoms other than the ones indicating intestinal malabsorption (abdominal pain, constipation) are diagnosed with non-classic type, most frequently monosymptomatic type of coeliac disease.

Asymptomatic type of coeliac disease gives no symptoms associated with CD, responsive to gluten-free diet. Patients with classic, non-classic and asymptomatic type of CD, despite diverse clinical manifestation, are found to have specific antibodies and mucosa damage in the small intestine, typical for CD. In the latter, rarely diagnosed type of potential coeliac disease associated with abnormal serological test results, the condition of the mucosa of the small intestine is normal [5].

Articles published in recent years indicate that in the pediatric population (unlike the population of adults), the classic type of CD is still the dominant one. However, the proportion of children with other types of the disease is constantly growing. According to Spanish authors, in a group of 974 patients with CD diagnosed before the age of 15 years, in years 2006-2007, as much as 70.9% of patients presented with the classic form of the disease, 21.9% with non-classic and 7% with asymptomatic [6].

Greek researchers compared epidemiological data from years 1978-1987 and 1998-2007 and concluded that first clinical symptoms of CD currently occur later (25 vs. 73 months of age) and the classic type of CD is diagnosed less frequently (75% vs. 38%) [7].

DIAGNOSIS OF CD

Diagnostic evaluation of CD currently involves serological and genetic tests, as well as small intestine biopsy. Modern serological diagnostic evaluation of CD includes tests detecting antibodies against tissue transglutaminase (tissue transglutaminase type 2 antibodies, anti-TG2 – sensitivity ≥90%; specificity ≥90%), antibodies against deamidated gliadin peptides (DGP – sensitivity 80.7%; specificity 95.1%) and endomysial antibodies (EMA – sensitivity ≥90%; specificity 98.2%) [8]. Patients with normal level of IgA in serum have serological tests performed, which measure the level of specific IgA antibodies. Patients with low level of IgA in serum (<0.2g/L) have specific IgG antibodies measured [1].

The gold standard in diagnostic evaluation of CD is endoscopic biopsy of the small intestine with collection of numerous bioptates of mucosa from the duodenal bulb and the second and third segment of the duodenum. In order to histologically confirm CD, it is necessary to detect increased number of intraepithelial lymphocytes (>25/100), crypt hyperplasia and partial, subtotal or total atrophy of intestinal villi. Abnormalities allowing for microscopic confirmation of CD currently also include intraepithelial lymphocytosis and crypt hyperplasia with no evident damage of the structure of intestinal villi [1].

Genetic tests, that have been used for several last years in the diagnostic evaluation of CD assess patient's genetic susceptibility for coeliac disease, associated with HLA DQ2 and/or DQ8 genes. Over 90% of patients with coeliac disease present with HLA-DQ2 in cis position, coded by HLA-DR3-DQA1*0501-DQB1*0201 or trans, coded by HLA-DR11-DQA1*0505 DQB10301/DR7-DQA1*0201 DQB1 0202. Majority of the rest of patients with CD have the HLA-DQ8 heterodimer, coded by gene DQA1*0301-DQB1*0302. Presence of HLA-DQ2 or HLA-DQ8 is not sufficient for the diagnosis of CD, since they widely occur in the healthy population as well (30-40% of Caucasian population). However, their absence practically excludes development of CD [1].

Treatment of CD is based on complete and permanent elimination of food products containing wheat, rye and barley from the patient's diet [1].

T1D AND CD EPIDEMIOLOGY

The relation between CD and autoimmune disease, including type 1 diabetes (T1D), is well-known. The proportion of patients with T1D suffering from CD is said by various authors to fall within a wide range of 1.6-12.3%. However, in studies including large groups of subjects (≥1000 patients), this prevalence was estimated at 3.3-9.7% [9]. In two independent studies by Italian researchers, conducted on groups of 4322 and 8717 children with T1D, coeliac disease was diagnosed in 6.8% and 6.3% of subjects respectively [10, 11]. In a group of 4379 children with T1D from Australia, CD was diagnosed in 4.2% of them [12]. Results of a study carried out on a group of 1000 patients indicate that CD occurs in 3.3% of adult British patients with T1D [13]. A meta-analysis, including 26 605 children and adults with T1D, confirmed that the prevalence of histopathologically proven CD amounts to 6% of patients (95% CI 5.0-6.9%) [9]. The relations described pertain to Caucasian race. Little is known about coincidence of both diseases in the African-American population [14].

The risk of CD and T1D depends on the patient's age and sex. A comparison of patients with T1D and CD with patients with T1D only shows that diagnosis of diabetes mellitus is made earlier in the first group. The risk of having both diseases is three-times higher in children with T1D diagnosed before the age of 4 years, than in those diagnosed after the age of 9 years (OR 3.27; 95% CI 2.20-4.85). Higher risk of having both diseases is also associated with female sex (OR 1.93; 95% CI 1.51-2.47) [10].

Frequent coincidence of CD and T1D results from similar genetic etiology, associated both with HLA and non-HLA genes. Increased risk of CD in patients with T1D results from addition of genetic predisposition towards occurrence of both diseases, mainly associated with presence of particular MHC class II genes.

In the Finnish population, HLA-DQ2 haplotype was present in approximately 90% of patients with CD and 55% of patients with T1D, while HLA-DQ8 haplotype – in 10% of patients with CD and 70% of patients with T1D [15]. Moreover, several non-HLA loci have been identified, some associated with the immune system (IL18RAP, CTLA4, CCR5, IL2, IL21, TAGAP, PRKCQ), which coincide in both diseases [16].

Factors affecting the development of autoimmune diseases include the ecosystem of gut microbiota called a microbiome, which not only prevents the intestine from being colonized by pathogens, but also affects the innate and acquired immunological response in the host.

Different qualities in the intestinal system of the microbiota than in a healthy population was confirmed both in patients with T1D and CD. Patients with T1D presented with significantly higher prevalence of Bacteroides and Firmicutes, while in patients with CD Proteobacteriae, Enterobacteriaceae, Staphylococcaceae are prevalent [17-19]. Potential enteric microbes can modulate immunological response by means of various mechanisms, such as changing the balance between Th1 and Th17 cells, which is observed in CD and possibly observed in T1D [20, 21].

Some infectious illnesses, especially viral ones, seem to increase the risk of T1D (enterovirus) and CD (rotavirus) [22, 23].

Gluten Introduction in Diet

Diabetogenic effect of gluten was indicated in the results of two cohort studies published in 2003. In the BABYDIAB study, it was concluded that early (before the age of 3 months) introduction of gluten into the diet of infant from high risk group (relative of the first degree with type 1 diabetes) increases the risk of occurrence of T1D-related islet auto-antibodies by a factor of four (HR 4.0; 95% CI 1.4-11.5, p = 0.01), compared to infants that for the first three months of life were fed naturally only. Introducing gluten into diet after the 6th month of life did not affect the risk of autoimmunization [24]. The DAISY study revealed that the risk of occurrence of antibodies against islets was four times higher in children, in whom gluten was introduced into diet during the first three months of life (HR 4.32; 95% CI 2.0-9.35) and five times higher in children, in whom gluten was introduced into diet after they were 6 months old (HR 5.36; 95% CI 2.08-13.8), compared to infants, who were given gluten between the age of 4 and 6 months [25]. Observation lasting for several years (mean 13.1 years) of children, who had gluten introduced early into their diet (<3rd month of life) confirmed higher risk of occurrence of auto-antibodies against islets and of T1D, compared to the group of children that for the first 3 months of life were breastfed only (HR 3.97; 95% CI 1.41-11.17 and HR 3.45; 95% CI 1.04-11.48, respectively), as well as compared to children in whom gluten was introduced between 3.1 and 6 month of life (HR 3.40; 95% CI 1.19-9.70 and HR 3.43; 95% CI 1.01-11.66, respectively). Late (after 6 months of age) introduction of gluten into the infants' diet did not increase the risk of development of specific antibodies or T1D [26]. Thus, it seems that avoidance of early (during the first 3 months of life) introduction of gluten into the diet of infants can be significantly protective against development of T1D.

Clinical Presentation

In a vast majority of patients with CD and T1D, diagnosis of T1D is made as the first one. It is estimated that only in 10-25% patients CD is diagnosed first [10]. The presented epidemiological data suggests that in patients genetically predisposed, one disease (most frequently non-diagnosed oligosymptomatic or asymptomatic coeliac disease) can predispose to occurrences of the second disease (most frequently type 1 diabetes or autoimmune thyroiditis). However, results of studies assessing the influence of gluten supply on the development of T1D are contradictory. It is necessary to carry out studies finally concluding whether gluten-free diet can prevent occurrence of T1D in genetically predisposed patients or stop progression of the disease in recently diagnosed patients [27].

The clinical picture of CD in children is diverse. In the group of children with coincident CD and T1D, the tendency towards change of the clinical picture of CD into non-classic and asymptomatic type is more evident than in the population of children with isolated CD. Currently, only 10% of patients (mainly infants and children at post-infantile age) suffer from classic CD with chronic diarrhea, loss of appetite, malnutrition, body weight loss, abdominal pain or abdominal distention. A vast majority of patients presents with asymptomatic type of CD or a non-classic one – most frequently manifested with short stature, delayed sexual puberty, chronic fatigue, symptoms of hypovitaminosis or anemia due to iron deficiency [28]. According to the results obtained in Italian multi-center studies, the prevalence of potential CD in pediatric patients with T1D is also high (12.2 vs. 8.4% in a control group) and in most cases CD is asymptomatic [29]. Results of recently published studies confirm that compared to children suffering from CD only, patients with CD and T1D present with milder CD phenotype with higher body mass and body height, higher body mass index (BMI Z score 0.81 vs. 0.01, p = 0.001), lower proportion of children with anemia (71% vs. 95%, p = 0.003) and decreased bone density (29% vs. 49%, p = 0.05) [30].

Screening

Currently, there are no uniform international recommendations specifying a set of tests for CD in patients with T1D. Lack of unanimity is observed not only among different countries but also associations within particular countries, which was confirmed in a recently published material based on questionnaires conducted among 514 dietitians and diabetes educators from endocrine departments, diabetology departments, private practice offices, community nutrition centers and specialist clinics in the United States and Canada [31].

The recommendations by the American Diabetes Association (ADA), published in 2011, indicate that it is justified to carry out serological tests for CD in all children, immediately after diagnosis of T1D. Seropositive patient should be provided with gastrological care in order to have endoscopic biopsy of the small intestine performed. Histopathological confirmation of CD should result in an introduction of gluten-free diet, monitored by a dietitian having experience in therapy of patients with T1D and CD. At the moment of diagnosis, a patient with no CD-specific antibodies should have serological tests repeated in case of growth disturbances, no expected weight gain or weight loss, chronic diarrhea, flatulence or abdominal pain, symptoms of malabsorption, as well as frequent episodes of hypoglycemia of unknown etiology or poor glycemic control [32].

According to the recommendations by the American College of Gastroenterology (ACG), serological screening for CD in patients with T1D should be carried out only if they present with gastrointestinal symptoms and symptoms or laboratory abnormalities suggestive of CD [33]. Similar recommendations were formulated by the Canadian Diabetes Association (CDA). According to this association, screening tests for CD should be carried out if in a patient with T1D we observe recurrent gastrological symptoms, no expected increase in body mass and height, chronic fatigue or anemia, as well as episodes of hypoglycemia of unknown etiology and poor metabolic control. Children with histopathologically confirmed CD should follow gluten-free diet. Asymptomatic patients should be informed that conducting screening tests for CD and carrying out treatment is controversial [34].

The National Institute for Health and Clinical Excellence (NICE) currently recommends to perform screening tests in every child, only at the moment of diagnosis of T1D but does not recommend repeating them in subsequent years [35]. Conduct according to the recommendations by NICE can result in a delay in the diagnosis of CD in half of patients with T1D, therefore it is suggested that these recommendations should be changed [36, 37]. In some countries, the diagnostic protocol presupposes performance of screening test for CD at the moment of diagnosis of T1D and then every year for at least next 4 years and every two years for the next 6 years [28].

According to the recommendations by the European Society of Pediatric Gastroenterology, Hepatology and Nutrition (ESPGHAN), tests for CD should be conducted in every child with T1D, however, the plan of diagnostic evaluation and monitoring of the patient should depend on the clinical picture of a particular case. In a child with T1D suffering from symptoms suggestive of CD, at first measuring the level of IgA in serum is recommended, as well as verifying the level of IgA-TG2, or – in case of IgA deficiency – IgG-TG2. Patients with abnormal results of serological test should be referred to a pediatric gastroenterologist.

In all cases, in order to obtain a certain diagnosis of CD, endoscopic biopsy of the small intestine should be performed and gluten-free diet should be applied if typical histopathological abnormalities are found in the mucosa of the small intestine. Only in symptomatic patients, in whom genetic predisposition towards CD (presence of HLA DQ2 and/or DQ8 haplotype) and presence of antyendomysial antibodies specific for CD and against tissue transglutaminase have been confirmed, resignation of biopsy of the small intestine can be considered, provided that the concentration of the latter antibodies must exceed the upper range limit by ten times at least [1]. Asymptomatic pediatric patients with T1D should first have genetic tests carried out. Negative result of genetic test rules out occurrence of CD and makes it unnecessary to repeat serological tests for CD.

Asymptomatic patients with confirmed genetic predisposition towards CD require measurement of anti-TG2 antibodies in serum, and biopsy of the small intestine (in order to confirm enteropathy). Asymptomatic patients with T1D, in whom the level of anti-TG2 antibodies is increased but does not exceed the upper normal range by three times, constitute a specific group. They require repeated serological tests and an individually made decision whether or not biopsy of the small intestine should be performed in their case [1]. In 30% of them, spontaneous normalization of the level of antibodies is observed during further observation, even though the diet continues to contain gluten [38].

The American Diabetes Association (ADA) recommends that in patients with T1D, alternative test of anti-TG2 antibodies or EMA should be carried out after confirmation of normal level of IgA in serum [32].

TREATMENT

With regard to the treatment, recommendations pertaining to patients with T1D and CD are the same as in the case of patients with CD only. Application of gluten-free diet is recommended only in children with histopathologically confirmed enteropathy [39]. In patients with the potential form, it is necessary to repeat diagnostic tests for CD, because within three years, 30% of them will develop abnormalities in the small intestine typical for CD [40]. Gluten-free diet is based on total elimination of products containing wheat, rye and barley from the patient's diet and must be followed for the rest of life. Additional dietary restrictions resulting from the necessity to exclude gluten from the diet of a patient suffering from T1D and CD and frequent asymptomatic or oligosymptomatic course of CD, as well as lack of direct effect of the gluten-free diet on the patient's general feeling result in the fact that strict gluten-free diet in this group of patients is followed less frequently than in the group of patients with coeliac disease only (59% vs. 78%). Another problem is high glycemic index of many gluten-free food products, since it can affect the glycemia, level of HBA1c, demand for insulin, lipid profile, and thus, probably also increase the risk of late complications of diabetes [39]. The recommendations point out to the need of supplementing the therapeutic process of a patient with T1D and CD with a dietitian experienced in the treatment of both of these diseases [32].

Data concerning the effect of gluten-free diet on health condition of patients with T1D and CD is not uniform. Without doubt, patients with clinically overt disease benefit from elimination of gluten from their diet due to remission of gastroenterological symptoms and non-gastrointestinal symptoms associated with CD. Dietary treatment also decreases the risk of late complications of untreated disease, especially the risk of neoplastic diseases (including small intestine lymphoma and non-Hodgkin's lymphoma) and increased mortality [41]. In recent years, in United States, an increase in the incidence of enteropathy-associated T-cell lymphoma (EATL) has been well documented [42].

Use of gluten-free diet in patients with T1D and asymptomatic CD can probably alleviate disturbances in bone metabolism, which are observed both in patients with one of the two diseases and both of them [30]. They are caused not only by endocrine mechanisms and malabsorption, but also better and better known disturbances in the RANKL/OPG system (receptor activator of nuclear factor-dB ligand/osteoprotegerin) [28].

Patients with T1D and CD who do not follow gluten-free diet suffer from osteopenia more frequently than patients strictly following the diet [43].

EFFECT OF TREATMENT AND COMPLICATIONS

A few previously published articles indicated that gluten-free diet in patients with CD and T1D does not affect the curve of growth and metabolic control of T1D measured by means of HBA1c level. They only suggested a possibility of body mass increase measured by an increase in body mass index. In their article published in 2012, Taler et al. showed no differences in the analyzed anthropometric (body mass index-standard deviation score, height-standard deviation score) and metabolic parameters (HBA1c level) in children with T1D and CD, compared to patients with only one of these diseases [44]. Similarly, Mackinder et al. did not find any differences in the anthropometric parameters that they analyzed (BMI z-score, height z-score, weight z-score) in a group of 23 pediatric patients with CD, 44 children with T1D and 23 children with both diseases, selected according to sex and age, within two years before and two years after the diagnosis of CD.

Also, they did not find any differences in the concentration of HBA1c between the studied groups [45]. The results of studies carried out within the last several years in adults are contradictory to the reports from pediatric centers. In a group of adult patients with adult type T1D, at the moment of diagnosis of CD, a higher level of HBA1c was observed (8.2% versus 7.5%, p = 0.05), lower level of total cholesterol (4.1 versus 4.9, p = 0.014) and HDL (1.1 versus 1.6, p = 0.017), as well as a higher proportion of patients with retinopathy (58.3% versus 25%, p = 0.02), nephropathy (41.6% versus 4.2%, p = 0.009) and peripheral neuropathy (41.6% versus 16.6%, p = 0.11), compared to patients with diabetes but without CD. Moreover, after introduction of gluten-free diet, improvement was observed as regards markers of nephropathy [46].

Similar results were presented by Rohrer et al., who carried out a study on 56 514 patients with T1D from 392 German and Austrian centers and showed that both retinopathy and microalbuminuria occur earlier in the group of patients with coincident CD, than in patients without CD (26.7 years vs. 33.7 years and 32.8 vs. 42.4 years, respectively). The risk rates of retinopathy (HR 1.263; 95% CI 1.078-1.481) and nephropathy (HR 1.359; 95% CI 1.228-1.504) were significantly higher in patients with T1D and CD, compared to patients with T1D only [47].

Results of a study of patients with CD registered in the Swedish National Patient Register indicate that in the case of coincident CD and T1D, the risk of death is higher than in patients with T1D only. This risk increases with duration of CD and is 2.8 times higher in patients having a diagnosis of CD for ≥15 years [48].

Data on the effect of CD diagnosis and additional dietary restrictions on the quality of life of patients with T1D are contradictory.

In a study by Leeds et al., conducted on a group of patients with T1D and CD, there were no differences observed in the life quality, compared to patients with only one of the two diseases [49]. In a comparison of patients with CD and T1D with patients suffering from T1D only, Bakker et al. observed a significantly lower quality of life in social aspects and higher anxiety associated with diabetes in the first group [50].

CONCLUSION

Analysis of the results of the material published so far leads to a conclusion that in asymptomatic patients with T1D, the diagnosis of CD and initiation of gluten-free diet does not affect the anthropometric parameters and probably does not improve metabolic control of diabetes. However, it can affect diabetic microangiopathy and macroangiopathy. This is why it is recommended to carry out further prospective studies in this area [47].

REFERENCES

[1] Husby, S., Koletzko, S., Korponay-Szabó, I. R., Mearin, M. L., Phillips, A., Shamir, R. et al. European Society for Pediatric Gastroenterology, Hepatology, and Nutrition Guidelines for the Diagnosis of Coeliac Disease. *JPGN* 2012; 54: 136-160.

[2] Catassi, C., Gatti, S., Fasano, A. The new epidemiology of coeliac disease. *JPGN* 2014; 59 Suppl. 1: S7-9.

[3] Mooney, P. D., Hadjivassiliou, M., Sanders, D. S. Coeliac disease. *BMJ* 2014; 348: g1561 doi: 1136.

[4] Kenrick, K., Day, A. S. Coeliac disease: where are we now in 2014? *Aust. Fam. Physician* 2014; 43(10): 674-678.

[5] Ludvigsson, J. F., Leffler, D. A., Bai, J. C., Biagi, F., Fasano, A., Green, P. H. et al. The Oslo definitions for coeliac disease and related terms. *Gut* 2013; 62: 43-52.

[6] Cilleruelo, M. L., Roman-Riechmann, E., Sanchez-Valverde, F., Donat, E., Manuel-Ramos, J., Martín-Orte, E. et al. Spanish national registry of coeliac disease: incidence and clinical presentation. *J. Pediatr. Gastroenterol. Nutr.* 2014; 59(4): 522-526.

[7] Roma, E., Panayiotou, J., Karantana, H., Constantinidou, C., Siakavellas, S. I., Krini, M. et al. Changing pattern in the clinical presentation of pediatric celiac disease; a 30 year study. *Digestion* 2009; 80: 185-191.

[8] Giersiepen, K., Lelgemann, M., Stuhldreher, N., Ronfani, L., Husby, S., Koletzko, S. et al. Accuracy of diagnostic tests for coeliac disease in children: summary of an evidence report. *J. Pediatr. Gastroenterol. Nutr.* 2012; 54: 229-241.

[9] Elfström, P., Sundström, J., Ludvigsson, J. F. Systematic review with meta-analysis: assciations between coeliac disease and type 1 diabetes. *Aliment. Pharmacol. Ther.* 2014; 40: 1123-1132.

[10] Cerutti, F., Bruno, G., Chiarelli, F., Lorini, R., Meschi, F., Sacchetti, C. Younger age at onset and sex predict celiac disease in children and adolescents with type 1 diabetes: an Italian multicenter study. *Diabetes Care* 2004; 27: 1294-8.

[11] Franzese, A., Iafusco, D., Spadaro, R., Cavaliere, O., Prisco, F., Auricchio, R. et al. Potential celiac disease in type 1 diabetes: a multicenter study. *Diabetes Res. Clin. Pract.* 2011; 92: 53-56.

[12] Pham-Short, A., Donaghue, K. C., Ambler, G., Chan, A. K., Craig, M. E. Coeliac disease in type 1 diabetes from 1990 to 2009: higher incidence in young children after long diabetes duration. *Diabet. Med.* 2012; 29: e286-9.

[13] Kurien, M., Leeds, J. S., Hopper, A. D., Wild, G., Egner, W., Tesfaye, S. et al. Serological testing for coeliac disease in type 1 diabetes mellitus: is immunoglobulin A level measurement necessary. *Diabet. Med.* 2013; 30: 840-845.

[14] Kaistha, A., Castells, S. Celiac disease in African American children with type 1 diabetes mellitus in inner city of Brooklyn. *Pediatr. Endocrinol. Rev*. 2008; 5 suppl. 4: 994-998.

[15] Hermann, R., Turpeinen, H., Laine, A. P., Veijola, R., Knip, M., Simell, O. et al. HLA DR-DQ-encoded genetic determinants of childhood-onset type 1 diabetes in Finland: an analysis of 622 nuclear families. *Tissue Antigens* 2003; 62: 162-169.

[16] Kumar, V., Wijmenga, C., Withoff, S. From genome-wide association studies to disease mechanisms: celiac disease as a model for autoimmune diseases. *Semin. Immunopathol.* 2010; 34: 567-580.

[17] de Goffau, M. C., Luopajärvi, K., Knip, M., Ilonen, J., Ruohtula, T., Härkönen, T. et al. Fecal microbiota composition differs between children with beta-cell autoimmunity and those without. *Diabetes* 2013; 62: 1238-1244.

[18] Murri, M., Leiva, I., Gomez-Zumaquero, J. M., Tinahones, F. J., Cardona, F., Soriguer, F. et al. Gut microbiota in children with type 1 diabetes differs from that in healthy children: a case-control study. *BMC Med.* 2013; 11: 46.

[19] Sánchez, E., Donat, E., Ribes-Koninckx, C., Fernández-Murga, M. L., Sanz, Y. Duodenal-mucosal bacteria associated with celiac disease in children. *Appl. Environ. Microbiol.* 2013; 79: 5472-5479.

[20] Fernández, S., Molina, I. J., Romero, P., González, R., Peña, J., Sánchez, F. et al. Characterization of gliadin-specific Th17 cells from the mucosa of celiac disease patients. *Am. J. Gastroenterol.* 2011; 106: 528-538.

[21] Sorini, C., Falcone, M. Shaping the (auto)immune response in the gut: the role of intestinal immune regulation in the prevention of type 1 diabetes. *Am. J. Clin. Exp. Immunol.* 2013; 2: 156-171.

[22] Yeung, W. C., Rawlinson, W. D., Craig, M. E. Enterovirus infection and type 1 diabetes mellitus: a systematic review and meta-analysis of observational studies. *BMJ* 2011; 342:d35.

[23] Dolcino, M., Zanoni, G., Bason, C., Tinazzi, E., Boccola, E., Valletta, E. et al. A subset of anti-rotavirus antibodies directed against the viral protein VP7 predicts the onset of celiac disease and induces typical features of the disease in the intestinal epithelial cell line T84. *Immunol. Res*. 2013; 56: 465-476.

[24] Ziegler, A. G., Schmid, S., Huber, D., Hummel, M., Bonifacio, E. Early infant feeding and risk of developing type 1 diabetes-associated antibodies. *JAMA* 2003; 290: 1721-1728.

[25] Norris, J. M., Barriga, K., Klingensmith, G., Hoffman, M., Eisenbarth, G. S., Erlich, H. A. et al. Timing of initial cereal exposure in infancy and risk of islet autoimmunity. *JAMA* 2003; 290: 1713-1720.

[26] Chmiel, R., Beyerlein, A., Knopff, A., Hummel, S., Ziegler, A. G., Winkler, C. Early infant feeding and risk of developing islet autoimmunity and type 1 diabetes. *Acta Diabet*. 2014.

[27] Antvorskov, J. C., Josefsen, K., Engkilde, K., Funda, D. P., Buschard, K. Dietary gluten and the development of type 1 diabetes. *Diabetologia* 2014; 57: 1770-1780.

[28] Camarca, M. E., Mozzillo, E., Nugnes, R., Zito, E., Falco, M., Fattorusso, V. et al. Celiac disease in type 1 diabetes mellitus. *Italian J. Pediatr.* 2012; 38: 10.

[29] Franzese, A., Iafusco, D., Spadaro, R., Cavaliere, O., Prisco, F., Auricchio, R. et al. Potential celiac disease in type 1 diabetes: A multicenter study. *Diabetes Res. Clin. Pract.* 2011; 92: 53-56.

[30] Tsouka, A., Mahmud, F. H., Marcon, M. A. Celiac disease associated with type 1 diabetes and celiac disease alone: are these patients different? *J. Pediatr. Gastroenterol. Nutr.* 2015 Mar. 19.

[31] Simpson, S. M., Ciaccio, E. J., Case, S., Jaffe, N., Mahadov, S., Lebwohl, B. et al. Celiac disease in patients with type 1 diabetes. Screening and Diagnostic Pracices. *Diabetes Educ.* 2013; 39: 532-540.

[32] Executive summary: Standards of Medical Care in Diabetes-2011. *American Diabetes Care* 2011; 34 Suppl. 1: S4-S10.

[33] Rubio-Tapia, A., Hill, I. D., Kelly, C. P., Calderwood, A. H., Murray, J. A.; American College of Gastroenterology. American College of Gastroenterology Clinical Guideline: diagnosis and management of celiac disease. *Am. J. Gastroenterol.* 2013; 108: 656-677.

[34] Canadian Diabetes Association Clinical Practice Guidelines Expert Committee. Type 1 diabetes in children and adolescents. *Can. J. Diabetes* 2013; 37: S153-162.

[35] National Institute for Clinical Excellence: Type 1 diabetes: diagnosis and management of type 1 diabetes in children, young people and adults. London: National Institute for Clinical Excellence (NICE): 2009.

[36] Jones, H. J., Warner, J. T. NICE clinical guideline 86. Coeliac disease: recognition and assessment of coeliac disease. *Arch. Dis. Child* 2010; 95: 312-313.

[37] Babiker, A., Morris, M. A., Datta, V. Coeliac disease and type 1 diabetes: 7 years experience versus NICE guideline 2009. *Arch. Dis. Child* 2010; 95: 1068-1069.

[38] Waisbourd-Zinman, O., Hojsak, I., Rosenbach, Y., Mozer-Glassberg, Y., Shalitin, S., Phillip, M. et al. Spontaneous normalisation of anti-tissue transglutaminase antibody levels is common in children with type 1 diabetes mellitus. *Dig. Dis. Sci.* 2012; 57: 1314-1320.

[39] Scaramuzza, A. E., Mantrgazza, C., Bosetti, A., Zuccotti, G. V. Type 1 diabetes and celiac disease: The effects of gluten free diet on metabolic control. *World J. Diabetes* 2013; 4: 130-134.

[40] Tosco, A., Salvati, V. M., Auricchio, R., Maglio, M., Borrelli, M., Coruzzo, A. et al. Natural history of potential celiac disease in children. *Clin. Gastroenterol. Hepatol.* 2011; 9(4): 320-325.

[41] Tio, M., Cox, M. R., Eslick, G. D. Meta-analysis: coeliac disease and the risk of all-cause mortality, and malignancy and lymphoid malignancy. *Aliment. Pharmac.* 2012; 35: 540-551.

[42] Sharaiha, R. Z., Lebwohl, B., Reimers, L., Bhagat, G., Green, P. H., Neugut, A. I. Increasing incidence of enteropathy-associated T-cell lymphoma in the United States, 1973-2008. *Cancer* 2012; 118: 3786-3792.

[43] Valerio, G., Spadaro, R., Iafusco, D., Lombardi, F., Del Puente, A., Esposito, A. et al. The influence of gluten-free diet in children and adolescents with type 1 diabetes mellitus and celiac disease. *Bone* 2008; 43: 322-326.

[44] Taler, I., Phillip, M., Lebenthal, Y., de Vries, L., Shamir, R., Shalitin, S. Growth and metabolic control in patients with type 1 diabetes and celiac disease; a longitudinal observational case-control study. *Pediatr. Diabetes* 2012; 13: 597-606.

[45] Mackinder, M., Allison, G., Svolos, V., Buchanan, E., Johnston, A., Cardigan, T. et al. Nutritional status, growth and disease management in children with single and dual diagnosis of type 1 diabetes mellitus and coeliac disease. *BMC Gastroenterol.* 2014; 14: 99.

[46] Leeds, J. S., Hopper, A. D., Hadjivassiliou, M., Tesfaye, S., Sanders, D. S. High prevalence of microvascular complications in adults with type 1 diabetes and newly diagnosed celiac disease. *Diabetes Care* 2011; 34: 2158-2163.

[47] Rohrer, T. R., Wolf, J., Liptay, S., Zimmer, K. P., Fröhlich-Reiterer, E., Scheuing, N. et al. Microvascular Complications in Childhood-Onset Type 1 Diabetes and Celiac Disease: A Multicenter Longitudinal Analysis of 56,514 Patients From the German-Austrian DPV Database. *Diabetes Care* 2015;38:801-807.

[48] Mollazadegan, K., Sanders, D. S., Ludvigsson, J., Ludvigsson, J. F. Long-term coeliac disease influences risk of death in patients with type 1 diabetes. *J. Intern. Med.* 2013; 274: 273-280.

[49] Leeds, J. S., Hopper, A. D., Hadjivassiliou, M., Tesfaye, S., Sanders, D. S. Potential coeliac disease in type 1 diabetes mellitus: does a positive antibody lead to increased complications? *Nutr. Metab. Cardiovasc. Dis.* 2014; 24: 378-83.

[50] Bakker, S. F., Pouwer, F., Tushuizen, M. E., Hoogma, R. P., Mulder, C. J., Simsek, S. Compromised quality of life in patients with both type 1 diabetes mellitus and coeliac disease. *Diabet. Met.* 2013; 30: 835-839.

In: Alterations in Glucose Homeostasis in Children
Editor: E. Otto-Buczkowska
ISBN: 978-1-63483-861-0

Chapter 16

SKIN LESIONS IN TYPE I DIABETES: AN UPDATE

K. Mędrek and A. Reich*
Department of Dermatology, Venereology and Allergology, Wroclaw Medical University, Wroclaw, Poland

ABSTRACT

Skin lesions are very common in diabetes, affecting up to 50% of all diabetics. Skin symptoms usually appear after the diagnosis of diabetes, sometimes even after many years of its duration, but on same occasions they may precede the development of this disease. The mechanism of many skin disorders associated with diabetes is still not fully elucidated, however, it is believed that the non-enzymatic glycation of proteins play a significant role at least in some of them. Based on the etiology, skin changes associated with diabetes may be divided into following categories: diseases associated with microangiopathy and degeneration of connective tissue fibres, infectious diseases of the skin, and cutaneous adverse effects related to insulin injections. Here, we have reviewed current data on skin problems in children and adolescents with diabetes.

INTRODUCTION

Skin lesions are common in diabetes – it is estimated that about half of all adult patients with diabetes present various abnormalities within the skin [1, 2]. In children and adolescents suffering from diabetes, skin diseases are also common and can affect about 20-50% of patients [3, 4]. In one recently published study on cutaneous lesions in young diabetic patients the most common abnormality were cutaneous reactions to insulin treatment observed in 45% of patients [5]. Such cutaneous adverse effects related to insulin injections include lipohypertrophy (41%), post-inflammatory hyperpigmentation (3%), lipoatrophy (0·6%) and acanthosis nigricans (0·4%).

* Corresponding author: Prof. Adam Reich, MD, PhD. E-mail: adam.reich@umed.wroc.pl.

Other skin lesions included joint mobility (16·8%), xerosis (15·8%) and scleroderma-like skin changes (10%) [5]. Excessive dryness was also reported by other authors and was observed in 22-48% of patients [3, 6]. Skin symptoms usually appear after the diagnosis of diabetes, in some cases after many years of its duration, albeit occasionally they can also precede the development of this disease [4].

Although the mechanism of many common skin disorders associated with diabetes is still not completely understood, it is believed that the non-enzymatic glycation of proteins plays a significant role in the development of peripheral circulatory disorders (microangiopathy), degeneration of collagen and elastic fibres, damage of the peripheral nervous system, as well as impaired cellular and humoral defence mechanisms [4].

As a result of the complex and multifaceted pathogenesis of diabetes, theskin thickens, becomes dry, increases its sensitivity to mechanical stress and physical stimuli, its elasticity decreases, and/or reparative and regenerative processes become impaired, causing an increase of the susceptibility to bacterial and fungal infections of the skin [4]. Based on the etiology, skin changes associated with diabetes may be divided into:

1. diseases associated with microangiopathy and degeneration of connective tissue fibres,
2. infectious diseases of the skin, and
3. the cutaneous adverse effects related to insulin injections [4].

Some other conditions, such as acanthosis nigricans or autoimmune diseases, or vitiligo and alopecia areata, are not directly related to diabetes but may coexist with this disease [7, 8]. On the other hand, certain skin diseases can occur less frequently in diabetic patients, for example atopic dermatitis [9].

It should also be noted that the assessment of the degree of protein glycation of the skin or the early changes in the microcirculation and/or innervation of the skin can be used as a predictor of the development of serious complications of diabetes [10-13].

Diseases Associated with Microangiopathy and Degeneration of Connective Tissue Fibres

Degeneration of small blood vessels within the skin belongs to the group of systemic diabetic microangiopathies. These changes are caused by poor glycaemic control that results in non-enzymatic protein glycation (evident by high HbA1c) and can lead to the development of several, relatively serious, conditions, e.g., scleroderma-like syndrome. It was shown that skin microcirculation has been already impaired in about 25% of paediatric patients suffering from type 1 diabetes mellitus (T1DM) for at least one year.

In the study on patients with T1DM exposed to the localized thermal hyperaemia with localized warming to maximum of 40°C and iontophoretic delivery of sodium nitroprusside (SNP), diabetic patients showed significantly reduced vasodilatative response to either physical or pharmacological stimulation with SNP [14].

Remarkably, endothelial-independent vasodilatation was better preserved in patients with an insulin pump compared to those on multiple daily injection (MDI) therapy ($p < 0.01$) [14].

Microangiopathy-dependent skin lesions often precede the development of serious organ impairments or coexist with them. Subsequent microangiopathy-associated lesions may occur during the developing age in T1DM: diabetic rubeosis, *necrobiosis lipoidica diabeticorum*, granuloma annulare, scleroderma-like syndrome, and nail abnormalities [4].

Diabetic Rubeosis

Diabetic rubeosis (*rubeosis diabeticorum*) arises from small, confluent telangiectasia and is characterized as a poorly demarcated erythema on the face (Figure 1), hands and, less often, on feet [4, 15]. Usually, skin lesions are not accompanied by any subjective symptoms. Diabetic rubeosis occurs in the early stages of T1DM in approximately 10% of patients, often in young subjects with poorly controlled diabetes. Patients should be advised to avoid sudden changes of temperature and exposure to the sun. They should also regularly use proper UV protection (wearing long sleeves and headgear, applying sunscreen creams with Sun Protection Factor – SPF >20). It is also important to align glucose metabolism. In more advanced cases, beneficial therapeutic effect can be achieved by laser therapy.

Necrobiosis Lipoidica Diabeticorum

Necrobiosis lipoidica diabeticorum (NLD) manifests as a well-demarcated, infiltrative, yellow to brown-coloured lesion or lesions.

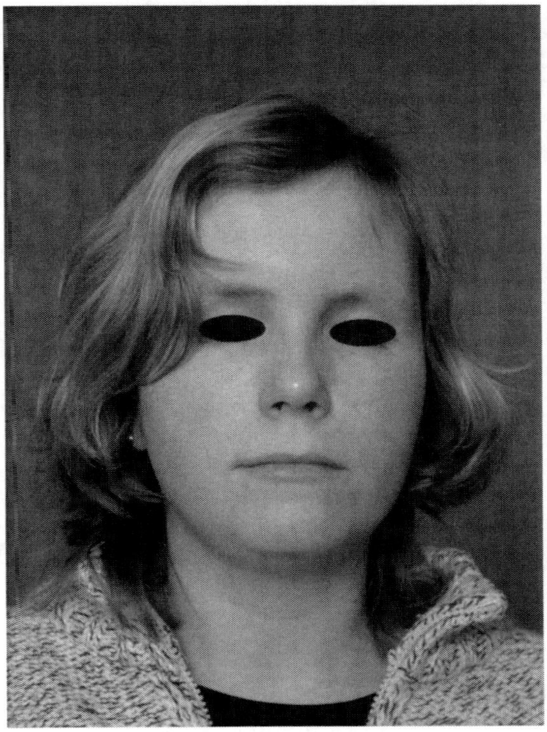

Figure 1. Diabetic rubeosis of the face.

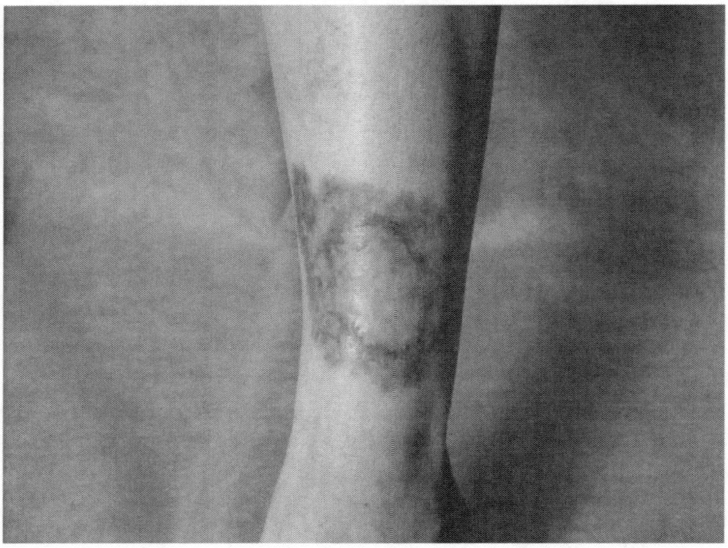

Figure 2. Necrobiosis lipoidica diabeticorum.

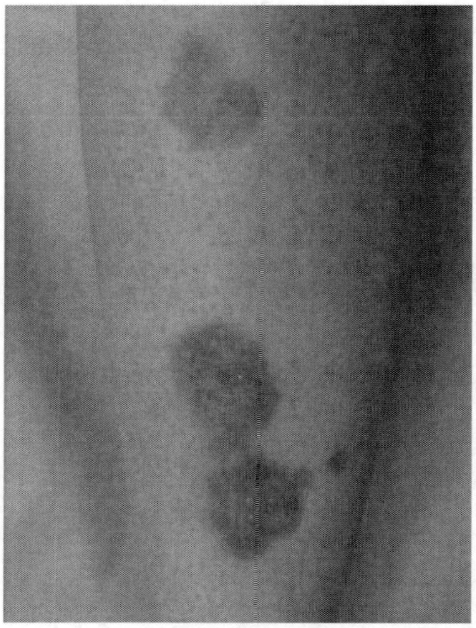

Figure 3. Necrobiosis lipoidica diabeticorum.

A translucent atrophy with characteristic “glazed-porcelain” sheen and arborizing pattern of dilated blood vessels is observed in the central part of skin lesions (Figure 2 and 3).

NLD most often is present on the extensor surfaces of lower legs, rarely on the forearms and hands, but it has also been reported on other, less typical placement, e.g., around the orbits [16]. Initially, skin lesions may be located unilaterally, but in the majority of patients skin changes are symmetrical. Sometimes, the lesions may ulcerate. It should also be kept in mind that the long-lasting NLD changes can be the starting point for squamous cell carcinoma [17, 18]. NLD affects about 0.3-1.2% of diabetic patients, the majority of whom are suffering from T1DM [19]. Interestingly, NLD seems to be even more prevalent in the paediatric population as compared to adults as it was found in 2.3% of paediatric patients [20].

Histological examination reveals degeneration and inflammation of small blood vessels, collagen necrobiosis, and the presence of lipoid bodies, mucin deposits and granulomatous reaction with histiocytes of palisade pattern at the edge of necrotic lesions.

Typically, there is also a presence of giant epithelioid cells. Although the NLD associated with diabetes is considered as skin disease linked with microangiopathy, recent studies have shown that blood flow within the affected skin is greater than in the nonlesional skin, which could argue in favour of an inflammatory nature of this disorder, and not an ischemic one as it is believed currently [21].

Opinions on the impact of blood glucose levels on the development of NLD lesions are divergent. Results of some studies have shown that glucose control does not appear to be correlated with NLD onset and severity, while other studies claimed that NLD is usually associated with poor glucose alignment. Moreover, it seems that a tighter glucose control might improve or even prevent development of this disorder [22].

Despite only about 40-60% of adult patients with NLD have diabetes, it is believed that in children and adolescents this condition is almost always associated with T1DM. NLD can precede the onset (in approximately 30% of patients), can be recognized simultaneously with the diabetes (in about 25% of patients), and can also occur during the disease course (in approximately 45% of patients) [4].

Spontaneous remission of NLD is observed in less than 20% of cases. Ulceration is the most frequent complication and affects about 25-33% of cases over time. Squamous cell carcinomas developing in areas of NLD have also been reported [24]. Treatment of NLD is difficult because there are currently no reliable management guidelines based on clinical trials and skin lesions are often resistant to employed therapeutic methods [23, 25]. Usually, it is recommended to use vascular drugs (pentoxifylline, extracts of horse chestnut), antiplatelet drugs (aspirin, ticlopidine) and potent topical corticosteroid formulations.

If such therapy fails, other drugs can be used. The effectiveness of antimalarials [26], thiazolidinediones (glitazones) [27, 28], thalidomide [29], esters of fumaric acid [30], cyclosporine A [31, 32], topically applied calcineurin inhibitors [33], or even tumour necrosis factor α (TNF-α) inhibitors [34] have been reported, albeit no controlled studies confirming their efficacy exist so far. For recalcitrant and extensive lesions, some authors recommend systemic corticosteroids [35]. In addition to pharmacological treatment, it is also important to protect the involved skin against injuries and secondary infections, as they may lead to chronic and difficult to treat ulcers.

Some importance in the treatment may also have photochemotherapy (PUVA) or UVA-1 therapy, particularly in cases with rapid progression [36-38]. Successful treatment of NLD has also been observed using photodynamic therapy with methylaminolevulinate, wherein the effectiveness of this method has only been reported in adults [39-41]. In refractory changes cosmetic camouflage can be recommended to patients.

Granuloma Annulare

Granuloma annulare is a rather mild condition with chronic, usually asymptomatic development prone to spontaneous regression. The disease can occur at any age, but usually skin lesions occur before the age of 15-years.

In children the disease usually manifests as a single, annular lesion, spreading peripherally with raised edges, and composed of cohesive, pearly-pink or flesh-coloured nodules, histologically characterised by a granuloma formation with centrally located necrosis or necrobiosis.

It is believed that in children this form of granuloma annulare is not associated with diabetes [42]. However, occasionally the disease may be diffused with numerous foci on the skin of the limbs, trunk and face (Figure 4). It is estimated that approximately 12-15% of adults with disseminated granuloma annulare have or will have diabetes. The connection of this form of granuloma annulare with diabetes in children remains unclear. However, in paediatric subjects diabetes may be associated with another rare form of granuloma annulare, namely subcutaneous granuloma annulare (SGA).

In a study on 34 children aged under 10-years with SGA, diabetes occurred in 2 (6%) patients [43]. In addition, there is one case report on SGA in a 2-year-old girl, who demonstrated skin changes before the diagnosis of IDDM (insulin dependent diabetes mellitus) and after achieving a control of glycaemia all skin lesions vanished [44].

Coexistence of necrobiosis lipoidica and granuloma annulare has also been reported in a 12-year-old girl, who developed diabetes two years after the appearance of first skin lesions [45]. Treatment of single lesions of granuloma annulare includes cryotherapy and potent topical corticosteroids, which are often applied under occlusion.

Because of the tendency toward spontaneous regression, some authors also recommend the 'to watch and wait' approach. In the diffused form of granuloma annulare in adults various systemic drugs were employed with some success (e.g., dapsone, short-term administration of systemic corticosteroids, isotretinoin, chloroquine, cyclosporine A) as well as photochemotherapy (PUVA) [4]. Anti-TNF-α drugs and topically applied calcineurin inhibitors also have been observed to be effective [46, 47]. However, in children aggressive treatment is usually not necessary [4].

Scleroderma-Like Syndrome

Scleroderma-like syndrome is characterized by thickening and hardening of the back skin of the hands, especially fingers, and limited mobility and contractures of the interphalangeal and metacarpophalangeal joints of the hands [4].

The pathogenesis of this syndrome has not been fully clarified, but it is believed that deposition of glycosylated proteins in the intercellular space and/or increased proliferation of fibroblasts play an important role. Skin thickening and stiffening of fingers usually begins with the pinky finger spreading towards the remaining fingers over time [4]. At first it comes to reduction of straightening in metacarpophalangeal and interphalangeal joints [9]. Cupped hands take on a distinctive prayer shape. In contrast to scleroderma, there is no evidence of skin atrophy and dyspigmentation in scleroderma-like syndrome; Raynaud's phenomenon is also absent. It is estimated that the scleroderma-like syndrome affects approximately 30% of patients with diabetes and is probably the earliest complication of the disease [6, 36, 48]. Its appearance is primarily associated with duration of diabetes and may indicate associated subclinical organ complications [7]. Treatment of this disorder is often unsatisfactory, but some improvement can be achieved through proper diabetes treatment, ensuring long periods of normoglycaemia [4].

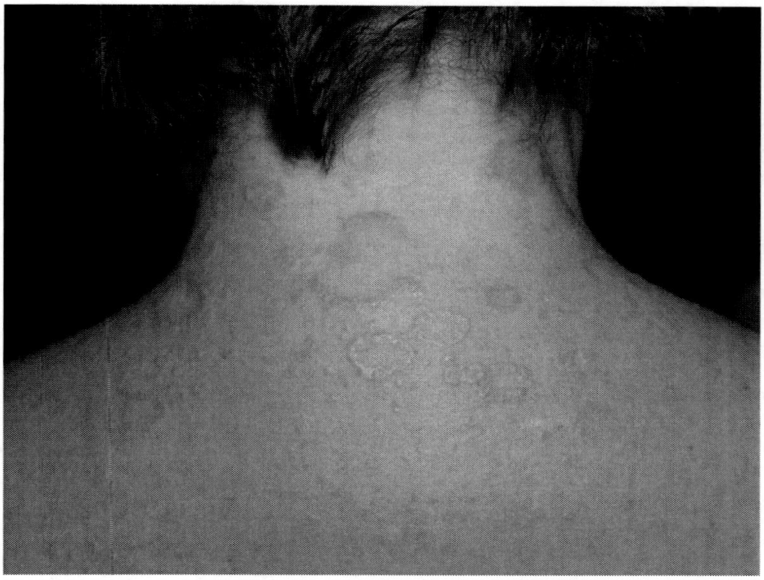

Figure 4. Disseminated granuloma annulare.

Diabetic Dermopathy

Diabetic dermopathy is one of the most frequent skin lesions in adults with diabetes and occurs in approximately 50% of patients. The causes of these changes are nutritional alterations of the skin as a result of microangiopathy.

Although diabetic dermopathy mostly relates to adults, on rare occasions it can also affect children with diabetes [7]. The disease manifests by the presence of yellowish or brown, small, round or oval, slightly depressed skin lesions, well delineated from the surrounding skin. They are most frequently located on the extensor surfaces of the lower legs. Individual lesions usually occur after a few to dozen months, leaving superficial, discoloured scars.

It is worth noting that patients with diabetic dermopathy were shown to possess significantly greater risk of developing diabetic retinopathy and other complications of diabetes compared to other patients with diabetes [49, 50].

Nail Changes

Proper growth of nails depends on adequate vascularity of the nail bed and nail folds. Thus, disorders of microvascular blood flow may lead to abnormal appearance of the nails. In diabetes, impaired nail growth, distortion, longitudinal or transverse crumbling of the nail plates, nail plate colour changes to yellow or brown, as well as subungual hyperkeratosis were observed. In addition, relatively frequently observed are nail fold lesions associated with yeast infection [51].

Most frequently, nail changes occur in type 2 diabetes; however, they have also been reported in patients with type 1 diabetes mellitus, including children [4, 51].

It was also shown that young patients with diabetes having nail fold changes demonstrate impaired feeling of vibration and more often suffer from subclinical neuropathy and microalbuminuria [51].

Skin Dryness

In diabetic patients the skin may be rough, stiff, fragile, and inelastic. Skin dryness is usually of mild to moderate severity. However, it may be accompanied by pruritus. Because the cutaneous barrier function is reduced in xerosis, the skin is more easily exposed to external hazards. Like in some other xerotic conditions, a greater susceptibility to irritation caused by soaps or detergents may be observed. For that reason, patients should be advised to avoid frequent hand washing and bathing in order to limit cumulative soap-induced irritation. In addition, daily moisturizing of the skin with emollients is highly recommended, as it may improve subjective and objective symptoms of dry skin. It is also worth mentioning that daily bathing should be short (maximum 10-15 minutes) in tepid to slightly warm water (37-38°C), ideally with bathoil addition. Importantly, shortly after bathing or showering the whole skin should be again moisturized with an emollient. When properly used, emollients reduce scaling and overall patient discomfort thus improving their quality of life.

INFECTIOUS SKIN DISEASES

Bacterial and fungal infections of the skin and mucous membranes are relatively frequent complications of diabetes. Their occurrence, course and response to applied therapy are closely linked with the state of metabolic control of diabetes.

Following parameters contribute to the development of infectious skin diseases: presence of other skin changes, high blood glucose level, acidosis, abnormal immune processes including impaired humoral and cell-mediated immunity, granulocyte dysfunction as well as abnormal microvascular circulation, collagen degeneration, impaired nutrition of the skin, sweating disorders, dry skin, and neuropathy. Moreover, infections may alter carbohydrate metabolism and increase the need for insulin [4].

Fungal Infections

Candida albicans seems to be the most closely associated pathogen with diabetes. In diabetic patients, yeasts often colonize the respiratory system and digestive tract. They more strongly adhere to the skin and epithelium of the mucous membranes of the mouth and genitals. In patients with diabetes *Candida* cells are weaker bonded, phagocytized and killed by granulocytes. *Candida sp.* can cause angular cheilitis and intertrigo, acute and chronic mucosal inflammation of the mouth and tongue, and inflammation of the nail folds. There is also a cause of chronic and recurrent vulvovaginitis and balanitis, which is often accompanied by persistent itching and burning. Sometimes candidiasis in infants and children may be the first sign of diabetes, leading to detection of the disease [52]. On the other hand, dermatophyte infections among children with diabetes occur with similar frequency compared with otherwise healthy children. However, they are characterized by a more chronic course and resistance to treatment, and may undergo bacterial superinfection which could be particularly dangerous within feet. For the treatment of fungal infections commonly available topical formulations (e.g., ketoconazole, miconazole, naftifine, ciclopirox) and systemic antifungal preparations (e.g., terbinafine, fluconazole) are recommended.

Bacterial Diseases

The most common bacterial infections in diabetic patients are suppurative diseases caused by *Staphylococcus sp.* and beta-haemolytic *Streptococcus sp.* In patients with diabetes recurrent furuncles, carbuncle, impetigo, ecthyma and erysipelas may occur (Figure 5).

A common phenomenon is also susceptibility to superinfection of wounds, intertrigo, excoriations and eczematous changes. Patients with diabetes are also at greater risk of developing serious cutaneous infections after surgery [53]. As in yeasts, bacterial cells are less phagocytized by neutrophils.

Topically applied antibiotics may be used for the treatment of mild bacterial infections of the skin (e.g., mupirocin, fusidic acid), whereas infection with accompanying systemic symptoms should be treated with systemic antibiotics, if possible, in accordance with the culture and the antibiogram.

Cutaneous Adverse Effects Related to Insulin Injections

Subcutaneous insulin injections are a daily routine in patients with IDDM. Local cutaneous adverse effects (LAE) may occur due to the drug itself but can also be caused by improper technique of insulin injection. LAE occur in patients using various methods of treatment, like multiple daily insulin injections (MDII) or continuous subcutaneous insulin infusions (CSII, pump therapy), with different frequencies. These groups of adverse effects can be effectively prevented by patient education and counselling about proper insulin injection practice such as rotation of injection sites, hygiene rules and frequent change of the needles.

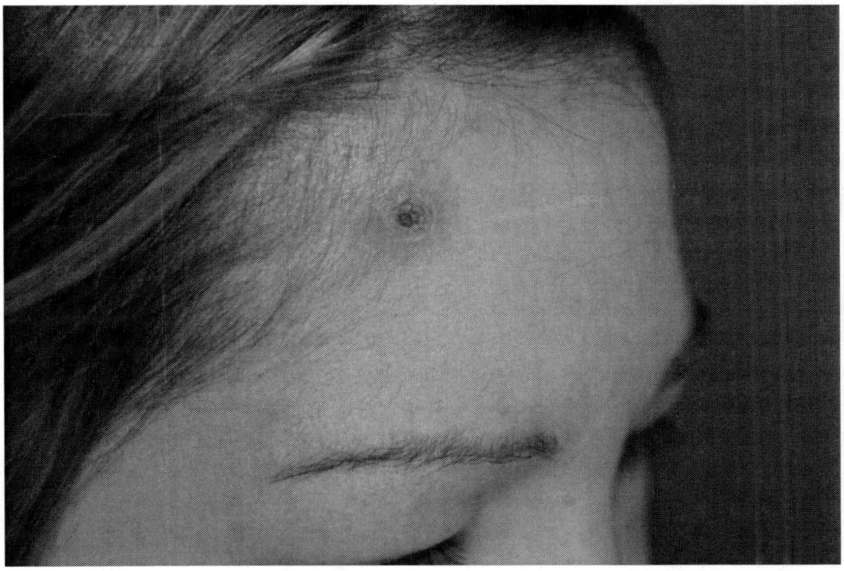

Figure 5. Furuncle in a diabetic girl.

Hypersensitivity Reactions

Cutaneous reactions to insulin were frequently observed before the introduction of modern, highly purified human insulin preparations. However, still a small number of patients may develop immediate or delayed-type local allergic reactions [36, 54].

Severe generalized allergic reactions such as urticaria or anaphylaxis are very rare. Recently, it has been pointed, that local allergic reactions may be observed to substances found in insulin pump sensor fixing patches [55]. Immediate local reactions are probably IgE-dependent reactions. Initially, the erythema is present, which evolves into a wheal within 30 minutes, and then regresses over an 1 hour.

In delayed-type hypersensitivity reactions at the site of insulin injections, approximately two weeks after the start of insulin treatment, itchy nodules appeared and persisted for several to a dozen days and healed, leaving scars and discoloration. Furthermore, additives used in insulin preparations (protamine, zinc, preservatives) or latex antigens contained in the insulin vial and/or insulin syringe can also cause local allergic reactions [56].

Lipoatrophy

Lipoatrophy is now rarely observed (in less than 1% of patients with diabetes), although there are still new reports published on that cutaneous complication [3, 57-61]. This problem currently concerns mainly patients treated with insulin pumps [57].

Lipoatrophy is characterized by the presence of demarcated regions of subcutaneous atrophy at the insulin injections site. The mechanism is not fully understood, but the importance of local abnormal immune reaction with hyperproduction of cytokines (including TNF-α) is postulated [54]. Lopez et al. [60] have also showed an increased number of mast cells within the area of lipoatrophy and have observed clinical improvement after using of cromolyn sodium. To prevent lipoatrophy it is recommended to change insulin administration site and insulin formulation to another one [61].

Lipohypertrophy

Similarly to lipoatrophy, lipohypertrophy (LH) was a rare insulin treatment complication (<2% of patients), however, in the era of insulin pumps use it continues to be a relatively common adverse effect [3, 58]. LH was the most frequent adverse effect observed in the study on 500 young patients (range 1-25 years) diagnosed with T1DM with a prevalence of 41% [5].

In LH the development of soft skin colour nodules is observed that arise probably under the influence of insulin lipogenic action. Occurrence of this condition is more likely related to the frequency of injections at a given site, purity of insulin and immunogenicity than to the dose or species of insulin [36, 62]. Another cause of LH may be the immunogenicity due to high titres of anti-insulin antibodies [63].

Moreover, it has been demonstrated that poor blood supply to this region leads to unpredictable (usually worse) absorption of insulin and the delay in insulin absorption is proportional to the extent of lipohypertrophy [64].

Post-Inflammatory Hyperpigmentation

Post-inflammatory hyperpigmentation (PIH) is usually transitory and resolves within 2 to 3 months, but in some patients it may take 6-12 months to fade. PIH appears as well-circumscribed light brown macules at injection sites. It is believed that the mechanism of PIH formation is linked with reusing of point blunt needle that cause more trauma while injecting with subsequent epidermal inflammatory response that leads to hypermelanosis [65]. Other possible causes include reuse of the non-sterile needles that may cause local skin infection and non-rotation of the injection site.

Acanthosis Nigricans

Acanthosis nigricans is a very rare (0.8-7.4% patients with IDDM) local cutaneous side-effect of insulin injection with very few reported cases [5, 6, 66]. The pathogenesis of acanthosis nigricans is not fully elucidated but is seems that insulin-like growth factor 1 receptor (IGF-1R), tyrosine kinase receptors such as epidermal growth factor receptor and fibroblast growth factor receptor might play a role in it [67].

Other Common Skin Conditions

Xerosis is the second most frequent skin symptom in patients with IDDM. It occurs in 15.8-22.2% of patients [3, 5]. Reduced hydration of the stratum corneum, decreased sebaceous and sweat gland activity, and autonomic peripheral C-fibre neuropathy are casual factors for skin xerosis [3, 6].

References

[1] Paron, N. G., Lambert, P. W. Cutaneous manifestations of diabetes mellitus. *Dermatology* 2000; 27: 371-383.

[2] Romano, G., Moretti, G., Di Benedetto, A. et al. Skin lesions in diabetes mellitus: prevalence and clinical correlations. *Diabetes Res. Clin. Pract.* 1998: 39: 101-106.

[3] Pavlovic, M. D., Milenkovic, T., Dinic, M. et al. The prevalence of cutaneous manifestations in young patients with type 1 diabetes. *Diabetes Care* 2007; 30: 1964-1967.

[4] Cisło, M., Reich, A. Skin lesions in diabetes of developmental age. In: *Diabetes of developmental age – what's new?* Otto-Buczkowska, E. (ed.), Cornetis, Wrocław 2009, 367-375.

[5] Sawatkar, G. U., Kanwar, A. J., Dogra, S., Bhadada, S. K., Dayal, D. Spectrum of cutaneous manifestations of type 1 diabetes mellitus in 500 South Asian patients. *Br. J. Dermatol.* 2014; 171: 1402-1406.

[6] Yosipovitch, G., Hodak, E., Vardi, P. et al. The prevalence of cutaneous manifestations in IDDM patients and their association with diabetes risk factors and microvascular complications. *Diabetes Care* 1998; 21: 506-509.

[7] Lucky, A. W., Powell, J. Cutaneous manifestations of endocrine, metabolic, and nutritional disorders. In: *Pediatric dermatology*, 3rd edition. Schachner, L. A., Hansen, R. C. (eds.), Mosby, Edinburgh, London, New York, Oxford, Philadelphia, St Louis, Sydney, Toronto 2003, 901-941.

[8] Hardin, D. S. Screening for type 2 diabetes in children with acanthosis nigricans. *Diabetes Educ*. 2006; 32: 547-552.

[9] Tzeng, S. T., Hsu, S. G., Fu, L. S., Chi, C. S. Prevalence of atopy in children with type 1 diabetes mellitus in central Taiwan. *J. Microbiol. Immunol. Infect.* 2007; 40: 74-78.

[10] Shun, C. T., Chang, Y. C., Wu, H. P. et al. Skin denervation in type 2 diabetes: correlations with diabetic duration and functional impairments. *Brain* 2004; 127: 1593-1605.

[11] Genuth, S., Sun, W., Cleary, P. et al. Glycation and carboxymethyllysine levels in skin collagen predict the risk of future 10-year progression of diabetic retinopathy and nephropathy in the diabetes control and complications trial and epidemiology of diabetes interventions and complications participants with type 1 diabetes. *Diabetes* 2005; 54: 3103-3111.

[12] Greenman, R. L., Panasyuk, S., Wang, X. et al. Early changes in the skin microcirculation and muscle metabolism of the diabetic foot. *Lancet* 2005; 366: 1711-1717.

[13] Ngo, B. T., Hayes, K. D., DiMiao, D. J., Srinivasan, S. K., Huerter, C. J., Rendell, M. S. Manifestations of cutaneous diabetic microangiopathy. *Am. J. Clin. Dermatol.* 2005; 6: 225-237.

[14] Heimhalt-El Hamriti, M., Schreiver, C., Noerenberg, A. et al. Impaired skin microcirculation in paediatric patients with type 1 diabetes mellitus. *Cardiovasc. Diabetol.* 2013; 12: 115.

[15] Serrao, R., Zirwas, M., English, J. C. Palmar erythema. *Am. J. Clin. Dermatol.* 2007; 8: 347-356.

[16] Sizmaz, S., Pelit, A., Bolat, F., Tuncer, I., Akova, Y. A. Periorbital necrobiosis lipoidica diabeticorum: case report. *Int. Ophthalmol.* 2008; 28: 307-309.

[17] Santos-Juanes, J., Galache, C., Curto, J. R., Carrasco, M. P., Ribas, A., Sánchez del Río, J. Squamous cell carcinoma arising in long-standing necrobiosis lipoidica. *J. Eur. Acad. Dermatol. Venereol.* 2004; 18: 199-200.

[18] Lim, C., Tschuchnigg, M., Lim, J. Squamous cell carcinoma arising in an area of long-standing necrobiosis lipoidica. *J. Cutan. Pathol.* 2006; 33: 581-583.

[19] Ahmed, I., Goldstein, B. Diabetes mellitus. *Clin. Dermatol.* 2006; 24: 237-46.

[20] Pavlović, M. D., Milenković, T., Dinić, M., Misović, M., Daković, D., Todorović, S., Daković, Z., Zecevi, R. D., Doder, R. The prevalence of cutaneous manifestations in young patients with type 1 diabetes. *Diabetes Care* 2007; 30:1964-7.

[21] Ngo, B., Wigington, G., Hayes, K., Huerter, C., Hillman, B., Adler, M., Rendell, M. Skin blood flow in necrobiosis lipoidica diabeticorum. *Int. J. Dermatol.* 2008; 47: 354-358.

[22] Cohen, O., Yaniv, R., Karasik, A., Trau, H. Necrobiosis lipoidica and diabetic control revisited. *Med. Hypotheses* 1996; 46: 348-50.

[23] Körber, A., Dissemond, J. Necrobiosis lipoidica diabeticorum. *Can. Med. Assoc. J.* 2007; 177: 1498.

[24] Uva, L., Freitas, J., Soares de Almeida, L. et al. Squamous cell carcinoma arising in ulcerated necrobiosis lipoidica diabeticorum. *Int. Wound. J.* 2013 Dec. 26.

[25] Hammami, H., Youssef, S., Jaber, K., Dhaoui, M. R., Doss, N. Perforating necrobiosis lipoidica in a girl with type 1 diabetes mellitus: a new case reported. *Dermatol. Online J.* 2008; 14: 11.

[26] Durupt, F., Dalle, S., Debarbieux, S., Balme, B., Ronger, S., Thomas, L. Successful treatment of necrobiosis lipoidica with antimalarial agents. *Arch. Dermatol.* 2008; 144: 118-119.

[27] Boyd, A. S. Treatment of necrobiosis lipoidica with pioglitazone. *J. Am. Acad. Dermatol.* 2007; 57: S120-S121.

[28] Boyd, A. S. Thiazolidinediones in dermatology. *Int. J. Dermatol.* 2007; 46: 557-563.

[29] Kukreja, T., Petersen, J. Thalidomide for the treatment of refractory necrobiosis lipoidica. *Arch. Dermatol.* 2006; 142: 20-22.

[30] Stanway, A., Rademaker, M., Newman, P. Healing of severe ulcerative necrobiosis lipoidica with cyclosporin. *Australas J. Dermatol.* 2004; 45: 119-122.

[31] Aslan, E., Körber, A., Grabbe, S., Dissemond, J. Successful therapy of ulcerated necrobiosis lipoidica non diabeticorum with cyclosporine A. *Hautarzt* 2007; 58: 684-688.

[32] Clayton, T. H., Harrison, P. V. Successful treatment of chronic ulcerated necrobiosis lipoidica with 0.1% topical tacrolimus ointment. *Br. J. Dermatol.* 2005: 152: 581-582.

[33] Kreuter, A., Knierim, C., Stücker, M. et al. Fumaric acid esters in necrobiosis lipoidica: results of a prospective noncontrolled study. *Br. J. Dermatol.* 2005; 153: 802-807.

[34] Zeichner, J. A., Stern, D. W., Lebwohl, M. Treatment of necrobiosis lipoidica with the tumor necrosis factor antagonist etanercept. *J. Am. Acad. Dermatol.* 2006; 54: 120-121.

[35] Tan, E., Patel, V., Berth-Jones, J. Systemic corticosteroids for the outpatient treatment of necrobiosis lipoidica in a diabetic patient. *J. Dermatolog. Treat.* 2007; 18: 246-248.

[36] Sreedevi, C., Car, N., Pavlic-Renar, I. Dermatologic lesions in diabetes mellitus. *Diabetol. Croat.* 2002; 31: 147-159.

[37] Narbutt, J., Torzecka, J. D., Sysa-Jedrzejowska, A., Zalewska, A. Long-term results of topical PUVA in necrobiosis lipoidica. *Clin. Exp. Dermatol.* 2006; 31: 65-67.

[38] Beattie, P. E., Dawe, R. S., Ibbotson, S. H., Ferguson, J. UVA1 phototherapy for treatment of necrobiosis lipoidica. *Clin. Exp. Dermatol.* 2006; 31: 235-238.

[39] Heidenheim, M., Jemec, G. B. Successful treatment of necrobiosis lipoidica diabeticorum with photodynamic therapy. *Arch. Dermatol.* 2006; 142: 1548-1550.

[40] Berking, C., Hegyi, J., Arenberger, P., Ruzicka, T., Jemec, G. B. Photodynamic therapy of necrobiosis lipoidica - a multicenter study of 18 patients. *Dermatology* 2009; 212: 136-139.

[41] De Giorgi, V., Buggiani, G., Rossi, R., Sestini, S., Grazzini, M., Lotti, T. Successful topical photodynamic treatment of refractory necrobiosis lipoidica. *Photodermatol. Photoimmunol. Photomed.* 2008; 24: 332-333.

[42] Martinón-Torres, F., Martinón-Sánchez, J. M., Martinón-Sánchez, F. Localized granuloma annulare in children: a review of 42 cases. *Eur. J. Pediatr.* 1999; 158: 866.

[43] Grogg, K. L., Nascimento, A. G.: Subcutaneous granuloma annulare in childhood: clinicopathologic features in 34 cases. *Pediatrics* 2002; 107: e42.

[44] Agrawal, A. K., Kammen, B. F., Guo, H., Donthineni, R. An unusual presentation of subcutaneous granuloma annulare in association with juvenile-onset diabetes: case report and literature review. *Pediatr. Dermatol.* 2012; 29: 202-205.

[45] Marchetti, F., Gerarduzzi, T., Longo, F., Faleshini, E., Ventura, A., Tonini, G. Maturity-onset diabetes of the young with necrobiosis lipoidica and granuloma annulare. *Pediatr. Dermatol.* 2006; 23: 247-250.

[46] Hertl, M. S., Haendle, I., Schuler, G., Hertl, M. Rapid improvement of recalcitrant disseminated granuloma annulare upon treatment with the tumour necrosis factor-alpha inhibitor, infliximab. *Br. J. Dermatol.* 2005; 152: 552-555.

[47] Harth, W., Linse, R. Topical tacrolimus in granuloma annulare and necrobiosis lipoidica. *Br. J. Dermatol.* 2004; 150: 792-794.

[48] Seibold, J. R. Digital sclerosis in children with insulin-dependent diabetes mellitus. *Arthritis Rheum.* 1982; 25: 1357-1361.

[49] Abdollahi, A., Daneshpazhooh, M., Amirchaghmaghi, E., Sheikhi, S., Eshrati, B., Bastanhagh, M. H. Dermopathy and retinopathy in diabetes: is there an association? *Dermatology* 2007; 214: 133-136.

[50] Morgan, A. J., Schwartz, R. A. Diabetic dermopathy: A subtle sign with grave implications. *J. Am. Acad. Dermatol.* 2008; 58: 447-451.

[51] Kapellen, T. M., Galler, A., Kiess, W. Higher frequency of paronychia (nail bed infections) in pediatric and adolescent patients with type 1 diabetes mellitus than in non-diabetic peers. *J. Pediatr. Endocrinol. Metab.* 2003; 16: 751-758.

[52] Williams, M. D., Sallee, D., Robinson, M. Diabetic ketoacidosis in toddler with a diaper rash. *Am. J. Emerg. Med.* 2008; 26: 834.e1-e2.

[53] Glynn, F., Skinner, L. J., Riley, N., Donnelly, M. Parapharyngeal abscess in an insulin dependant diabetic patient following an elective tonsillectomy. *J. Laryngol. Otol.* 2007; 121: e16.

[54] Beltrand, J., Guilmin-Crepon, S., Castanet, M., Peuchmaur, M., Czernichow, P., Levy-Marchal, C. Insulin allergy and extensive lipoatrophy in child with type 1 diabetes. *Horm. Res.* 2006; 65: 253-260.

[55] Diabetes Research in Children Network (DirecNet) Study Group, Buckingham, B., Beck, R. W., Tamborlane, W. V., Xing, D., Kollman, C., Fiallo-Scharer, R., Mauras, N., Ruedy, K. J., Tansey, M., Weinzimer, S. A., Wysocki, T. Continuous glucose monitoring in children with type 1 diabetes. *J. Pediatr.*, 2007; 151: 388-393.e1-e2.

[56] Towse, A., O'Brien, M., Twarog, F. J., Braimon, J., Moses, A. C. Local reaction secondary to insulin injection. A potential role for latex antigens in insulin vials and syringes. *Diabetes Care* 1995; 18: 1195-1197.

[57] Radermecker, R. P., Piérard, G. E., Scheen, A. J. Lipodystrophy reactions to insulin: effects of continuous insulin infusion and new insulin analogs. *Am. J. Clin. Dermatol.* 2007; 8: 21-28.

[58] Conwell, L. S., Pope, E., Artiles, A. M., Mohanta, A., Daneman, A., Daneman, D. Dermatological complications of continuous subcutaneous insulin infusion in children and adolescents. *J. Pediatr.* 2008; 152: 622-628.

[59] Del Olmo, M. I., Campos, V., Abellán, P., Merino-Torres, J. F., Piñón, F. A case of lipoatrophy with insulin detemir. *Diabetes Res. Clin. Pract.* 2008; 80: e20-e21.

[60] Lopez, X., Castells, M., Ricker, A., Velazquez, E. F., Mun, E., Goldfine, A. B. Human insulin analog-induced lipoatrophy. *Diabetes Care* 2008; 31: 442-444.

[61] Szypowska, A., Skórka, A., Pańkowska, E. Lipoatrophy associated with rapid-acting insulin analogues in young patients with type 1 diabetes mellitus. *Pediatr. Endocrinol. Diabetes Metab*. 2008;14:117-118.

[62] Mahadevan, S., Bhadada, S. K., Velayutham, P., Bhansali, A. Bovine insulin therapy and lipohypertrophy. *Arch. Dis. Child*. 2004; 89: 624.

[63] Raile, K., Noelle, V., Landgraf, R., Schwarz, H. P. Insulin antibodies are associated with lipoatrophy but also with lipohypertrophy in children and adolescents with type 1 diabetes. *Exp. Clin. Endocrinol. Diabetes* 2001; 109: 393-396.

[64] Young, R. J., Hannan, W. J., Frier, B. M. et al. Diabetic lipohypertrophy delays insulin absorption. *Diabetes Care* 1984; 7: 479-480.

[65] Lacz, N. L., Vafaie, J., Kihiczak, N. I., Schwartz, R. A. Postinflammatory hyperpigmentation: a common but troubling condition. *Int. J. Dermatol*. 2004; 43: 362-365.

[66] Sawatkar, G. U., Dogra, S., Bhadada, S. K., Kanwar, A. J. Acanthosis nigricans – an uncommon cutaneous adverse effect of a common medication: report of two cases. *Indian J. Dermatol. Venereol. Leprol*. 2013; 79: 553.

[67] Higgins, S. P., Freemark, M., Prose, N. S. Acanthosis nigricans: a practical approach to evaluation and management. *Dermatol. Online J*. 2008; 14: 2.

INDEX

E

F

G

H

I

L

M

N

O

P

R

S

T

V

W